INDUSTRIAL
ROBOTICS

KEITH DINWIDDIE

❋ Cengage

Australia • Brazil • Canada • Mexico • Singapore • United Kingdom • United States

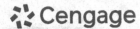

Industrial Robotics, **First Edition**
Keith Dinwiddie

SVP, GM Skills: Jonathan Lau

Product Director: Matthew Seeley

Senior Product Manager: Katie McGuire

Executive Director of Development:
Marah Bellegarde

Content Developer: Jenn Alverson

Product Assistant: Mara Ciacelli

Vice President, Strategic Marketing Services:
Jennifer Ann Baker

Marketing Manager: Andrew Ouimet

Project Manager, Production Vendor
Management: James Zayicek

Content Project Management:
Lumina Datamatics, Inc.

Art Designer: Angela Sheehan

Cover Image Credit: SvedOliver/Shutterstock.com;
TUM2282/Shutterstock.com; bogdanhoda/
Shutterstock.com; Rainer Plendl/Shutterstock.com

For product information and technology assistance, contact us at
**Cengage Customer & Sales Support, 1-800-354-9706
or support.cengage.com.**

For permission to use material from this text or product, submit all
requests online at **www.copyright.com.**

Library of Congress Control Number: 2017959358

ISBN: 978-1-133-61099-1

Cengage
200 Pier 4 Boulevard
Boston, MA 02210
USA

Cengage is a leading provider of customized learning solutions
with employees residing in nearly 40 different countries and sales in more
than 125 countries around the world. Find your local representative at:
www.cengage.com.

To learn more about Cengage platforms and services, register or access
your online learning solution, or purchase materials for your course,
visit **www.cengage.com.**

Notice to the Reader

Publisher does not warrant or guarantee any of the products described herein or perform any independent analysis in connection with any of
the product information contained herein. Publisher does not assume, and expressly disclaims, any obligation to obtain and include information
other than that provided to it by the manufacturer. The reader is expressly warned to consider and adopt all safety precautions that might
be indicated by the activities described herein and to avoid all potential hazards. By following the instructions contained herein, the reader
willingly assumes all risks in connection with such instructions. The publisher makes no representations or warranties of any kind, including but
not limited to, the warranties of fitness for particular purpose or merchantability, nor are any such representations implied with respect to the
material set forth herein, and the publisher takes no responsibility with respect to such material. The publisher shall not be liable for any special,
consequential, or exemplary damages resulting, in whole or part, from the readers' use of, or reliance upon, this material.

Printed in the United States of America
Print Number: 02   Print Year: 2022

Contents

Preface

This is my second textbook on robotics. Unlike my first book, which covered the basics of robotics in general, this book focuses on the industrial side of things. Inside, you, the reader, will find all the basic knowledge needed to start your exploration of the world of industrial robotics, written in a straightforward, conversational style that many find refreshing. The jargon used by industry is explained along the way and highlighted so you can find it as needed later on. In the pages of this book you will find tales from industry, hard-won knowledge presented in an easy-to-understand format, and meticulously researched historical facts. This book is the culmination of my many years learning about, working with, and teaching robotics. While many texts seem to be written for top-level engineers or industry experts, this book is designed for those who are new to robotics and eager to learn. It is my sincere hope that you enjoy reading this book as much as, if not more than, I did writing it!

As you dig into the text, you will find that Chapter 1 covers all the basic safety concerns that one might encounter in any technical environment. Here you benefit from my many years of teaching safety and first aid. In Chapter 2, you will learn what a robot is, how this technology evolved, and why we use them. The history behind robotics is longer and more intricate than you might imagine. Chapter 3 introduces the various parts of the robot and explains what each part does for the system. Notably, this chapter describes how to number the different axes of the robot for different configurations—coverage often missing in other books or addressed only for the arm type. Chapter 4 discusses how we group robots and describes the different work envelope geometries used by robots. This chapter ends with the ISO classifications utilized in many industries. Chapter 5 finishes up our exploration of the core robotic components with a look at the world of robotic tooling. Without proper tooling, most robots are fairly useless—so it is important to understand which options are available and what they are used for. As part of this chapter's coverage, we even consider ways to use multiple tools on one robot.

Chapter 6 explores the process of programming the robot and the basic steps that go into this process. This chapter gives the reader the knowledge base needed to create working programs for any robot, once they learn the specifics of the robot they are using. Chapter 7 examines the various sensors used in robotic work cells and other places in industry. Since the robot knows only what we tell it, it is crucial to understand the types of information that sensors can provide. Chapter 8 deals with robotic vision, an exciting addition to any robotic system that can help it see the world in a whole new light. Lighting is one of the primary concerns with any vision system, and you can find information on that topic in this chapter.

Chapter 9 presents an overview of networking and integration, which have become crucial skills in the modern factory. Gone are the days of dumb machines isolated and alone. Today, most (if not all) of the equipment in the plant is part of the plant's network and accessible from remote terminals. Chapter 10 covers the basics of programmable logic controllers (PLCs) and human–machine interfaces (HMIs). In this chapter, I share some hard-won knowledge about how the PLC works as well as how to think like a PLC. Most often, technicians have trouble with PLCs because of the way these devices' operation was explained to them. In Chapter 10, you will learn several crucial points that can prevent this kind of problem from happening to you!

Everything eventually breaks, and Chapter 11 provides tips on how to keep the robot running as well as how to fix it when something goes wrong. This chapter includes a section on arc flash, which is a major concern when engaging in any electrical work. It also has some tips for crash recovery, for those times when the robot runs into something it should not have. Chapter 12 finishes things off with a look at robots versus human labor and the return on investment from robots. When the topic of robotics comes up, it often leads to a discussion of whether robots will take all the jobs away from people. If you want to learn more about this debate, check out Chapter 12.

By the end of this book, you will have a deeper knowledge of robotics and will be set to explore the field deeper. There are many fields to go into once you have read this book and completed the course that introduced you to it. For those who want to program or design robots, several engineering fields might be appropriate career avenues. If you like getting your hands dirty and fixing broken things, a career as a repair technician could be a good fit. If you like to find solutions to production needs/problems, the integrator field—that is, the area where you find the right robot and get everything set up—may be your calling. No matter where your journey takes you, this text will provide a solid foundation so you can dig deeper in the field of your calling.

SUPPLEMENTS

Instructor Companion Site

Everything you need for your course in one place! This collection of book-specific lecture and class tools is available online via www.cengage.com/login. Access and download PowerPoint presentations, images, and an instructor's manual.

Cengage Learning Testing Powered by Cognero

Cengage Learning Testing Powered by Cognero is a flexible, online system that allows you to:

- Author, edit, and manage test bank content from multiple Cengage Learning solutions
- Create multiple test versions in an instant
- Deliver tests from your LMS, your classroom, or wherever you want

MindTap for Industrial Robotics

MindTap is a personalized teaching experience with relevant assignments that guide students to analyze, apply, and improve thinking, allowing you to measure skills and outcomes with ease.

- *Personalized teaching*: MindTap becomes your own when you make available a learning path that is built from key student objectives. Control what students see and when they see it; match your syllabus exactly by hiding, rearranging, or adding your own content.
- *Guide students*: MindTap goes beyond the traditional "lift and shift" model by creating a unique learning path of relevant readings, multimedia, and activities that move students up the learning taxonomy from basic knowledge and comprehension to analysis and application.
- *Measure skills and outcomes*: Analytics and reports provide a snapshot of class progress, time on task, engagement, and completion rates.

ACKNOWLEDGMENTS

I would like to thank Katie McGuire from Cengage for giving me the chance to work on another textbook and Jennifer Alverson from Cengage for working with me throughout the project. Without their help and trust in me, you would not be enjoying this book. Special thanks go out to Jerry Guignon and Matthew Morris for helping with the ABB photos. Also, thank you to all the generous people from Yaskawa America and Panasonic-Miller, Kiel Vedrode from FANUC, and Kelly Fair from Schunk who helped me track down the images used in the text—this book would not be the same without these great images. And thank you, the reader, for taking the time to explore my book.

About the Author

Keith Dinwiddie has been teaching at Ozarks Technical Community College, based in Springfield, Missouri, for 10 years full time and 12 years total as of the writing of this book. In the past, Keith has worked as a maintenance technician in industry; he also did a stint in the army working on the Huey helicopter. Keith has loved all things robotic from a very young age and has had the chance to work with Panasonic, FANUC, Mitsubishi, and UR3 industrial robots as well as NAO, WowWee, and hobby robotic systems. Keith is a FANUC C.E.R.T (Certified Education Robotic Training) instructor for handling pro and vision, which means he can give his students a certificate that carries the same weight as going to the FANUC classes. This is Keith's second textbook but he has also authored various articles for Balluff or AZO.

DEDICATION

This book is dedicated to my loving wife, Lucia Dinwiddie. Thank you for becoming a book widow once more, and for all the support you have given me through the writing process. It is appreciated more than you know!

CHAPTER 1

Safety

WHAT YOU WILL LEARN

- How to work safely with robots

- The three conditions that will stop a robot

- The three zones around a robot

- How to ensure the safety of people in the danger zone

- Operation of several common safety sensors

- The dangers of electricity

- How to lock out or tag out equipment

- How to deal with emergencies

- Some basic first-aid procedures

OVERVIEW

Industrial robots are incredible technological marvels that have the ability to move heavy materials, perform machining functions, fuse metals, deposit various substances, and move much faster than humans. The same functions that make a robot valuable are also the biggest dangers of working with a robot. From the person who works around a robot, to those responsible for its maintenance, to the creators of robots—safety needs to be a primary concern at all times. To help you survive working with and around industrial robots, we will cover the following topics in this chapter:

- Robots require respect
- Danger zones
- Guarding
- Safety devices
- Electricity and you
- LOTO
- Handling emergencies

 FOOD FOR THOUGHT

1-1

The following story comes from one of my students and is a great example of just how dangerous it can be to work around a robot without the proper safety precautions. In truth, my student is very lucky to be alive today.

The company where this event happened produces large lead-acid batteries. It utilizes robots to move the batteries from one station to the next as different operations are performed. In the cage where the accident happened, there were three different stations where the robot would pick up the battery and move it to another process. The robot had been stopped and my student was inside the **work envelope**—that is, the area that the robot could reach. The student was performing some routine cleaning operations, while another employee outside the protective cage was engaged in similar activities.

While my student was busy inside the work envelope, an engineer for the company came up and started to do something with the robot controller. While my student never fully understood what the engineer was trying to do, he did find out painfully what the engineer had done. One minute he was working; the next minute he was lying on the ground trying to figure out what happened. Somehow the engineer had started the robot with my student inside the cage and it had knocked my student down.

As he laid there trying to figure out what had happened, my student heard someone holler, "Don't get up, it's coming back!" A moment later he saw the robot sweep over his head at full speed, going about its normal programmed duties. Someone finally stopped the robot. After being trapped on the floor for what seemed like an eternity, my student was able to move once more. As he started to get up and take stock of his injuries, he realized that his head was only inches away from anchor bolts left in the floor from a previous machine that had been moved, and the severity of what had just happened started to sink in.

My student was lucky to walk away from this event with only some bumps and bruises. If he had been standing where the robot picked up the batteries instead of just in its movement path, the robot could have easily clamped down on him and thrown him around the work envelope. Considering that the parts the robot moved weighed as much as 150 pounds, it is very reasonable to think the robot could have picked up my student. If he had landed a few inches over, his head would have likely hit the bolts sticking up, which could have been fatal. If he had fallen onto some of the other equipment instead of hitting the floor and getting out of the robot's path, he could have been crushed. The possible ways he could have been severely injured or killed go on and on.

This incident raises some big questions—namely, what went wrong and what could my student have done to prevent this incident? First, one should never enter the work envelope of a robot without access to an emergency stop. Indeed, it is for this very reason that industrial robots have an E-stop on the teach pendant. It is also why OSHA requires you to have the teach pendant with you at all times while inside the reach of the robot. Second, no one should try to make alterations to the robot or controller while people are in harm's way. This is just asking for trouble—and in this case, trouble was found. Third, when working with industrial robots, we must always be aware of what is going on around us. The moment the engineer walked up and started to mess with the controller, my student should have exited the work envelope of the robot and waited until the engineer was finished. Of course, he also should have had the teach pendant with him so he could have hit the E-stop should the need arise, as it did in this case. We can also fault the engineer for working with the system while someone was inside the work envelope. Even those people who have tons of experience with robots can accidently hit the wrong button or switch from time to time.

ROBOTS REQUIRE RESPECT

Fatalities involving robotic systems are rare, but they do happen. The Occupational Safety and Health Association (OSHA) cites 27 fatalities involving robots from 1984 to 2013, an average of less than one death per year, but these are the numbers for just the United States. In 2015, two fatalities brought robotic safety to the forefront of concern—one in India, where a worker was impaled by a welding robot, and another in Germany at Volkswagen, where a worker was grabbed and crushed against a steel plate. In 2016, a young woman in Alabama was crushed inside a robotic station while trying to clear a fault. These deaths serve as a tragic reminder that industrial robots still have the ability to kill in spite of the safety equipment and regulations in use today.

When it comes to safety, you will often hear about the three Rs of robotics: *Robots Require Respect*. If safety equipment and regulations were enough, injuries and fatalities involving robots would be a thing of the past. Unfortunately, that is not the case. Some of the cases of injury or death involving robots trace back to faulty equipment, and others are the result of improper training, but many have a direct correlation to complacency. When we work with robots day after day, we become accustomed to their methodical nature and forget the inherent dangers of their functionality. When people stop respecting the robot as a powerful piece of equipment, they often start to take risks that could literally cost life or limb.

A robot performs its actions via programming, direct control, or some combination of the two. That is it. Robots do not have feelings, they do not have moods, they do not have intuitive thought, they do not think as we do, and they do not have their own agenda. While it is true we are working on artificial intelligence (AI) programs and have given robots the ability to deal with complex situations in which there may not be a clear-cut right answer, these machines are still performing only as programmed. Because of this functionality, there are only three conditions that stop a robot:

- *The program/driven action is finished.* The program or direct control is used to control robots. Thus, once the robot has run its program or we have stopped sending action signals, the robot simply stops and waits for the next command. A sensor or other system may initiate the next command/program, which explains why robots sometimes seem to start unexpectedly, but this is the robot working under program control.
- *There is an alarm condition.* Almost all modern robotic systems have some sort of alarm system.

In many cases, this system monitors such things as safety sensors, E-stops (emergency stops), load on the motors, vision systems, and other available devices that give the robot information about the world around it or its internal systems. The alarm function stops the robot in an effort to prevent or minimize harm to people, other equipment, and/or the robot.

- *Some type of mechanical failure occurs.* Robots are mechanical systems, and like any other machine they are susceptible to breakdown. Motors fail, bolts break, air hoses rupture, wiring shorts out, and connections work loose, just to name a few of the potential failures. Anything of this nature can cause a robot to stop. In a worst-case scenario, the robot would keep operating, but perform its tasks erratically or unpredictably.

If you have the misfortune to be in the robot's path when it starts up either automatically or via your direct control, there is no amount of pleading, no bribe you can offer, and no reasoning with the system to halt its operation—nothing but one of the three conditions mentioned previously will stop the robot. If you do not respect the robot, then you, too, might learn this hard lesson.

DANGER ZONES

At this point, you may be wondering if it is ever safe to work around a robot. The answer is an absolute yes. Every day we use thousands of robots safely and effectively in many facets of the modern world. One of the first things we do to create this safe working condition is to determine the various zones around a robot. Each zone has a risk level and requires a certain level of awareness based on that risk. For our purposes, we will focus on three major zones: the safe zone, the cautionary zone, and the danger zone (Figure 1-1).

Safe Zone

The safe zone is where a person can pass near the robot without having to worry about making contact with the system. This area is outside of the reach of the robotic system as well as beyond the area the robot can affect. The distance from the robot to the safe zone depends on the type of robot, the maximum force of the system, and the actions performed by the robot. The more powerful the robotic system, the farther away you will need to be to remain in the safe zone. An example can help shed some light on what a safe zone truly is.

Suppose a group of new hires are on a tour of the plant to get an overview of where everything is and which

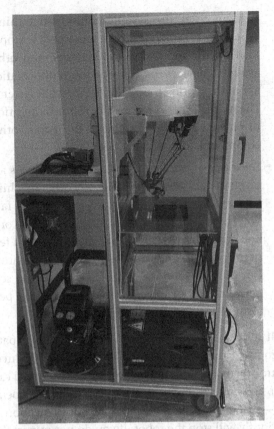

FIGURE 1-1 This is one of the trainers my students use. This system has a completely encapsulating cage that creates two zones: a danger zone inside the Plexiglas and a safe zone outside the safety glass as long as the door remains closed. If the door opens, it creates a cautionary zone directly in front of the robot. In the upper-right corner above the door handle, notice the red door switch: It senses when the door is open and keeps the robot from running in automatic mode in that situation.

dangers are associated with the various departments. The group walks through a production area where several robots are moving parts and performing welding operations. For many of the new employees, this is their first look at an industrial robot in action, but they can watch it safely because they stay in the taped-off pedestrian traffic path. This path is routed so that people are safe from a majority of the hazards of the area. In this safe zone, anyone could walk through the area without needing to know anything specific about the robot to remain safe.

Cautionary Zone

The **cautionary zone** is the area where one is close to the robot, but still outside of the work envelope or reach of the system. While the robot cannot reach you in this area, there could be danger during operation from things such as chips, sparks, thrown parts, high-pressure leaks, crashes, overspray, or flash from welding. Since this is often the area from which operators perform their tasks,

they must understand the potential hazards involved to work safely. Because of the need for understanding of the system, this is not an area for just anyone.

A good example of this area is the operator station for an industrial robot that performs a **pick and place operation**—that is, picking up items from one area and placing them in another. Often this type of system will pick up a raw part from a conveyor or container, remove a finished part from a machine, place the raw part into the machine for processing, deposit the finished part on a conveyor or something similar, and then wait to start the process all over again. The operator is usually responsible for such tasks as loading raw parts onto the conveyor, checking the dimensions and quality of the finished parts, making corrections to the process as needed, and any other tasks needed to complete this portion of the production process. The requirements of this job often place the operator in close proximity to not only the robot but also other production equipment. Workers spend millions of hours in these areas each year without injury or incident, though accidents do happen on occasion. Because of this potential for accidents, workers in the cautionary zone need to be aware of all the dangers and know how to handle any situations that might arise. Because this is the normal space for the operator, the cautionary zone will contain stop buttons, emergency stops (E-stops), the controller for the robot, operator interfaces such as the teach pendant, and other ways to stop or control the system as needed.

Danger Zone

The **danger zone** is the area the robot can reach or the work envelope; it is where all the robotic action takes place. The various axes of the robot and the design of the system define the work envelope, so you need to be familiar with the robot to enter the danger zone. Each robot has its own danger zone, and you must exercise extreme caution in this area because it has the highest potential for injury or death. When in the danger zone you have to watch out for the robot, any tooling used by the robot, and any place where the robot could trap you against something solid, known as a **pinch point**.

In industrial settings, we have to mitigate the hazards associated with the danger zone in some way, so that when people enter this area the system either slows to a safe velocity with extra sensitivity for impacts or stops automatic operation. A popular method to achieve this kind of protection is to place metallic fencing around the robot, creating a cage that keeps people out of the danger zone while providing one or two entrances for necessary repair, cleanup, tool changes,

FIGURE 1-2 The teach pendant for the FANUC trainer. Note the big red E-stop on the upper-right corner of the teach pendant as well as the E-stop on the controller directly below the teach pendant.

or other normal job requirements. These entrances have sensors in place such that when they are opened, the robot stops automatic operations or in some other way renders itself safe for humans to be near it.

When we put ourselves in the danger zone, we should always have some way to stop or shut down the robot. If you enter the danger zone with no way of stopping the robot, you are asking for trouble—like

the student in the Food for Thought box. In industry, an OSHA requirement states that anyone entering danger zone must take the teach pendant along. The **teach pendant** is a handheld device, usually attached to a fairly long cord, that allows people to edit or create programs and control various operations of the robot (Figure 1-2). It also contains an E-stop should the need for it arise.

Let us return to our industrial pick and place robot to explore the three zones in greater detail. This robot works with several machines and moves parts around in a defined area enclosed by a metal mesh cage. The operator has a workstation where he or she monitors operations, checks parts, and makes adjustments as needed. There is a clearly marked main isle near where the operator works, but it is 15 feet away from the process. The isle can be used by anyone and is considered a safe path for people—that is, a safe zone. The operator station is in the cautionary zone, as there is some potential for injury, but this area is safe for people to be as long as they have the proper training and safety equipment. Anywhere inside the cage is in the danger zone, exposed to all the dangers of the system and where access to an E-stop for the system is an OSHA requirement (Figure 1-3). To meet the E-stop requirement, anyone entering the danger zone should bring the teach pendant and have the ability to control the robot and prevent bad things from happening in the first place (Figure 1-4).

As another example, consider a robot that is responsible for dipping molten aluminum from a large cauldron and then pouring it into a feed tube for an injection molding machine. When I saw this actual robot in operation, the taped-off walkway used for my tour would be the safe zone. An operator for this

FOOD FOR THOUGHT

1-2

The Occupational Safety and Health Administration (OSHA) was officially formed in 1971 due to an act signed into law by President Richard Nixon on December 29, 1970. OSHA's sole mission is to assure a safe and healthful workplace for every working person. It does so by inspecting worksites and factories to ensure they are following the guidelines that OSHA has written or mandated by reference. Many of the rules OSHA enforces are derived from sources such as the National Electrical Code (NEC) or rules established by the National Fire Protection Association (NFPA).

OSHA enforces these rules by going to factories and work sites to conduct inspections, in which its personnel look for violations or things that do not conform to the rules. Companies are fined for violations and then given a set amount of time to fix the problem or face steeper fines and penalties. OSHA averages approximately 40,000 to 50,000 of these inspections each year, many of which occur in response to written complaints from workers.

If you would like to know more about OSHA or how to report safety violations, check out www.OSHA.gov.

Image courtesy of ABB Inc.

FIGURE 1-3 These three robots are handling parts inside a caged area. Notice how much floor space is considered in the danger zone, as denoted by the white metal cage behind the robots.

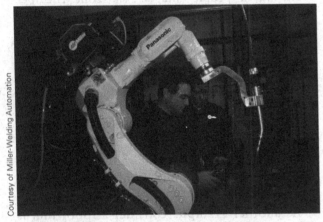

Courtesy of Miller-Welding Automation

FIGURE 1-4 Here you can see a couple of operators in the danger zone working on the program with teach pendant in hand.

machine monitored the process as well as dealt with the finished pieces. He worked in the cautionary zone, but had to take some extra care as this robot worked with molten metal. From my safe observation spot, one of the things that caught my attention was the amount of aluminum that had spilled between where the robot collected it and where the robot filled the machine. One expects drips and spills at the collection and fill points, but not so much as to create a path of spilled aluminum lining the way between the two. In addition, aluminum could be observed *outside* the reach of the robot. Thus, the danger zone in this case extended past the reach of the robot to the area where the liquid aluminum splashes could reach.

These examples emphasize that each robotic system is unique and, therefore, requires individual assessment of both the risks it poses and the locations of the boundary zones. Remember, the more powerful the system and the more dangerous its function, the greater the potential for injury or death. Failing to respect the requirements of the cautionary zone or the danger zone is a quick way to experience the horror of being at the nonexistent mercy of a robot.

GUARDING

In our discussion of the danger zones, we briefly mentioned means of guarding robotic systems. **Guards,** for our purposes, are devices designed to protect us from the dangers of a system. The two types we will explore here are guards installed directly on the equipment and guards placed a set distance from the device. No matter the placement, a guard's main purpose is to keep people safe; only rarely do they improve the operation of the equipment. In fact, there are many situations where guarding limits the operation of the equipment and/or makes it more difficult to work with.

When it comes to questions about robotic safety systems such as "What are the min/max perimeter guard dimensions for a robot?" or "How is it possible for the Baxter robot to run without a cage?", ANSI/RIA R15.06-2012 provides the answers. This safety standard for robotics was developed by the **Robotic Industries Association (RIA)**, which has been working to further the use of robotic technology in United States for years; it was also adopted by the **American National Standards Institute (ANSI)**, which administers and coordinates private-sector voluntary standardization systems. This revision is an update to the 1999 standard and took full effect on January 1, 2015. While not referenced by OSHA and thus not legally binding, it is considered the proper way to make sure that workers and those around robots are properly protected. Any organization using this standard correctly to guard its robots would have a firm legal leg to stand on should anything go wrong.

The guarding that protects people from moving parts such as chains, pulleys, belts, or gears is mounted directly on the robot and is a part of the system (Figure 1-5). The manufacturer or designer of the robot supplies this guarding, which usually forms the outer structure of the robot. Often this guarding is made of a sturdy plastic or light metal that allows for its removal during repairs and when preventive maintenance is required. Depending on how the guarding fits together, it may be necessary to move the robot into specific positions to remove certain pieces of guarding. If you are having trouble removing a piece

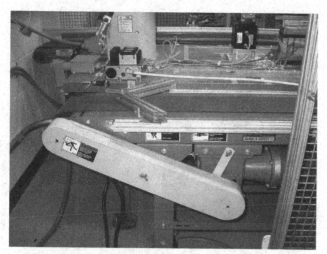

FIGURE 1-5 A guard to protect operators from the belt or chain underneath.

FIGURE 1-7 This small work cell is protected by metal mesh, the yellow fence, and two light curtains at the front opening.

of guarding, look for hidden screws or other pieces of guarding that may be holding it in place. Trying to force a piece of guarding off is a good way to break the guarding and possibly damage parts underneath. Always repair or replace damaged guarding to ensure proper operation of the system and the safety of those who work around it.

Guarding that encloses the work area of a robot, as mentioned earlier, can be made of various materials, with expanded metal or metal mesh being favorites. Expanded metal guarding is metal that is perforated and stretched to create diamond-shaped holes with 0.25-inch pieces of metal around it (Figure 1-6). Metal mesh consists of thick wire welded and/or woven together to create a strong barrier that is easy to see through (Figure 1-7). Metal mesh may be welded into metallic frames, usually angle iron pieces, that make up the panels of the robot cage. This creates a robust guarding system that is easy to see through, but strong enough to resist thrown parts, robot impacts, and people falling into or leaning on it. Add a few sensing devices (a topic covered later in this

FIGURE 1-6 An example of expanded metal guarding rotating parts.

chapter), and you have an OSHA-approved system to ensure the safety of workers in the cautionary zone.

Metal cages are not the only way we can guard the work area of the robot. Another guarding option that is gaining popularity is a camera-based system mounted on the ceiling above the robot that detects when people pass into the danger zone, triggering the robot to respond accordingly. These systems often have a projector that defines the monitored area with clearly visible white lines, so operators can see the danger zone. With this camera-based system, it is easy to adjust the danger zone and increase the area protected as needed, while avoiding the costs and down time associated with moving and fabricating metal cages. The tradeoff is that this system does nothing to stop flying parts or anything physically entering the danger zone.

The Baxter robot, which debuted in 2012, performs danger zone guarding by using a 360° camera system in its head that detects when humans are in the danger zone (Figure 1-8). When it senses someone within range, the robot slows to what is considered a safe movement speed by the RIA, OSHA, and International Standards Organization (ISO); an organization that creates standards for any industry in the world to use), while also monitoring the sensitive collision detection system that will stop all movement of the robot if an impact is detected. This combination of safety features allows Baxter to work outside of the cage that restricts so many robots. Rethink Robots, which created Baxter, also created a more traditional style of robotic arm that uses the same technology. It is not alone in the charge to free robots, as Universal Robots put the cage-free UR5 into production at a Volkswagen plant

FIGURE 1-8 Baxter operates in an industrial setting without any cage around it. This kind of machine is leading the charge to free the industrial robot from confinement.

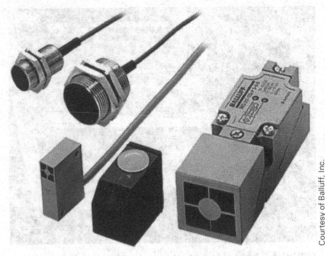

FIGURE 1-9 Examples of various proximity switches.

in 2013. While the cage-free robot has not taken over industry as of yet, the longer these systems run without incident, the better the chances for adoption and evolution of this technology.

We have not covered every form of guarding available by any stretch of the imagination, but you should have a good idea of what guarding is and why we use it. Advancements in safety technology have given us options far beyond the traditional cage, even though the cage remains the most popular choice of today, especially in areas where the cage acts as a physical barrier to any parts that might go flying or similar dangers. As the newer technology proves its worth, more robots will likely be freed from the cage in cases where this barrier exists solely to prevent contact with humans.

SAFETY DEVICES

So far, we have looked at the broad picture of robot safety, ways to work with robots in general, and the threat levels associated with various areas around the robot. This discussion is just the tip of the iceberg, as the modern robot uses a multitude of sensors and devices to ensure human safety. Without these devices, many of the tasks we perform with and around robots would have a greatly increased risk of injury or death. This section introduces some of the devices used to help ensure the robot remains a benefit in the workplace and avoids the detriment of increased danger to workers.

Proximity switches are devices that are widely used to ensure the safety of those who work around robots as well as to direct robot operation (Figure 1-9). A **proximity switch** is a device that generates an

electromagnetic field and senses the presence of various materials based on changes in this field (instead of physical contact). Such switches are used to sense parts, determine machinery position, and track items on conveyors, among many other applications. When it comes to safety, we tend to use the proximity switch to ensure something is in a specific position before an operation takes place—for example, to ensure parts are in position, monitor the entrance to robot danger zones, verify the machine door is open before loading or is closed before the process starts, prevent the robot from rotating too far in a given direction, and verify that the robot's tooling is in the correct configuration or position before an action occurs (Figure 1-10). In short, the proximity switch is a great tool for answering simple yes/no-type questions related to safety or operation.

FIGURE 1-10 This proximity switch is guarding the door to the robot's cage. It consists of the two yellow pieces on the black metal near the top of the cage and over the light-yellow piece.

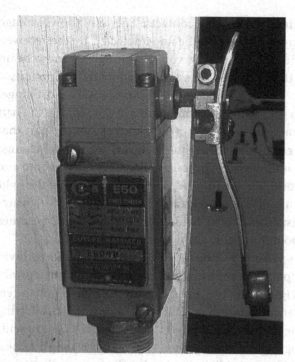

FIGURE 1-11 A limit switch.

A close relative of the proximity switch is the limit switch (Figure 1-11). The **limit switch** senses the presence or absence of a material by contact with a movable element attached to the end of the unit. Limit switches are used in much the same way as proximity switches, but the main difference between the two is that the limit switch actually makes contact with whatever it monitors. Because this device makes physical contact instead of depending on a sensing field, we can extend the range of a limit switch by simply extending the actuator mechanism. The physical contact aspect of the limit switch helps it overcome dirty environments and gives it a larger sensing distance compared to most proximity switches. The downside is that the limit switch experiences more wear due to the physical contact and requires more maintenance than a properly selected and mounted proximity switch.

Many times, we use proximity and limit switches as safety interlocks. A **safety interlock** is a system in which all the safety switches must be closed or "made" for the equipment to run automatically. If at any point during the operation of the equipment these switches open or lose connection, the system automatically drops into a manual or alarmed condition, with many systems doing both. Safety interlocks are often used to guard the doors of the robot cage: Whenever an entrance is opened, the robot automatically enters a safe mode and stops all automatic operation. If the

system includes easy-to-remove covers, we can use the same trick to make sure they are in place before the robot runs in automatic mode. The downside to using interlocks on cage entrances is that someone could potentially open the door, step in, close the door, and then use the teach pendant to reset any alarms and put the robot back in automatic mode. To prevent this situation, we can use **presence sensors**—sensors that detect when a person is inside an area. The camera system used by the Baxter robot, pressure mats, and light curtains are examples of presence sensors that we can tie into the safety interlock system to add another layer of protection and prevent automatic operation of the robot with people nearby.

A **pressure sensor** detects the presence or absence of a set level of force. For safety purposes, these devices are placed in mats that can detect the weight of the operator and respond accordingly. They also contain safety circuitry that can detect when the pressure mat is malfunctioning, thereby preventing the dangerous situation in which someone is on the mat yet the system thinks everything is clear. In some cases, pressure mats are used to verify the operator is standing in a specific location, but more typically they are used to make sure everyone is clear of a danger zone. Small pressure sensors have also been used to cover robot grippers or other tooling so the robot can gather information on force exerted as well as information about the shape of the part.

A relatively new development in presence sensing for robots is robotic "skin." The skin in this instance is a covering of the robot that allows the system to sense people or objects and respond accordingly. Bosch unveiled its first version of this technology in 2014 on the APAS robot, which was covered in leather with embedded tactile sensors to detect impact. The 2017 version of this robot uses a capacitive system to detect humans and stop before any contact occurs. As of the writing of this text, Bosch was the main company working on this technology. As it works the bugs out, however, other companies may add this type of protection to their robots.

Many robots can tell when a motor has encountered something unexpected by sensing the increased power drawn. As a motor encounters resistance, it naturally starts to use more energy in an attempt to overcome this force, thereby increasing the amount of amperage used. When this happens, the part of the robot that drives the motor recognizes the additional power draw and shuts down the robot while sending an alarm to the teach pendant, letting the operator know that something is wrong. In the past, robots primarily used this technology to detect major collisions, but now it is part of the safety system for robots like ABB's YUMI and ROBERTA, freeing them from the traditional cage. With the older technology, it took a fair amount of resistance to create a large enough difference in the power draw to prompt the system to respond. With the new force torque sensors and other mechanical sensing means, robots can now detect impact and abnormal forces at lower levels and respond more quickly. Some systems can even move the affected joint of the robot in the opposite direction, absorbing the impact of the blow via motion and greatly reducing damage to whatever was hit.

A wide range of safety devices are also available that rely on interrupted or reflected light, known as photo eyes. The **photo eye** emits an infrared beam that strikes a shiny surface or reflector, which then reflects the light back to a receiver in the unit. These devices

often include two contacts: one that allows voltage through when the beam is detected by the receiver, and one that allows voltage through when no beam is detected. This gives the user a broad set of options.

The downside to the photo eye is its short range. To get around this limitation, we can separate the part that sends the beam (the emitter) and the part that receives the beam (the receiver), or we can change from infrared light to laser light. The **laser photo eye** works like the infrared photo eye except that the light emitted is a concentrated beam, allowing for greater distance of travel before the returned signal becomes too diffused for sensing purposes (Figure 1-12).

When we house the emitter and the receiver separately, we often call the system a **light curtain** (Figure 1-13). A light curtain may have only a few transmitters and receivers or a large number, allowing them to cover and protect a large area via a line of sensing. Light curtains protect workers by creating a vertical or horizontal light barrier that detects any disruption.

FIGURE 1-12 A standard photo eye and its reflector.

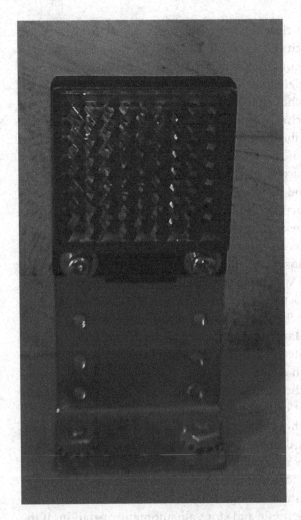

Courtesy of Miller-Welding Automation

FIGURE 1-13 The two yellow bars on either side of the open space are light curtains. In this case, when the robot is not running, the operator can freely enter the work area and change parts out (or do whatever else is needed). If the operator crosses the invisible barrier while the system is in automatic mode, the robot immediately stops operation.

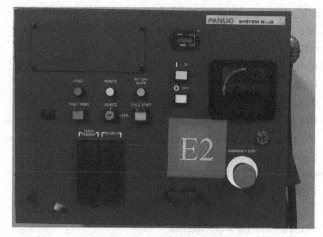

FIGURE 1-14 This FANUC control panel includes a big red E-stop on the lower-right side. It also has an on, off, cycle start, and fault reset button.

Often these systems stop the equipment only when its actions present a danger to the worker, allowing workers to change out parts and do other necessary activities safely in close proximity to automatic equipment. The downside to the light curtain is the need for precision alignment of the emitter and receiver array. A very small amount of error in this alignment can result in the light traveling to one side or the other of the receiver such that it never creates a complete sensing path, which the machine interprets as something being in the danger zone.

Stops, pauses, and E-stops are another crucial part of the safety circuit (Figure 1-14). The stop or pause button provides us with the ability to stop the system during normal operation and gives us control over when the system runs. In many modern systems, stop and pause are options on the teach pendant. Sometimes, however, we need a faster, more decisive response: For these situations, we have the E-stop. E-stops are wired into multiple systems on the robot and provide a more immediate and system-wide response than the stop button. When we press the E-stop, the robot will halt motion as quickly as possible, and will often shut down many of the power and drive systems in an effort to prevent or minimize damage and injury. The E-stop is one of the ways we can tell the robot to stop, cease and desist, freeze, quit, do no more, or however you prefer to think of it. The stop button tells the robot to stop, but generally lets it finish the current motion or slow to a stop. The E-stop, by comparison, is like yelling "Freeze!" Industrial robots

that have teach pendants will have an E-stop, and OSHA regulations require workers to take the teach pendant into the danger zone so they have access to an E-stop at all times.

While this is by no means a complete listing of the safety devices available, this discussion should give you a good idea of what it takes to keep us safe. We will dig deeper into the operation of many of these devices in Chapter 6 as well as describe how other sensor systems factor into safety. In the meantime, I encourage you to do your own research on the subject of safety devices and see where it leads.

ELECTRICITY AND YOU

No matter which power source a robot uses, you must pay it proper respect. Hydraulic power can generate bone-crushing force as well as release hot fluid at high pressure. Pneumatic power can generate enough noise to damage the ears, blow chips and dust into your eyes, or whip around busted hoses with great ferocity. Electricity can paralyze muscles, stop your breathing or heart, and literally burn you inside and out. We will explore the dangers of hydraulic and pneumatic power later in this book, but you should take special note of the dangers of electricity. There is a high probability that any robot experiments you perform in conjunction with this course will involve electrical systems, and it is my sincere hope that you will pay attention to this section and work carefully around all electrical systems.

Electricity is something we live with every day and is a crucial part of the modern world. We use it to heat and cool our homes, bring light to the darkness, cook and preserve food, and power many of our modern entertainments. Whenever the power goes out, we are reminded how truly reliant we are on electricity for many facets of our daily lives. Most robots are just as dependent on electricity as we are, given that it powers the control systems, drives, sensors, and peripheral systems of the robot. If you plan to work with an industrial robot, you are almost guaranteed to be working with and around electricity and all of the dangers that it presents.

Before diving into the safety side of electricity, we need to define a few terms:

- **Voltage**: This force drives electricity through the system: The greater the voltage, the greater the driving force. We measure it in volts (V).

- **Amperage**: Amperage is a measurement of how many electrons, or how much electricity, is flowing through a system. We measure it in amps (A).

- **Resistance**: Resistance is a measurement of how much force is working against, or resisting, the flow of electrons. We measure it in ohms, represented by the Greek letter Ω (omega).

- **Electricity**: Electricity is a flow of electrons from a point with more electrons to a point with fewer electrons. We control this flow and harness its power to do meaningful work or tasks.

In typical electrical systems, we control the flow of electrons using insulated wire and proper connections to route this force through specific systems and perform the desired work. Under normal circumstances, electricity is a safe and reliable power source, as our everyday lives have shown us. However, when we become careless or something goes wrong, we can become a part of the **circuit** (the path through which the electrons flow), which is where the danger lies. In cases where a person gets an electric **shock** or becomes a part of the circuit, the electricity enters the body at the point of contact with the electrical system, electrons pass through the body, and then the electricity exits at a **grounded point** (a point somehow connected to the earth). Three main factors determine the severity of an electrical shock:

- The amount of current that passes through the body

- The path the electricity takes through the body

- The duration of the shock

The amount of current that passes through the body determines how much damage the affected area suffers. Current is the number of electrons passing through the system, so the amount of current will depend on the voltage that is providing the driving force and the resistance of the material it is passing through. The formula that represents this relationship is known as **Ohm's law**, named after the German physicist Georg Simon Ohm (1787–1854). Ohm's law states that the current through a conductor between two points is directly proportional to the potential difference across the two points or $I = V/R$; in words, this means current (I) is equal to voltage (V) divided by resistance (R). We can also manipulate this formula to find voltage or resistance using algebraic rules, giving us $V = I \times R$ or $R = V/I$.

Example 1

How much current could flow in a system where there is 100 ohms of resistance and we power it with 120 volts?

$$I = V/R$$
$$I = 120/100 = 1.2 \text{ A}$$

Example 2

How many volts would it take to push 3 A through 100 ohms of resistance?

$$E = I \times R$$
$$E = 3 \times 100 = 300 \text{ V}$$

Example 3

What is the resistance of a system that allows 4 A to pass through when powered by 120 volts?

$$R = V/I$$
$$R = 120/4 = 30 \text{ } \Omega$$

When a person is shocked, the more current that passes through the person's body, the worse the injury will be. If you look at Table 1-1, you can see that 0.003 to 0.010 A of current can be a painful shock, on par with shocking yourself on a doorknob after walking across a carpeted room. However, if the shock increases to 0.100 to 0.200 A, there is a high possibility of **ventricular fibrillation**, a condition in which the heart quivers instead of actually pumping blood. In other words, just a portion of an amp can stop your heart from working! A current of 2 to 4 A can stop your heart altogether, damage your organs, and possibly cause irreversible damage to your body. Many industrial robotic systems use voltages from 220 V to 480 V

TABLE 1-1 Electrical Effects on the Human Body

Current in Milliamperes	Effect on the Body
1 to 3	Ranges from unnoticed to mild sensation
3 to 10	Painful shock
10 to 30	Muscle contractions and breathing difficulty begins, with loss of muscle control possible
30 to 100	Severe shock with high possibility of respiratory paralysis
100 to 200	Ventricular fibrillation highly possible
200 to 300	Severe burns and breathing stops
2000 to 4000 (2 A to 4 A)	Heart stops beating, internal organ damage occurs, irreversible bodily damage possible

Note: 1 Milliamp is equal to .001 A.

and have amperages ranging from 30 A to 100 A. Most 110-V plugs, like the ones found in classrooms and homes, have a 15-A or 20-A breaker that protects the system. Thus, any of these systems have more than enough current available to do severe and irreparable damage to you, should you become part of the circuit.

The second main factor in the severity of a shock is *the path it takes through the body*. If current or amperage enters your body at the tip of your finger and then exits through the center of the same hand's palm, your heart and other internal organs would be safe. Burns to your finger or hand may occur, and you may lose a finger and part of your hand to the event, but you will more than likely survive. If that same shock comes in through your finger or hand, travels through your body, and then exits through your foot (a common path), then your lungs, heart, and any other organs the current passes through are in danger of electrical burns and failure, which is life-threatening.

 FOOD FOR THOUGHT

1-3

Metal is a great conductor of electricity. So any metal in contact with your skin—such as the frame of your glasses, necklaces, rings, or piercings—provides a great initial path for a shock. Having metal in direct contact with or through the skin greatly reduces the skin resistance factors noted in Table 1-2.

Imagine for a moment the horrific effects of having your metal frame glasses or facial piercing bring electricity into your body, making the initial point of entry your head. Either case creates the potential for

electricity to pass through the brain, damaging the control center for the body. While this is not a common occurrence, this path is more dangerous than that going through the entire body, as shown by Table 1-2.

If you have piercings or other metal parts in your body (e.g., an implant) that you cannot remove, you will need to take extra precautions when working around electricity. You may need to cover your piercings with some type of insulating material or make sure to keep that part of your body far away from anything that has electricity flowing through it. Failure to do so could lead to an unwanted electrical lobotomy.

TABLE 1-2 Material Resistance and Amperage Flow Chart

Material	Resistance in Ohms	Current When 120-V AC Is Applied
Dry wood, 1 inch thick	200,000 to 200,000,000	0.0006 to 0.0000006 A
Wet wood, 1 inch thick	2000 to 100,000	0.06 to 0.0012 A
1000 foot of 10 AWG copper wire	1	120 A
Dry skin	100,000 to 600,000	0.0012 to 0.0002 A
Damp skin	As low as 1000	Up to 0.12 A
Wet skin	As low as 150	Up to 0.8 A
Hand to foot (inside body)	400 to 600	0.3 to 0.2 A
Ear to ear (inside body)	100	1.2 A

The last factor in the severity of a shock is *the duration of the voltage passing through the person.* It does not take long for damage to occur, as just 0.1 A passing through the heart for one-third of a second can cause ventricular fibrillation. Nevertheless, the longer the current is passing through you, the more damage it will do. In the worst-case scenario, a person becomes part of the circuit and loses muscular control so the victim cannot free himself or herself, but there is not enough total current flowing through the fusing devices to kill the circuit. In this situation, the damage continues until a muscle spasm of the victim breaks the connection, the damage to the body destroys the conductive path, someone frees the victim, or the amperage is finally great enough that protection devices open the circuit. Unless someone is nearby or responds immediately, this scenario means almost certain death.

Table 1-2 lists some common resistance values and the current that can pass through a material when exposed to 120 V. For instance, 1000 foot of copper wire will have 120 A passing through it ($E = IR$ or 120 volts = 120 A $\times$ 1 ohm). Notice that dry skin barely allows any current to pass, but damp skin can let enough through to be dangerous. Damp skin includes lightly sweaty skin, washed but not fully dried hands, and so on. Wet skin is a huge risk; it is considered to be present whenever the skin has a noticeable layer of sweat or water on it. In Table 1-2, we can see that one of the worst-case scenarios is when current passes from ear to ear, with 120 volts pushing 1.2 A through, doing damage directly to the brain.

No matter the voltage and amperage level present, you should always treat electricity with respect. Many times, tragedy strikes when we get in a hurry or no longer consider the dangers of working with electricity. In the section on handling emergencies, we will discuss how to deal with the situation in which someone gets shocked or electrocuted. First, though, we describe how to work safely with and around electrical equipment.

LOTO

In 1982, OSHA began work on the LOTO standard; it continued to work on this standard until September 1, 1989, when OSHA issued its final rule. On January 2, 1990, this final rule took effect. LOTO, which stands for lock out/tag out, means that all equipment must have a means of shutting off and locking out the various power sources of the machinery. OSHA's LOTO standard covers the 3 million-plus workers who service equipment in some fashion and are at great risk from electrical shocks. In accidents involving unintentional power release, the events that do not kill workers outright average 24 days of recovery before the person returns to work. Not all of those who recover get back full functionality of the body parts affected. The LOTO standard is estimated to prevent approximately 120 fatalities and 50,000 injuries each year when properly implemented, making it a crucial part of any safety and health program.

LOTO is a methodical process of removing all power from equipment so there is no active or latent power, also known as a zero-energy state, making it safe for people to work on and around the equipment. To get to a zero-energy state, it is necessary to turn off all electrical, hydraulic, and pneumatic power coming to the machine; bleed off any residual pressure; discharge any capacitors; and either lower portions of the equipment that could fall or block them in place (Figure 1-15). As part of the LOTO process, specialized devices called lockouts are used to hold the power source in the blocked or de-energized state. When the

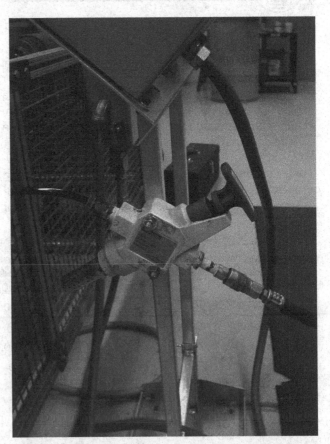

FIGURE 1-15 This disconnect for an air system would require attention during the LOTO process.

FIGURE 1-16 A lockout hasp on the power for a robot, complete with a lock to keep someone safe. Don't forget to put your name on your lock so everyone knows you are working on the equipment.

LOTO standard is enforced, no matter how you turn the power source off, there is a lockout designed to keep it off and each lockout has a place for a lock or hasp that can hold multiple locks. Each person who works on the equipment must put his or her own uniquely keyed lock on each lockout for the machine (Figure 1-16). This is the only way to ensure that someone else does not turn on the power while you are working on the equipment. Only two people can legally unlock that lock—the person who applied the lock and his or her supervisor. Before a supervisor can remove a lock, the supervisor must talk with that person to verify that he or she is not in the danger zone.

Once you believe that you have all the power locked out, bled off, and accounted for, you should try to turn the system on. This step helps to ensure that the machine is in a zero-energy state before you are exposed to danger and verifies that you have done everything right. Make sure you check any gauges on the system for fluid power pressure and use a multimeter to verify that no electricity is present. Once you have performed all these checks and proved there is no power left in the system (neither kinetic nor potential), you can safely begin work.

As noted earlier, each person working on the robot must have his or her own lock on the lockout devices, to ensure everyone's safety. Any necessary tags should be in place as well. Common warnings on such tags include "DO NOT RUN" or "Maintenance working on machine." Make sure your lock has your name on it, so everyone knows you are working on the machine.

A lock with no name makes it very difficult to verify the owner's safety.

The following checklist provides a step-by-step procedure for the LOTO process:

1. Notify affected individuals you are about to shut the machine down. This includes operators, employees nearby, and management.

2. Stop the machine cycle, if necessary.

3. Turn off or remove all external power supplies and lock them in the off position using lockout devices and a lock with your name on it.

4. Place appropriate information tags on the equipment, such as "Do Not Run, Under Maintenance."

5. Verify a zero-energy state. Make sure to account for capacitors, compressed springs, items that could fall, stored fluid pressures, or other potential energy sources.

6. Perform the necessary repairs.

7. Once the repairs are finished, remove all tools and any blocking devices or other items you added to the machine for safety reasons.

8. Once everything is clear, each person working on the robot should remove his or her own lock. The last person can return power to the equipment.

Failure to follow this procedure can put not only your life but also the lives of your coworkers in danger. History is full of tales of injury or death from improper lockout, so this issue is not something to take lightly. In fact, you can perform all the steps of LOTO and still end up in a life-or-death situation. The nearby Food for Thought feature recounts a near miss that happened to my best friend. Always be aware of your surroundings when working on equipment, especially any signs that someone is powering up the system. LOTO protects you from harm, but it does not mean you can check your brain and situational awareness at the door.

HANDLING EMERGENCIES

Unfortunately, the modern world we live in tends to create emergencies. An **emergency** is a set of circumstances or a situation that requires immediate action and often involves the potential for or events that have caused injury to people and/or severe damage to

FOOD FOR THOUGHT

1-4 LOTO Awareness

This LOTO incident happened to my best friend James E. Stone, who passed away in 2012 after losing his battle with cancer. It happened during his maintenance technician years, while he was working at an injection molding company. In the injection molding process, the machine presses two halves of a mold together, injects either molten plastic or metal into the mold, and then rapidly cools the part before the two halves separate, exposing the newly created part to the world. The machines used in this process range from bulky to massive; the one in this story was the massive type for metal parts. It was so large, in fact, that my friend Jim could easily fit between the two halves when the machine was open.

On the day in question, Jim had some maintenance to do on the molds, so he locked out the piece of equipment, verified the zero-energy state, and crawled into the mold area with tools in hand to get to work. He had been working on the machine for a while when an operator showed up and decided to start the machine. After hitting the start button and getting no response, the operator noticed the main electrical disconnect was turned off and someone has stuck an aluminum tag thing (Jim's lockout) through a hole and put a lock on it. The operator studied the device long enough to determine that it would have to be pried off, but apparently did not get the meaning of the tag "Maintenance Working on Machine." Most lockout devices are made of thick plastic or thin aluminum, so they are not indestructible; that is, they are more of a hassle to remove than impossible.

After a bit of work, the operator bent and warped the lockout device in such a way that the operator could remove it and throw the arm on the side of the disconnect, returning power to the injection-molding machine. This is where the real problem began for Jim. During the several minutes the operator was making his bone-headed changes, my friend had been working in the molds and had no idea what was going on. Jim's first clue was when he heard the hydraulics for the machine

cycling up. Knowing the piece of equipment all too well, Jim recognized that part of the power-up process for this machine was to close the two halves of the mold and then open them back up, ensuring everything was working correctly. The halves of the mold for this machine closed with tons of force, and Jim was right in the danger zone.

It was only a combination of fast reflexes, panic, and odd luck that saved Jim's life that day. As he scrambled to get clear of the machine, he happened to reach up and grab a large bar that ran from one side of the mold to the other. The bar he grabbed was the main power bus for the two halves of the mold and was energized with 480 V of three-phase power. His grab was a panicked reflex, the fact it was the bus bar was the odd luck. As soon as Jim grabbed the bus bar, the 480-V three-phase power electrocuted him and caused such a violent muscle contraction that it threw him clear of the machine just as the halves closed. In fact, the timing was so close that part of his pant leg was caught between the two halves of the mold. Jim was literally saved by an event that most maintenance technicians work their entire careers to avoid!

So there Jim was, lying on the ground, feeling the effects of a severe electrical shock, bleeding from the ears and nose, and riding out the fight-or-flight adrenaline reaction. Once he realized he was alive and functional, he had a "discussion" with the operator of a rather physical nature. Jim then informed the management of the facility that if they were going to hire that caliber of employee, he was going to quit. His supervisor did try to talk him into staying, but Jim had had enough of that place.

This event just goes to show that you can do everything right and still find yourself in a life-or-death situation. Learn from Jim's experience and keep your wits about you at all times when working on or repairing equipment! This advice has saved me from injury more than once over my years of repairing equipment and robots, making the difference between near misses and life-altering accidents.

---

property. Most readers of this book have experienced some kind of emergency in the past and can relate to how intense these situations can be. Usually we have to react as fast as possible to these situations; we have little time to sit back and think about our actions. In this section, we look at some general rules for dealing with emergencies and then drill down into some specifics for various situations.

General Rules for Emergencies

Rule 1: Remain Calm

If you allow your emotions—especially fear—to get the better of you, your odds of dealing effectively with any situation are dramatically reduced. Fear clouds judgment and blocks logical thought patterns. Fear can prevent a person from thinking, speaking, or acting.

Fear is also contagious. If someone is injured and that person senses your fear, it will only add to the victim's own negative reactions. For instance, which of the following statements do you think would be more beneficial: "I have training in first aid and I am here to help you" or "You're bleeding a lot! What should I do?!" The second statement will just make the situation worse and focus the victim's mind on the negatives of the situation.

Rule 2: Assess the Situation

Just because someone has been hurt and needs help, it does *not* mean that the situation is safe for you to help the person. Many times, rescuers must first take care of the dangers linked to a situation before they can worry about helping the injured victims. Blindly rushing in is a good way to become another victim or, even worse, lose your life because of another person's mistake. Unfortunately, history is full of cases in which people were injured or killed while trying to help others. Assessing the situation is another part of being calm; you have to think before you act so that you will act wisely and not impulsively, irrationally, or emotionally. What if you are the only person there who could call for medical help but instead you become the second victim? Who will call for help then?

Rule 3: Perform to the Level of Your Training

When responding to an emergency, you need to make things better or, at the very least, no worse. If you are unsure what to do, contact someone who has a higher level of training and can advise you. A wealth of resources are available by calling 911, contacting the Centers for Disease Control and Prevention's (CDC) poison hotline, or reaching first responders, doctors, and many more by other means. Someone in your facility or nearby may also be able to help you deal with the situation. *Do not* try something you saw in a movie once or heard about randomly on the Internet! Often this kind of uninformed action will only make the situation worse, may put your life in danger, and could open you up to legal liability.

Rule 4: After It Is Over, Talk It Out

Emergencies are high-stress, high-emotion situations. Even if no one is hurt, there is a good chance the emergency got your adrenaline flowing. It may have scared or upset you, your heart may be pounding, and you may find that it has troubled you on a deeper level. These are common reactions and nothing to be worried about, but you should talk with someone about what happened. The who, what, where, and why of the emergency will determine the best person(s) to reach out to, but you should take the time to talk about your reaction with whomever is appropriate. For mild events, friends and family are good groups to use as sounding boards. For severe events, you may want to talk with other people who were involved or a counselor. For accidents in the workplace, protocol often details whom you follow up with to ensure that victim information remains private while providing you with support.

Specifics of Emergency Responses
Bleeding

There are a multitude of ways to cut, damage, or break the skin, all of which can result in bleeding. Because of this high risk, we shall start here with our deeper look at emergencies.

For minor cuts and abrasions, the procedure for dealing with bleeding is simple: Clean the wound, apply some kind of antiseptic ointment, and cover the wound with a nonstick bandage. For serious bleeding, the first step is to stop blood loss. To do so, you should take a clean bandage, cloth, or gauze and apply firm pressure directly to the wound. This will likely hurt, but almost all bleeding can be stopped in 5 to 15 minutes with this method. If the material you are using on the wound becomes soaked with blood, you will need to apply more material on top of it *without* removing the soaked material. Removing the original covering has the potential to rip open any clotting that has occurred and allow the wound to flow freely again. When possible, elevate the injured portion of the body above the heart, as this will help to reduce the pressure on the wound and accelerate clotting to stop the bleeding.

In cases where direct pressure does not stop the bleeding, a tourniquet may be required. A **tourniquet** is a tightened band that restricts arterial blood flow to wounds of the arms or legs in an effort to stop severe bleeding. Because of how the tourniquet works—that is, by stopping blood flow below where it is located—a tourniquet can also damage the tissue of the area it affects. In many cases, application of a tourniquet has led to severe tissue damage and ultimately amputation. Tourniquets can save lives, but they are a last resort when all else has failed and should be applied only by those persons who are trained in their proper use. If you would like to know more about tourniquets, you should take a first-aid course offered through a reputable source.

Burns

Burns are another common injury and thus worthy of a deeper look. A degree scale is used to classify burns, with first-degree burns being the most minor and fourth-degree burns being the most severe.

- A first-degree burn is easily recognizable by the reddened, dry appearance of the skin. This kind of burn is painful, but heals in about a week.

- A second-degree burn is more severe and often includes blistering of the skin along with a reddened or whitish appearance. The burn can look dry or wet, and some sensation of the area may be lost. A second-degree burn can take up to three weeks to heal with the most severe requiring medical assistance.

- A third-degree burn is a very severe burn and is often blackened or ash white in appearance. The skin may be leathery, and open wounds are often present. Such burns, which require immediate medical attention, usually hurt less at first because all the nerves are destroyed. They take months to heal.

- A fourth-degree burn extends past the skin into the muscle and bone of the victim. It has the same characteristics as a third-degree burn.

When dealing with burns not associated with open wounds (e.g., first-degree and minor second-degree burns), submerge the area in cool water for 10 to 15 minutes or until the pain subsides, and then wrap the area with a dry, nonstick, sterile bandage. Do not pop blisters should they appear, and seek medical help if the burned area is a sensitive region or if pain persists.

For burns that are accompanied by open wounds, such as severe second-degree, third-degree, or fourth-degree burns, *do not* place the wounds in water and do not try to remove any clothing that may be stuck in the burned area. Cover the burn with a cool, moist, sterile, nonstick bandage or cloth and seek immediate medical help. Infection is the number one enemy when dealing with burns, so it is paramount to let medical professionals deal with cases involving broken skin.

Blunt-Force Trauma

In the case of blunt-force trauma, referring to an impact that does not penetrate the skin, the proper action depends on the level of the injury. Severity usually depends on how large the impacting object is, how much force is behind the impact, and where it hits. For minor impact injuries, you may want to apply an ice pack and monitor for continued swelling, discomfort that does not fade, or other signs of serious injury. For major impact injuries, apply an ice pack and seek medical help, as there is an increased chance of internal injury that is not outwardly visible.

In the case of broken bones, immobilize the limb as best you can with a splint device from the first-aid kit. If a splint is not available, you can use materials such as wood, rolled-up magazines, or anything rigid placed on either side of the broken bone and a cloth wrap or similar to hold it in place. The key point is to immobilize the injured limb for transport to medical help.

In case of head trauma or internal injuries, keep the victim calm and get medical help as quickly as possible.

Electrocution

The last type of emergency we will address here is electrocution. You learned earlier what can happen when electricity passes through a person's body, but not how to deal with this situation. If the person is still being shocked, you will need to either cut the power to the circuit or use a nonconductive item like a wooden broom handle or dry rope to get the person free of the circuit. *Do not* touch the victim! If you do, you will become part of the circuit and get shocked as well. If the victim is already clear of the circuit, make sure there is no chance you will be electrocuted before rendering aid.

Once the victim is clear of the circuit and it is safe for you to do so, check for a response and a pulse. If the person responds, get that individual to a medical professional as soon as possible. If there is no response, call for help and have someone call 911. If you know cardiopulmonary resuscitation (CPR), perform the steps and respond accordingly. Otherwise, try to find someone who has CPR training, as often electrical shock will stop the heart or cause ventricular fibrillation.

Obviously, this is not the entire list of emergencies you might face in a setting with industrial robots, nor do we have the time to cover them all. Nevertheless, if you remember the general rules from this section and keep your wits about you, there is a good chance you can deal with most emergencies you encounter. If you are worried about your ability to perform first aid or want to get a deeper knowledge of any of these topics, then seek out additional training. Some employers have programs in which employees can volunteer for first-aid training, so that might be an avenue to learn more. The more training you have in this area, the better you can deal with any emergencies you encounter.

REVIEW

By now, you should have a deeper understanding of what it takes to work safely with robotics as well as how to handle any emergencies that might arise. Safety is an ever-changing environment as we adopt new devices and new standards to ensure we can work safely with robots. As you proceed along your chosen path in robotics, you will learn more about the specifics of how safety applies to your field. Here is a quick overview of the topics we covered in this chapter:

- **Robots require respect.** We discussed what happens when you no longer fear the robot as well as how to work safely with robots.
- **Danger zones.** This section outlined the various zones around a robot as well as who could be in those areas.
- **Guarding.** This section was all about how we keep people out of the danger zone.
- **Safety devices.** We discussed some of the devices used to ensure people's safety around robots.
- **Electricity and you.** This section described the effects electricity can have on the body and explained how to work safely with electricity.
- **LOTO.** We covered how to safely power down equipment so we can work on or around it.
- **Handling emergencies.** We introduced some general rules for dealing with emergencies and then drilled down into the first aid for some of the injuries you might encounter.

KEY TERMS

American National Standards Institute (ANSI)
Amperage
Blunt-force trauma
Cautionary zone
Circuit
Danger zone
Electricity
Emergency
E-stop
Expanded metal guarding

Grounded point
Guards
International Standards Organization (ISO)
Laser photo eye
Light curtain
Limit switch
Lockouts
Lock out/tag out (LOTO)
Metal mesh

Occupational Safety and Health Association (OSHA)
Ohm's law
Photo eye
Pick and place operation
Pinch point
Presence sensors
Pressure sensor
Proximity switch
Resistance

Robotic Industries Association (RIA)
Safe zone
Safety interlock
Shock
Teach pendant
Tourniquet
Ventricular fibrillation
Voltage
Work envelope
Zero-energy state

REVIEW QUESTIONS

1. What are the three Rs of robotics?

2. What are the three conditions that can stop a robot?

3. What are the safe zone, the cautionary zone, and the danger zone as they relate to a robotic system?

4. What is a common way to keep people out of a robot's danger zone?

5. Which organization made it a requirement that whenever you enter a robot's work envelope you must take the teach pendant with you? What is one of the main benefits of having the teach pendant?

6. What are some of the tasks for which proximity switches are used?

7. Which purposes do pressure sensors serve in relation to robotic safety?

8. What is the difference between a safety interlock and a presence sensor?

9. Describe what happens when a person is shocked.

10. What are the three factors that determine the severity of a shock?

11. What is the formula for Ohm's law?

12. What is ventricular fibrillation, and at which amperage passing through the body does it become a high possibility?

13. In terms of industrial robotic systems, what are the common ranges for voltage and amperage?

14. What are the steps of LOTO?

15. What are the general rules for dealing with emergencies?

16. How would you stop serious bleeding?

17. How would you treat a minor burn?

18. How would you treat a severe burn?

19. What would you do for a broken bone?

20. When a person is being electrocuted, what is the first thing you must do?

CHAPTER 2

Introduction to Industrial Robotics

WHAT YOU WILL LEARN

- What an industrial robot is
- Some of the history behind modern technologies

- Key events leading to the modern industrial robot
- The four Ds of robotics
- Where and why we use industrial robots

OVERVIEW

The realm of robotics is wide and diverse, with many interesting divisions and utilizations that one can investigate to learn their specifics. This book, however, is designed to delve deeper into the world of the industrial robot. From the days of the earliest **automata**—that is, devices that worked under their own power—many people envisioned mechanical devices capable of being humans' equal in functionality, design, and purpose, which could take over dangerous work, complete superhuman tasks, or take on menial chores so we could pursue avenues requiring our intelligence and creativity. Over the centuries, the automata of old evolved into robotic systems that could follow voice commands, write words, play instruments, simulate breathing, emulate facial expressions, and perform other tasks showcasing the abilities of the technology. While these early units were impressive, some began to call robotics "a technology in search of an application." Ultimately, industry would give the field of robotics that application, along with the focus needed to drive the technology and create the truly impressive machines we have today.

In this chapter, we will look at a broad overview of the world of industrial robotics before we drill down into the nuts and bolts of things in later chapters. To that end, we will cover the following topics:

- What is an industrial robot?
- A history of technology
- The rise of the industrial robot
- Why use a robot?

WHAT IS AN INDUSTRIAL ROBOT?

The word *robot* was first used by Karel Capek in 1921 in his play *R.U.R. (Rossum's Universal Robots)*. It comes from the Czech word *robota*, which means "drudgery or slave-like labor." In Capek's play, a factory creates robots that look like people, are capable of performing complex tasks, and can even think on their own (Figure 2-1).

In 1927, Fritz Lang released his movie *Metropolis*, which brought the term *robot* to a larger audience. In Lang's film, a robot takes the place of one of the main characters and wreaks some havoc in her name.

Since its debut, we have used the term (RIA, 2017) *robot* to describe fictional human-like machines, various toys, people who display little or no emotion, programs that search out information on the web, and a multitude of systems, ranging from ones that can complete

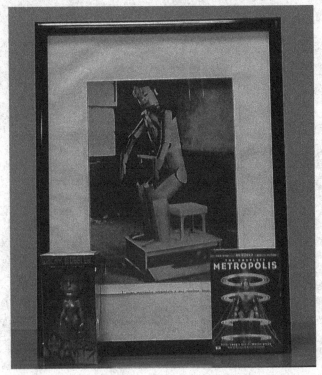

FIGURE 2-1 One of the actors suited up for the *R.U.R.* play, along with a poster for Lang's film *Metropolis*. The film was rereleased a few years ago as an anniversary edition.

tasks based off programming to systems controlled directly by people. Because of this diversity of usage of the term *robot*, we need to start our exploration of the industrial robot with a definition agreed on by industry.

The **Robotic Industries Association (RIA)** is an organization with the goal of driving innovation, growth, and safety in manufacturing and service industries through education, promotion, and advancement of robotics, related automation technologies, and companies delivering integrated solutions (RIA, 2017). It has been the champion of all things robotic in the United States since its inception and is one of the co-creators of the ANSI/RIA 15.06-2012 safety standard, which is considered the guiding principle for robotic safety in industry. RIA (2017) defines an **industrial robot** as an automatically controlled, reprogrammable, multipurpose manipulator programmable in three or more axes, which may be either fixed in place or mobile for use in industrial automation applications (Figure 2-2). Notably, under the new safety standard, the tooling that the robot uses to complete its tasks is *not* considered part of the robot. Given that this tooling is interchangeable and often purchased from companies other than the supplier of the robotic system, it makes sense to consider it separately from the robot.

Other definitions for the term *robot* are also used, and you will likely encounter several of them along your journey of robotic learning. In this book, we will stick to the RIA's definition of a robot. While the definition

FIGURE 2-2 This is a perfect illustration of what RIA considers an industrial robot. In the upper-right corner, you can see the red Lincoln welder unit on the robot that supplies power and wire to the welding gun that the robot is using.

we are using is fairly concise, the concepts and information contained in this book are applicable to all realms of robotics in varying degrees and will give you a good knowledge base for your future learning endeavors. The parts of the industrial robot, the way we interact with the robot, the way we program it, and the various types of motions and mechanical configurations all share commonalities with nonindustrial robots.

Because the RIA definition is concise, it rules out many of the robotic systems you may have heard of or possibly had some experience with. None of the systems from *Robot Wars* or *Robot Combat League* would be considered an industrial robot, as they are controlled directly by humans. The same goes for the da Vinci medical robot, first introduced in 1999 by Intuitive Surgical Inc., as it is controlled by doctors. You can purchase robots from iRobot and several other sources to clean your floors, gutters, pool, and mow your lawn, but none of these systems is designed for reprogramming or multipurpose usage, so they also do not qualify as industrial robots. WowWee makes multiple robotic systems, ranging from the Robosapien line to various stand-alone robotic toys that can be reprogrammed and do have multiple axes, but they fall short on the multipurpose manipulator portion of the RIA definition.

In short, the industrial robot is a highly automated piece of equipment designed to do the things that humans either cannot, will not, or should not. In addition, the industrial robot has the capacity to change its primary function with little more than a change in

tooling and updated programming, making it highly versatile and thus a huge benefit in many production applications.

A HISTORY OF TECHNOLOGY

When we think of technology such as the industrial robot, we often believe that its origins trace back only to the late twentieth century. In truth, the modern robot is the result of millennia of human advancements in science and math, which we have built on for generation after generation. In 420 BCE, Archytas of Tarentum inspired his peers with a wooden pigeon that could fly using steam or compressed air. From 285 to 222 BCE, Ctesibius of Alexandria worked out many of the basic principles of pneumatics, earning him the title of "Father of Pneumatics"; this technology continues to be heavily utilized by robotics today. In 1206 CE, Al-Jazari published a book on mechanical and automated devices, allowing others to learn about gears, timing, sequential function, and other such concepts.

In 1495, Leonardo da Vinci created a suit of armor that could move its arms, head, and visor, all under clockwork-type power and automation. In 1525, Hans Bullman created humanoid automata so lifelike that many credit them as being the first androids—that is, a synthetic organism designed to imitate a human. Several others would follow in Bullman's footsteps and create impressive clockwork systems that emulated nature.

In 1620, William Oughtred created the first slide rule, a 2-foot-long ruler that includes a logarithmic scale and makes complex math much easier to perform. This technology would survive well into the 1970s, when the scientific calculator finally made this technology obsolete (Figure 2-3). Wilhelm Schickard invented the first four-function calculator, known as the Rechenuhr, in 1623. His work came centuries before the invention of the microchip and modern computers that utilize the binary system of arithmetic developed by Gottfried Wilhelm von Leibniz in 1679.

One could argue that without the factory, there is no industrial robot. Like the robot, the factory is not really a modern concept. In fact, the first true factory was set up in England, next to the Derwent River in Cromford, Derbyshire, by Richard Arkwright, Jedediah Strutt, and Samuel Need in 1771, more than two centuries ago. The Industrial Revolution first began in Britain and is considered by many historians to encompass the period from 1760 to 1840. When most of us think about the Industrial Revolution, we are actually thinking about the second Industrial Revolution,

FIGURE 2-3 This slide rule is positioned next to the device that made it obsolete—the scientific calculator.

which happened in the nineteenth and twentieth centuries and brought us assembly lines, large factories, **automation** (machines that work largely on their own, performing tasks in industry), and all the wonderful products we enjoy today.

But surely programmable control is a modern concept, one might protest. In 1772, Pierre Jaquet-Droz built the *Writer*, an automata that resembled a child, could write with spacing and punctuation, and was controlled by a programmable mechanical computing device of impressive complexity. In 1804, Joseph-Marie Jacquard invented the Jacquard loom, which utilized punch cards, was designed to fit on looms for weaving fabric, and created intricate patterns that could only be done by hand before its introduction. **Punch cards** are rigid cards that have holes in a specific pattern that can control the functionality of equipment; this technology would also survive until the late 1970s. In 1810, Friedrich Kaufmann created a mechanical trumpet player controlled by a **stepped drum**—that is, a cylinder with high and low spots of varying lengths around the outside to trigger switches during the drum's rotation. Some of the early industrial robotic systems would use this same technology to time their movements. The operator could change the equipment's functionality by adding, removing, or changing the location of pegs inserted into the drum. The longer the drum, the more sequences it could control.

Computers and programs also have an impressive history, though we tend to think of them as emerging only in the mid-twentieth century. While the digital computer is a modern invention, development of the first computer is credited to Charles Babbage in 1835. He designed an analytical engine complete with processor and memory function, though he did not actually build the machine. In 1843, Augusta Ada King published her notes on Babbage's analytical engine, primarily on how

to control it, and went down in history for writing the first computer program. A few years later in 1847, Gorge Boole published *Mathematical Analysis of Logic*; he would later expand his theory into Boolean algebra, a mathematical system used widely in the computer field. Lord Kelvin, also known as William Thomson, developed an analog computer that could predict tides in 1873.

In 1889, Herman Hollerith patented his punch-card-driven tabulation machine for tabulating data collected in the 1890 census. Seven years later, Hollerith started the Tabulating Machine Company, which would ultimately become International Business Machines Corporation (IBM). IBM continues to thrive today and is still heavily involved with computers and programming.

The first electronic computer was the brainchild of Thomas H. Flowers, who, with his colleagues, created *Colossus* in 1943. The early electronic computers were controlled by relay logic and took up rooms' worth of space. Reprogramming these behemoths required several days and a large amount of labor to physically change the machines' wiring. (See Food for Thought 2-1 for more on relay logic.) The primary function of the early electronic computers was to complete complex calculations for the National Aeronautics and Space Administration (NASA) or other scientific endeavors.

Staying on the topic of computers and related technology, 1958 was a big year for industry on the computer side of things, with two key events occurring that proved crucial for the development of the modern robot. The first event was the creation of the microchip by Jacky Kirby. Microchips are a necessity of the digital age, and we use the descendants of Kirby's microchips in modern robotics, electronics, and many other aspects of modern life. These tiny chips allow for Boolean algebra logic flow, power manipulation and control, data storage, and many other complicated tasks necessary for modern computation and control of devices. The microchip is the invention that allowed computers to become small portable units instead of remaining room-size behemoths.

The second seminal event that occurred in 1958 was the FANUC Corporation's shipping of its first **numerically controlled (NC)** machine. NC machines used punch cards (like the Jacquard loom) or magnetic tapes (similar to VHS or cassette tapes), with position and sequence information being used to control the motions and actions of the machine. Eventually **computer numerical control (CNC)** would supersede numerical control as computers using microchips became strong enough and small enough to use in industry, thereby replacing the punch cards and magnetic tapes of old with programs and computer code.

FOOD FOR THOUGHT

2-1

Relay logic is a control system that utilizes relays to create logic gates for function control. By activating certain relays at certain times, the system creates logic gates for sorting information. The common gates are AND, OR, NOR, NAND, and XOR.

- **AND**: This logic function requires two or more separate events or data states to occur before the output of the function occurs.
- **OR**: If at least one of two or more events happens, then the output of the function occurs. If more than one condition is true, the output of this logic command engages as well.
- **NOR**: With this logic filter, all input conditions must be false before the output occurs.
- **XOR or Exclusive OR**: This works like the OR command, with the exception that only one of the conditions can be true. If more than one condition is true, the output does not occur.
- **Not or NAND**: This is the opposite of the AND command, in that all the input conditions must be false before the output is triggered.

A **relay** is a mechanical or solid-state device that, when control power is applied, opens or closes internal connections. **Mechanical relays** use a coil of copper wire to generate a small magnetic field that moves parts inside the relay to open and close contacts. Usually this type of relay will have an even number of contacts, half of which conduct power between two points without power to the control coil, and the other half of which conduct power only when the coil is energized. The normal position for a relay is considered the state of the contacts when no power is flowing through the control coil. **NC**, which stands for "normally closed," indicates that there is connection between two points when there is no power to the coil. **NO**, which stands for "normally open," indicates that there is no connection between the two points when there is no power to the coil. Each set of NO and NC contacts has a common terminal point, which is where we connect whatever it is we are switching. With no power to the coil, whatever we connect to the common connection flows through the relay and out the NC contact terminal tied

to that specific common. When we energize the coil, the relay mechanically breaks the NC connections and connects the common connection to the NO contact associated with it, applying the power tied to the common to the terminal associated with the NO contact.

Solid state is a term describing those components that control power without moving parts. Solid-state relays work the same way as mechanical relays, except that no physical movement occurs inside the relay to make and break connections. Because nothing moves within solid-state components, they are often rated for millions of operations before they are expected to wear out. The down side is that, even in an open state in which nothing should be flowing out of a connection, there will be a trickle of power within the solid-state relay. Often this flow is only a volt or two and not enough to be a problem, but it is enough to measure with a meter and may sometimes be mistaken for a signal when used for an input by something else. Solid-state components are also vulnerable to vibration and voltage spikes, which can damage the internal structure and, therefore, the component's ability to control electricity.

Luckily for us, relay logic is pretty much a thing of the past: It is *not* the way we control modern industrial robots. For the most part, relay logic has been replaced with some form of controller that is much easier and faster to reprogram. With relay logic, the only way to change the machine's function is to physically change the machine's wiring, which involves many hours of human labor and creates the opportunity for incorrect wiring that could damage or destroy not only the relays but also the equipment they control. On top of the time it takes to change the wiring, the relays generate a large amount of heat that may cause problems with the systems around them. Lastly, these control systems are expensive from a component standpoint. Each relay has a cost and will wear out at some point, even if nothing is wired up incorrectly. There is cost in the wire to connect everything. There also has to be some place to mount all these relays, which means larger (and more expensive) control boxes or enclosures.

In the chapter on programming, we will explore how we control today's industrial robots and give you an idea of what you will find both in the field and in your classroom.

Electricity, which represents the main power source for the modern world, has to be a twentieth-century accomplishment, because otherwise we would have made even greater advancements by now, right? This one is a bit of yes-and-no. What appears to be chemical batteries and similar devices have been found in the

pyramids in Egypt and other archeological sites, but our scientific exploration of electricity really kicked off only in the middle of the eighteenth century. In 1752, Benjamin Franklin carried out his famous kite experiment (though he was not really holding the string) and sparked interest in what electricity could do.

In 1746, Pieter van Musschenbroek, a Dutch physicist, accidentally discovered the **Leyden jar**, a device for storing static charge and the predecessor of the modern capacitor. This jar would be used by Luigi Galvani in 1786 to cause muscle contractions in the leg of a frog as part of his effort to understand how electricity affected animals. In 1800, Alessandro Volta created the **voltaic pile**, a battery consisting of alternating disks of zinc and silver or copper and pewter separated by paper or cloth soaked in either salt water or sodium hydroxide (*Britannica*, 2017). Volta's invention was the first simple and reliable source of electrical current that did not require recharging like the Leyden jar, and it became a crucial piece of equipment for early explorations of electricity.

In 1827, Georg Simon Ohm, a German physicist, gave the world Ohm's law, a mathematical representation of how voltage, current, and resistance all interact. This understanding was a crucial step in conquering electricity, as we now had a way to mathematically predict how electricity would flow. In 1831, Michael Faraday discovered electromagnetic induction, the principle behind power generation as we know it today and the first consistent way to generate power via a means other than chemical interactions or static charges. Sir William Grove created the first fuel cell in 1839, a technology that only now is being explored with any depth. In 1841, James Prescott Joule stated Joule's law, which we commonly know as the power formula today; it represented another big piece of the mathematical puzzle needed to control electricity predictably.

Of course, no discussion about the history of electricity is complete without mentioning Nikola Tesla, the father of AC power and other driving forces of the modern world. Some of Tesla's most famous inventions include AC power, the transformer, the AC induction motor (patented as the rotary transformer), the Tesla coil, and radio-controlled boats. When Tesla built the world's first AC power plant at Niagara Falls in 1895, his work started us down the road to where we are today. He faced fierce competition from his former employer Thomas Edison, who was a strong proponent of DC power. Many historical accounts describe their public and harsh battles for which type of power would run the United States. Luckily for us, Tesla won that battle, and AC power is the go-to power source of the modern world. When AC voltage is stepped into the thousands of volts range, electricity can be transmitted over miles of wire with only a minimal loss of power, allowing power plants to transmit electricity over long distances.

By the 1930s, most urban areas in the United States had electrical power available, with many homes wired up to take advantage of that electricity. Unfortunately, it would be several more years before rural Americans would enjoy the same access to the electrical grid. In 1944, approximately 55% of U.S. farms lacked electricity, but that percentage dropped quickly, to 5% by 1956. Nevertheless, some places in the world today do not have access to commercial electricity, forcing them to either generate their own power or do without. For the most part, these are either remote locations or areas in countries that have infrastructure problems of some kind. From this standpoint, one could argue that electricity is a twentieth-century technology, though the groundwork for its current role was laid in eighteenth and nineteenth centuries.

This whirlwind tour has provided just a quick glance at some of the key events leading up to today's technology utilized by robots and industry the world over. In later chapters, we will dig deeper into these areas as they relate to industrial robotics and expand your understanding of key areas. You might want to do some digging of your own if you would like to learn more about historical figures and their discoveries that shaped our world.

THE RISE OF THE INDUSTRIAL ROBOT

While the path of robotics in general is a bit dicey to plot due to differences of opinion and definition, the evolution of the industrial robot is easier to track due to our concise definition. The first advance of interest for the industrial robot was a crane with a gripper used primarily to move metal billets; it was designed by Seward Babbitt and Henry Aiken in 1892. This was the first real **Gantry**-type system, which many early robotic materials-handling systems would mimic years later. (Gantry machines are typically simple, two- or three-axis machines designed to pick up parts from one area and place them in another area.)

In 1941, the DeVilbiss Company built what many consider the first industrial robot, based on the patented parallel robot design of Willard L. G. Pollard, Jr., bringing robots into the world of industry. Harold Roselund directed the creation of this robot for spray painting applications to ensure even, consistent coats of paint while minimizing waste. Ultimately, this application for robots found its way into the automotive industry and began what has proved to be a beneficial partnership. The automotive industry is currently one of the top users of robotic technology, with some companies, such as General Motors (GM) and Honda, actually helping to design advanced robotic systems.

In 1957, Planet Corporation exhibited its five-axis, hydraulically powered, polar-coordinate, robotic arm called *PLANETBOT* at an international trade fair, giving many industry members their first look at a truly useful robot. In 1960, American Machine and Foundry shipped its first VERSATRAN programmable robotic arm to an American customer; in 1967, the same company shipped its first robot to Japan. In 1961, George Devol patented his industrial robot arm, which became one of Unimate's robotic models used by GM. In 1968, Kawasaki Robotics started producing hydraulically powered robots for Unimation, beginning that company's rise in the robotics field.

In 1969, Victor Scheinman designed the Stanford Arm, an all-electric, computer-controlled robotic arm. He used this arm to assemble parts during various experiments, proving that robots were capable of performing many tasks that his contemporaries thought impossible. In 1974, Scheinman created the VICARM Company to promote and sell his robotic arm design. In 1977, he sold VICARM to Unimation, which ultimately led to the development of the Programmable Universal Machine for Assembly (PUMA) robot. The PUMA robot was the primary reason the Staubli Group bought Unimation from Westinghouse in 1989, using it as the foundation of a new Staubli robotics division.

Richard Hohn designed the T3 robot for Cincinnati Milacron; this robot, which was released in 1973, is recognized as the first commercially available robot controlled by a microcomputer. The T3 utilized a microprocessor, or computer on a chip, that the Intel Corporation developed in 1971. The microprocessor reduced the size of the robot controller, increased the system's overall computational ability, and enabled the rise of modern robotics. Also in 1973, KUKA, a German company, joined the robotics industry with the IR 600, which went into Europe's first welding transfer line in a Daimler-Benz plant. In 1976, KUKA built FAMULUS, its first six-axis industrial robot. KUKA survived the turbulent 1980s to take a place in today's world robotics market. Cincinnati Milacron survives today as an industrial machinery company, but sold its robotics line to ABB in 1990.

In 1974, ASEA (now known as ABB, short for ASEA Brown Boveri—a firm created by the merger of ASEA with BBC, formerly known as Brown Boveri, in 1988) delivered the first fully electric, microprocessor-controlled robot to the Magnussons Company in Sweden. Forty years later, this robot was still in service handling and polishing parts for Magnussons! There

is a high probability that this robot is still producing parts today, demonstrating how, with proper maintenance, industrial robots can stand the test of time (Figure 2-4). By 1977, ASEA's line of robotics had grown to include two sizes of electrically powered robots. ASEA built its first welding robot in 1979, and in 1982 it began moving into Japanese markets. Today ABB is one of the largest producers of robots worldwide, with its newest models utilizing the latest technology.

Image courtesy of ABB Inc.

FIGURE 2-4 While these robots are not from the Magnussons Company, they are similar to that company's robot and are a common sight in many industrial facilities.

In 1974, FANUC Ltd. of Japan began to develop and install robots in its factories in Japan (Figure 2-5). The company established the FANUC Corporation in the United States in 1977, and in 1982 entered into a joint venture with General Motors known as GMFanuc. This led to the development and installation of robots in GM plants throughout the United States. In 1992, General Motors sold its share of the robotics

Photo Courtesy of FANUC America Corp. www.www.fanucamerica.com

FIGURE 2-5 FANUC's larger robots can handle a car body with ease.

company due to finical issues, and GMFanuc became FANUC Robotics Corporation, a wholly owned subsidiary of FANUC Ltd. FANUC Robotics is unique in that its line of automotive paint robots are manufactured at its U.S. headquarters in Rochester Hills, Michigan—most other robotics companies manufacture their robots overseas, primarily in Europe and Japan.

In 1976, Motoman started a robotic welding company in Europe that would survive the harsh competition of the 1980s and emerge as one of the key players in modern robotic systems. In 1978, Hiroshi Makino developed the Selective Compliant Articulated Robot Arm (SCARA) in Japan (Figure 2-6). This robot holds the honor of being the first Japanese industrial robot as well as another design that proved sound for use in industry. Today companies like KUKA and Motoman (now called Yaskawa Motoman) sell this type of robot for use in the precise handling of small components for assembly, fastening, and soldering as well as for tasks requiring a large amount of downward force. The SCARA robot has become a favorite among various electronics companies for the rapid placement of components on circuit boards.

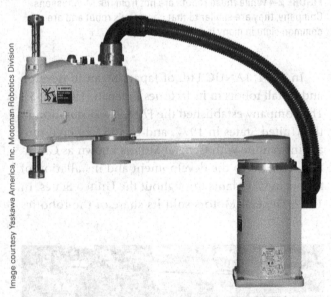

Image courtesy Yaskawa America, Inc. Motoman Robotics Division

FIGURE 2-6 This photo depicts one of Motoman's SCARA-type robots. The popularity of this design led several robotics companies to invest in and create this type of robot.

In the early 1970s, labor was fairly cheap, so that the total cost of goods remained at acceptable levels for many companies. During the same time frame, robots were costly due to their small production runs, the need to work out the kinks in the technology, limited sales, and a high hourly cost to run the robots. Most industries in the United States that were approached during the early 1970s about buying a robot countered these appeals with an economic argument: "Why would I buy a robot for *x* amount when I can hire workers for less?" Starting in 1975 and extending into the early 1980s, however, several changes in the American manufacturing environment would change this balance. A major factor was increasing wages and benefits for U.S. workers, especially in heavily unionized industries such as the automotive and steel industries. Roger Smith, chairman for the board of GM in the 1970s, explained: "Every time the cost of labor goes up a dollar an hour, a thousand more robots become economical" (Rehg, 2003). Another big cost increase came with the several years of double-digit inflation during the 1970s, which raised the costs of everything from labor to raw materials to shipping of products. This increase in their costs required companies to either hire fewer workers, raise the prices of the products they sold, find some way to cut costs, take a cut in profits, or use bits and pieces of several approaches in an effort to stay in business.

By this time, many of the bugs in the early robotic systems had been worked out, making the technology easier to work with and more stable. As demand for their devices started to increase, the robotics companies of the time ramped up production and produced more systems for sale, which helped to drive down their per-unit costs without sacrificing their profits. Recognizing that robots could save on raw materials in many applications, had a history of improving product quality, and do not take sick days or vacations, many manufacturers changed their tune from "Why would I use a robot?" to "Which one would you recommend?" during the early 1980s.

This sudden increase in demand created a boom of new robotics manufacturers in the 1980s, as start-ups and small companies scrabbled to claim a share of the robotics market. At the height of the boom, a new robotics company entered the market every month, but many of these companies ultimately lasted only a year or two. Some simply did not have the expertise to compete in the field or could not capture the market share they needed to survive. Others were successful enough to draw the attention (or ire) of the larger, more well-established robotics companies. If a successful small company had an impressive product or a significant market share, the larger companies would often try to buy it out or collaborate with it in some way. If that kind of olive branch failed or the small company did not have technology they wanted, the larger companies would simply work to take market share from the smaller company and force it out of business, thereby eliminating their competition. While the names of these small companies are not important for our purposes, it is important to note that some of these systems are still doing their jobs in industry. The ABB robot at Magnussons, for example, proves robots can survive for decades with proper maintenance, and some manufacturers are reluctant to buy a brand new robotic system when the one they have is still working.

The boom of the early 1980s might have continued throughout the decade if not for a few issues that emerged in the middle of the decade. One of the key factors was labor costs, which began to level off from the drastic increases that began about a decade before in 1975. Labor costs continued to rise, as they do today, but at a more reasonable rate of 2% to 5% for the most part versus the double-digit increases that had occurred earlier.

Another factor that slowed growth of the robotics market was other forms of **industrial automation**, or equipment that completes processes with minimal human assistance, which became readily available and competitive in the latter part of the 1980s. In applications where the same task is performed hundreds or thousands of times a day, it is often better, easier, and cheaper to use a machine designed for that specific task than to adapt a robot for that purpose. The same technology that made robots smarter, faster, and smaller also created similar advancements for machinery. The NC technology of old became CNC technology, with microprocessors and programs replacing punch cards and magnetic tapes. Relay logic and mechanical timing systems were replaced by **programmable logic controllers (PLC)**—specialized systems that run one type of code very efficiently in industrial environments to monitor signals coming in, filter information through instructions entered by the user, and then send power out to activate various pieces of equipment. Another advance that was not strictly a technology, but rather a change in work flow, was **work cells** (or just **cells**), which are a logical arrangement of equipment to increase workflow via reduced part travel and wasted motion. These cells increased production output while reducing production costs and, best of all, utilized equipment the company already had.

By the late 1980s, most of the dust had settled in the robotics field, as smaller companies realized

they could not compete with the more aggressive, larger companies, and as the demand for robots tapered off to some extent. The companies that survived the 1980s had managed to carve out a portion of the robotics market large enough to sustain their businesses, with sales in Europe, Japan, and the United States being crucial for these companies to remain viable. The continued quest for market share has been a driving force behind the advancement of industrial robots from the large, clunky systems of old working behind a cage to the modern collaborative robots of today. **Collaborative robots** are designed to work *with* humans, instead of separated away from humans, and have safety systems that limit the danger to humans by carefully monitoring their surroundings and often slowing down when humans are nearby (Figure 2-7). Many of these robots have outer skins that are rounded in shape and made of materials that can help absorb some of the force should an impact occur.

Courtesy of Rethink Robotics

FIGURE 2-7 Rethink Robotics led the collaborative robots charge in 2012 with the release of the Baxter robot. Here you can see Baxter working to help someone in industry.

This turbulent decade for robotics also saw some new technology emerge. In 1982, COGNEX released DataMan, the brainchild of Dr. Robert J. Shillman. This system was capable of **character recognition**, meaning the ability to read written or printed letters, numbers, and symbols. It allowed manufacturers to track parts and process written data as well as giving robotic systems the ability to "see." In 1983, Adept Technology began producing industrial robots, machine vision systems, and other automation, starting a robotics company that remains alive and well today. In 1984, Takeo Kanade and Haruhiko Asada patented their direct-drive robotic system, in which the motor and the joint connect directly, instead of using gears, pulleys and belts, or sprockets and chains to create motion. This resulted in increased speed and accuracy of the robot while reducing maintenance requirements—which is why the majority of modern systems use direct drive. In 1985, Kawasaki ended its collaboration with Unimation, which had begun in 1968, and began its own global business in robotics. Like Adept, Kawasaki is one of the major suppliers of robotics to today's industries.

The 1990s was a time of maturation for robotics technology, with the advancements of the previous decades being refined and exploited in new ways. In 1991, advances in display technologies and new microelectronics improved robots and revived interest in the field, helping to stimulate sales figures that had begun to decrease toward the end of the 1980s. The early part of the 1990s saw a few major players bow out of the field, with ABB buying Cincinnati Milacron's robotic line and GM selling off its share of FANUC Robotics. In 1994, record numbers of robots were sold in the United States—a total of 4355 robotic units worth $383.5 million. Of that amount, $200 million went to ABB and $120 million went to FANUC; ABB and FANUC remain two of the main suppliers in today's robotics market. Also in 1994, Motoman became Yaskawa Motoman when Yaskawa Electric America bought out Hobart Brothers, with which they had originally gone into business in a 50/50 joint venture. That same year, Yaskawa Motoman made the world's first controller capable of controlling two robots at once, with the systems working together in a complementary fashion. The result was the ability to perform large, complex tasks without the need to use multiple controllers with networked communications between them.

The new millennium ushered in some truly fantastic innovations in the robotics world as well as signaled the rise of the robotic integrator. **Robotic integrators** are companies, usually other than the robotic system manufacturer, that specialize in selecting, adapting, installing, and programming robots for whatever application(s) their clients need. Many of these companies have zero involvement with the development or manufacture of the robotic systems they use; instead, their focus is utilization of the robot. Some integrators decide to work with only one robotics company, while others utilize any available system(s) to meet customer needs. Sometimes they work completely at the client's facility, but at other times they set up the system at their home base—often miles, states, or even continents away from the client—so that they can work the bugs out first and then ship everything out to the client. Robotic integrators often bill themselves as a "one-stop shop," as their role is to start with the client's needs and keep working until the system is up and running to the customer's satisfaction. These projects can take from a few months to a few years to complete, depending on their complexity. If you like varying challenges and want the opportunity to work with new robotic systems as they come out, this is a great field to investigate.

In 2000, NASA created the first version of the Robonaut, a robotic torso designed to perform the functions of an astronaut that included a tracked vehicle base for movement (Figure 2-8). This development would eventually lead to a partnership with GM in 2007. While NASA's interest is development of an aid for astronauts, GM is looking for a robotic system that would have greater benefits in the industrial world as well as a chance to get back into the robotics game.

FIGURE 2-8 NASA's Robonaut 1 ultimately sparked GM's interest enough to persuade that company to become involved in building the second generation of the system.

In 2004, Aaron Edsinger-Gonzales and Jeff Weber created *Domo* for Massachusetts Institute of Technology (MIT). This robotic research platform for advancing human–robot interactions represented one of the first focused efforts to bring the robotic system out of isolation and place it side by side with human workers, using a unique feedback system specifically designed with human safety in mind. The seeds planted by this research would begin to bear fruit in 2012, as industry bought into the concept of robots and humans working together, instead of isolated from each other, and Baxter (discussed later in this section) was released.

In 2006, Yaskawa Motoman gave industry a dual-arm robot that resembled a human torso without a head. The whole point was to mimic the motions of human workers and give the robot two arms instead of the standard single arm. Another innovation of this system was that all of the cabling and air hoses run internally, thereby preventing damage from contact, impact, or friction—all too common problems with many systems. In 2009, Yaskawa Motoman released a robot controller that managed up to 8 robots or 72 different axes of movement, opening up exciting new motion options and raising the bar once more. These continued innovations in controller capability helped to keep Yaskawa Motoman in the game, even though companies like ABB and FANUC continue to control a large chunk of the market.

Robonaut 2, the result of collaborative work by NASA and GM that began in 2007, completed the first human–robot handshake in space during 2012 (Figure 2-9). While this might not seem like much of an achievement, it represents the culmination of sensor, feedback, and safety controls necessary for robots to work not just near people but directly with them. This feat came eight years after *Domo* started the focused research into human–robot direct interaction for industry. In the same spirit of human–robot interaction, Kawada Industries tasked Tecnalia, one of its top engineering partners, with adapting its Hiro robots to work with workers in Europe instead of behind cages the same year. Late in 2012, Rethink released the Baxter robot, which works in the industrial environment without the need for a cage. Baxter has a human torso, a tablet that doubles as both interface and head for the robot, force sensors, a 360° camera system, and artificial intelligence (AI) software that helps to make the robot safe and easy to use (Figure 2-10). Baxter can adjust to changes in its work environment and slow to safe levels when humans are detected close by.

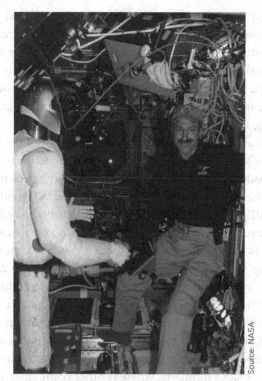

Source: NASA

FIGURE 2-9 The same technology that allows robots to work beside humans in industry was used to make the Robonaut 2 safe enough to shake hands with an astronaut, completing the first human–robot handshake in space back in 2012.

Courtesy of Rethink Robotics

FIGURE 2-10 Rethink Robotics' original robot Baxter, alongside the company's 2015 release for industry, Sawyer.

Thus, 2012 proved to be the year when industry truly began to seek and invest in robotic systems that can work with employees instead of behind a fence. By 2017, companies had at least 19 different choices for collaborative robots produced by companies such as ABB, BOSCH, FANUC, Rethink Robotics, and Universal Robots.

By 2013, several manufacturers had rolled out collaborative robots to capitalize on this new area of robotic use. Yaskawa Motoman offered Dexter Bot, a dual-arm, 15-axis collaborative robot with a six-figure price tag. KUKA introduced LBR iiwa (intelligent industrial work assistant), which was originally designed for use in space, through an agreement with the German Aerospace Center. Universal Robots had the UR, a single-arm robot with six axes that was first introduced in September 2012. ABB had a prototype dual-arm concept with 14 axes that was designed for small-part assembly. Toyota also developed a collaborative robot, but like most of the company's robots, it was used in house; thus it was not given a specific name to capture sales interest. Another trend in 2013 was greater interest in dual-arm robots, as industry warmed to the idea of a robot that is built more like a human to do human jobs. Yaskawa Motoman and ABB both decided to mix dual-arm functionality with collaborative safety protocols in hopes of achieving higher sales numbers, whereas other companies such as Seiko Epson, Nachi, and Kawada focused on just the dual-arm concept.

While its work is not directly related to industry, NASA's Johnson Space Center Engineering Directorate created the R5, commonly known as the Valkyrie robot, to compete in the 2013 DARPA Robotics Challenge Trials. Valkyrie capitalizes on many of the lessons learned in building the two Robonaut series, but has a key difference: It has legs for a base and is designed to walk like a person (Figure 2-11). Some other innovations relative to the Robonaut series include stronger motors for higher payloads, highly modular construction for easy repair or component swapping, additional sensors, and cameras on the forearm to give operators a better view of the hands. While NASA will ultimately send this technology into space if it can get all the bugs worked out satisfactorily, it is quite reasonable to believe similar technology may find its way into industry alongside workers.

In 2014, Yaskawa Motoman celebrated its 25th year of being in the robotics industry—an achievement shared with only a few other companies. That same year, it also had a chance to work with NASA on a teleoperation demonstration. One of Yaskawa Motoman's SIA50 seven-axis robots was used for the test. It was located at the Kennedy Space Center but controlled from Goddard Space Flight Center in Greenbelt, Maryland, more than 800 miles away. This is yet another tie between NASA and industrial technology that could bear fruit down the road.

Source: NASA

FIGURE 2-11 NASA's R5 robot, known as Valkyrie.

During 2014, Yaskawa Motoman's dual-arm robot was in its fifth generation of development (Figure 2-12). In the same year, ABB's dual-arm concept prototype was introduced as YuMi® ("you and me"), and it was announced that YuMi® would be commercially available on April 13, 2015. Universal Robots doubled its robotics sales from 1000 units in 2013 to 2000 units in 2014 to companies such as BMW and Volkswagen.

The use of various robotic systems for welding applications also expanded in 2014. It takes a large amount of time—usually measured in years—to train a person to weld consistently given all the various materials and weld configurations used by industry. As the numbers of available and/or interested workers in the welding field began to dwindle, companies turned to welding robots to fill the gap. With welding robots, they need only a few individuals with deeper knowledge of welding to create programs and set up the system for new runs. Once the setup is done, employees can learn the operation of the system in a few days and start turning out consistent welds, thanks to the robot.

The start of 2015 was marked by a deeper focus on sensors, software, and tooling for the robot, rather than by advances in the core robotic system. Many companies worked to refine the advances they had

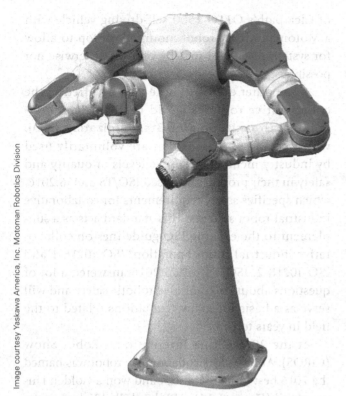

Image courtesy Yaskawa America, Inc. Motoman Robotics Division

FIGURE 2-12 This photo depicts one of Yaskawa Motoman's dual-arm robotic systems. Some companies place something on top of the central torso to make the robot look more like a person and usually add camera vision, but either way they get the job done.

made while integrating equipment such as soft grippers that would not mar the parts of 3D printed specialized tooling. This was also a time of greater use of robots to complete **3D printing** projects—that is, projects where a three-dimensional design is created layer by layer in some medium such as ABS (acrylonitrile butadiene styrene) plastic, hardened resin, glued powder, heat-fused metal powder, or even concrete, to name a few possibilities.

In March 2015, Rethink Robotics released Sawyer, an arm-type robot that utilized the collaborative technology of the Baxter robot but was designed as a better fit for industrial applications to open up new utilization markets (Figure 2-13). Sawyer, which is still in production as of the writing of this book, is smaller, faster, stronger, and more precise than Baxter.

Around the same time Yaskawa Motoman partnered with ARGO, an international leader in exoskeleton technology and the manufacturer of the ReWalk exoskeleton. **Exoskeletons**, in this case, are wearable frames with motors and electronics embedded that allow people with reduced or lost mobility to walk

Courtesy of Rethink Robotics

FIGURE 2-13 Sawyer at work in industry, cage free.

again. The Motoman robotics division of Yaskawa entered this partnership to pursue open innovation as well as to explore healthcare applications of its robotic systems.

Several robotics companies were purchased by other industrial interests in 2015. Teradyne bought Universal Robotics, famous for its collaborative robots, for $285 million, with a promise of another $65 million being paid if set performance goals are accomplished through 2018. Universal Robotics was found in Denmark in 2005 and had $38 million in profits the year before the buyout. Teradyne sells automatic test equipment for semiconductors, wireless products, and electronics; it had no robotics division before the acquisition of Universal Robotics. That same year, Omron purchased Adept through stock acquisition. Omron specializes in sensing and control technology, but had also dabbled some in robotics before this purchase. Adept had restructured in 2013 in an attempt to improve its financial standing, and it entered into the arrangement with Omron with hopes of increased international sales.

The tail end of 2015 was FANUC's time to shine, as it announced both a collaboration with Cisco and the company's best performance in sales and manufacturing to date. In October 2015, FANUC announced a partnership with Cisco in which Cisco devices will be embedded directly into FANUC's robots for data collection and down-time prediction. This agreement came just after a yearlong project known as ZDT (Zero Down Time), in which the company collected data directly from the robotic process in an effort to predict failures before they happen and prevent down time of the system. In December 2015, FANUC announced it had produced and sold more than 400,000 robots worldwide during the year. This record catapulted FANUC to the position of number one robotic supplier at the time and was a strong indicator of interest in robotic systems the world over.

In 2016, FANUC continued its trend of connectivity and data usage started with the Cisco merger to create a Cloud-based system at GM's Lake Orion facility. Using what it learned with the ZDT pilot, with the Cisco technology, and from working with Rockwell Automation, FANUC set up a Cloud server to data mine operations and roll out the new ZDT program with style. GM's global director of manufacturing engineering was quoted as saying the following about the system: "The robot calls in and says, 'I've got an issue in one of my motors in one of my joints' or it may say it's about to get sick in a few seconds" (Vanian, 2017). FANUC also released the Fanuc Intelligent Edge Link and Drive (FIELD) system in 2016 to allow other suppliers to download new applications to FANUC robots in the field. The FIELD system works similarly to the app store for your favorite smartphone, but software developers who wish to use this system will have to pay to play. Around the end of 2016, FANUC partnered with Nvidia to add in graphics processing units to work with artificial intelligence software to improve robot operation. The process starts with video of the robot doing what it is programmed to do; this video is then run through software that analyzes the motions and changes the program to improve the process. The result is a new program tweaked to optimize efficiency that can be created in hours, instead of the days or weeks that it might take an engineer to identify and fix all the various spots for improvement.

A new wave of collaborative robots gained traction as Clearpath Robotics' OTTO robot, a self-driving vehicle designed to transport heavy material in industrial/warehouse areas, gained enough popularity to warrant splitting off its production into its own division. In 2016, OTTO Motors was born as a separate division in Clearpath Robotics, which had started out specializing in unmanned vehicles. Clearpath unveiled the OTTO in 2015, and in just two years demand for and the potential of this mobile collaborative robot was deemed high enough to support its own division. Shortly after the OTTO Motors division announcement, Yaskawa Motoman announced its partnership with Clearpath to develop a mobile machine tending and material movement solution. This system consists of Clearpath's OTTO 1500 self-driving vehicle with a Motoman MG12 robot mounted on top to allow for system reach and tooling options otherwise not possible.

A few other events of note also occurred in the collaborative robotics world in 2016. The **International Organization for Standardization (ISO)**, which creates standards that are voluntarily used by industry members to prove levels of quality and safety in their products, released ISO/TS 15066:2016, which specifies safety requirements for collaborative industrial robot systems. This standard acts as a supplement to the existing ISO guidelines on collaborative industrial robot operation, ISO 10218-1 and ISO 10218-2. ISO/TS 15066:2016 answered a lot of questions about collaborative robotic safety and will serve as a basis for many regulations related to the field in years to come.

At the 2016 China International Robot Show (CIROS), ABB's YuMi® collaborative robot was named the 2016 best industrial robot and won a Golden Finger award (Figure 2-14). ABB has billed YuMi® as the world's first truly collaborative robot since its release, and it would seem that the CIROS organization agreed. Given CIROS is one of the three largest robot technology yearly events in the world, this is no small win for ABB.

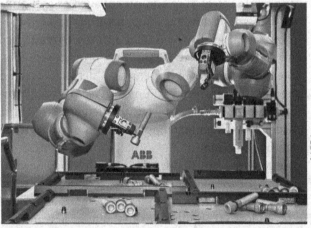

Image courtesy of ABB Inc.

FIGURE 2-14 The award-winning YuMi® robot, ABB's premier collaborative robot.

This brief summary by no means includes all the various events that were necessary for the creation of today's industrial robots, but it does give you an idea of the various innovations and technology that came together in the modern robot. As you dig deeper into the robotics field, you will discover more information on topics of interest to your chosen focus and expand your knowledge. If you would like to do some digging on your own, I recommend starting with the various robotic manufacturers, as they often have a history and/or news section on their websites that contains a wealth of information.

WHY USE A ROBOT?

So now we know *what* an industrial robot is and *how* it got here, but what really sells robots is the *why*. Where do we use robots? Why a robot instead of another piece of automation? What is the robot good for? In this section, we answer these questions and many others pertaining to industrial robots. Luckily for us, it is easy to answer these questions for industrial robots because of their specific focus.

In the industrial setting, we typically use robots for situations that are Dull, Dirty, Difficult, or Dangerous (the **four Ds** of robotics) or for things that are Hot, Heavy, Hazardous, and Humble (the **four Hs**). Both of these sets of conditions, which are very similar, are things people either should do or would rather not do, and they were a driving force behind robots' acceptance by both management and workers in industry. In addition to the four Ds of robotics, we will also explore precision, performance, and cost of operations, all of which play an important role in the "why" behind industrial robots.

The first D of robotics is **Dull**, which is similar to Humble. As the name implies, these tasks are repetitive in nature and often require little or no thought. The danger to the human worker in these situations is often damage to the body due to doing the same thing over and over and over again for weeks, months, or years on end. These injuries tend to involve the worker's joints or back and can range from sprains or carpal tunnel syndrome to injuries requiring complete joint replacement or back surgery. Another danger with dull tasks is that the worker may become bored and not pay attention to what he or she is doing, sometimes known as running on autopilot. While this inattention could lead to part or machine damage, the main concern is that the person might become injured by the moving parts of the machine due to carelessness. This type of injury can range from cuts, bruises, and deep lacerations, to lost fingers, complete amputations, and death. Given the dangers involved and the fact that these types of tasks rarely change, this is a perfect job for robots. A robot is able

to perform thousands, if not millions, of repetitive actions with only minimal maintenance and no loss of focus. This also frees up the operator to do important things such as checking part quality, making machine adjustments, or other things better suited to humans.

The second D on the list is **Dirty**, which may well involve the Hot and a bit of the Hazardous designation from the four Hs. Dirty tasks in industry involve processes that produce dust, grease, grime, sludge, or other substances that people would rather not get on them. They can present health risks in the form of allergic reactions or irritation of the skin as well as slips, trips, or falls, but for the most part these are simply jobs in which people end up covered in "something" by day's end. While these types of jobs do not bother some people, many of us would rather avoid the mess when given a choice. The dirtiness of the job may affect worker satisfaction, which in turn can affect productivity, product quality, and the **turnover rate** (how often workers quit a job or leave an employer). Dirty jobs are another great place for robots: They do not care how dirty they get, nor do they have skin that can be irritated. When using a robot in a dirty environment, the only real concern is damage to the mechanical parts of the robot. To prevent this kind of damage, robots may be sealed in plastic or designed specifically for the dirty environments where they work. To do the same for a human worker often takes suits that are hot to wear, respirators that require special training and medical clearance, and other protective equipment that adds to the difficulty of the job, not to mention the potential discomfort for the wearer.

The third D is **Difficult**, where we find the Heavy tasks. This consideration includes all the tasks in industry that humans struggle to perform (Figure 2-15). The construction of various machines and the processes involved in production at times require people to bend, twist, and move in ways that are difficult for the human body. While the rotational joint of a robot may have 270° of movement, the human elbow is limited to approximately 90°—and no human can rotate their wrist 360°. Sometimes the difficulty arises due to a combination of position and weight; that is, the person may be able to move as needed, but the added weight of the part makes it difficult due to the added strain on joints and muscles. Sometimes it is simply difficult for workers to reach the point needed due to the physical limitations of the length of their arms or their overall build. Instead of trying to find a seven-foot-tall contortionist, it makes sense to use a robot instead, especially since we can specify things such as reach, payload, and number of axes.

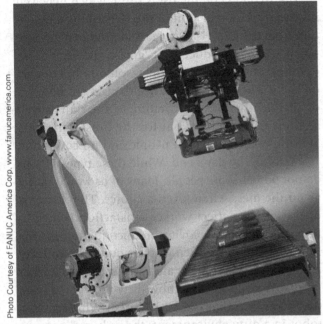

FIGURE 2-15 If a person were to move sacks of dog food all day, the wear on the individual's body would be immense. By comparison, a robot can do this task for years at a time with ease.

The last of the Ds is **Dangerous**, which includes all things Hazardous. These environments can be excessively hot, contain toxic fumes or radiation, involve working with unguarded machinery, or include any other condition that has a high risk of injury or illness for humans (Figure 2-16). This is a perfect job for a robot because we can either repair or replace a damaged robot, but a person may require months or years of medical assistance following an injury or illness and even then, never attain the quality of life he or she once had. In industry, there is no profit margin worth risking the worker, especially when a robot could do the job instead. A great example of this is the robotic systems used to clean up the Chernobyl nuclear facility. These robots are able to work in a radioactive environment that would be harmful or fatal to humans while cleaning up radioactive waste for proper disposal. When the area is clear or the robot is no longer functioning, it goes to the same facility as the rest of the nuclear waste for storage until the radiation levels are safe once more.

FIGURE 2-16 This robot is picking up a red-hot piece of metal—a dangerous task for which robots are well suited. Notice in the background how chipped and worn the paint is on the robot as well as the debris on the tooling. This is dangerous and dirty all in one!

Beyond the four Ds of robotics, another good reason for using robots is precision. **Precision** is performing tasks accurately or exactly within given quality guidelines. With human workers, differences in their vision, body construction, and experience all contribute to varying levels of possible precision. In industry, this variance can translate into increased production times or parts that do not function properly. Robots, by contrast, have the capacity to perform tasks repeatedly with levels of precision that are difficult, if not impossible, for humans to match. In fact, many robots have tolerances ranging from 0.35 mm to 0.06 mm (0.014 in. to 0.002 in.)! Their precision has made robots especially useful in the production of electronics, aerospace components, and other precisely manufactured parts.

Robots' precision leads to consistency and repeatability that is difficult for their human counterparts to match. **Consistency** is the ability to produce the same results or quality each time, while **repeatability** is the ability to perform the same motions within a set tolerance. A multitude of factors affect a human worker's performance, such as illness, injury, fatigue, distraction, emotions, job satisfaction, temperature, and ability, to name just a few. Any of these considerations can affect a worker's quality or quantity of production. Since industrial robots do not have emotions and they are not organic systems, their work does not suffer from these kinds of human ailments. With a robot, you have to worry about the program controlling the system, the mechanical and electrical components of the system, and the validity of the signals coming and going from the robot. As long as these are in good order, the robot will perform in the same way, no matter whether it has been working for 1 hour or 48 hours, whether it is Monday morning or Friday afternoon, regardless of the company hiring or laying off employees, or any other condition that might affect human workers. This consistency translates into cost savings through less materials used, higher part quality, and consistent manufacturing times that help with determining production schedules (Figure 2-17).

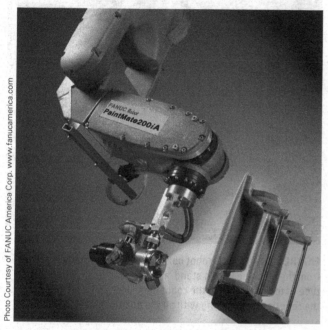

Photo Courtesy of FANUC America Corp. www.fanucamerica.com

FIGURE 2-17 The FANUC robot can paint parts consistently, over and over again. Painting was the application of the first industrial robot mentioned in this chapter.

Cost savings is a factor that has helped to drive the integration of robotics into industry and another "why" for the industrial robot. In industry, workers are generally paid a set per-hour or per-piece wage, along with **fringe benefits**—things such as health and dental insurance, retirement plans, life insurance, Social Security contributions, and anything else for which the company pays part or all of the cost on behalf of the employee. With a robot, the company must pay for the initial cost of the system, the cost of parts and maintenance, and the cost of electricity or fuel to run the system. In many cases, a robot can pay off the initial costs in two years or less; from that point forward, it operates at the cost of replacement parts, consumed energy, and maintenance, which for some systems can be as low as $0.42 per hour! These cost savings are on top of any time or material savings that come about

due to the robot's precision and speed. For instance, the welding of a car frame by human workers used to take 4 to 6 hours, but now a team of robotic systems does the same task in 90 minutes. In addition to the time saved, the robot's precision allows smaller welds to have the tensile strength of larger human welds, thereby saving the company money in the form of fewer materials used and fewer rejected parts. It also frees up human workers to do important tasks such as quality inspections, process modifications, and other tasks that require judgment-based decisions and that are difficult, if not almost impossible, to automate.

Another big draw when using robots in industry relates to technology like the collaborative robots that can work with people and the vision systems that can measure images to capture quality data. Each new trick learned by the robot that makes it easier to work with or gives it an ability humans lack increases the number of units sold. Universal Robotics focused heavily on the collaborative market to carve out its piece of the industrial robotics pie, while the major robotics companies, including FANUC, ABB, and Yaskawa Motoman, all developed a collaborative system or two of their own to make sure they did not get left out. Collaborative robots in many applications are more like "cheap workers" rather than another piece of industrial equipment.

While a person may be able to look at something and determine it is out of alignment, robots utilize cameras that not only detect the deviation but also measure it; they then use these data to update program positions for that cycle of the program or to determine whether a part should be rejected. Robots can use devices that can see heat, detect various specific gases, or even pick up sound waves humans cannot hear to find leaks, check quality, or perform other inspections on each part/system faster than the average human could. This allows inspection of everything coming down the production line, instead of only one in five units, one in ten units, or even more, as is common when humans perform the task. As we continue to expand robots' capabilities through improvements in sensors and tooling, robots will continue to find new jobs in industry.

The last area we will explore in this section is robot flexibility. Robots are one of the few systems in industry that boast such a wide range of functionality, which can be achieved with nothing more than tooling changes and a new program (Figure 2-18). An automated machine or system is often designed to perform a specific task or tasks efficiently over and over but requires major reworking to perform a different type of task. Robots, by contrast, come with a set of specifications such as reach, payload, and speed, but the rest is generally up to the user. If you want to use the robot to paint, you simply attach a spray gun to the end and write the program. Rather have it weld? Add a welding gun and power supply, along with a few program changes, and the robot is ready to go. Need a boring operation done, but don't want to buy another machine? Add either a stand-alone station or new robot tooling to spin the drill and make a few modifications to the program. We can even swap the robot tooling during operation and have the robot perform multiple functions in each cycle of the production process it works with.

In later chapters, we will explore the systems of the robot in great detail and learn more about many of the topics raised in this section. While the specifics of the four Ds change from industry to industry and from year to year, they are still as valid in today's manufacturing facility as they were in the plants of the 1970s and 1980s. To sum up this section, industry uses robots to solve a need or problem. Sometimes those needs are related to the human side of things (the four Ds, for example), while other times they reflect the need for a system with more flexibility than is offered by the standard automated piece of equipment.

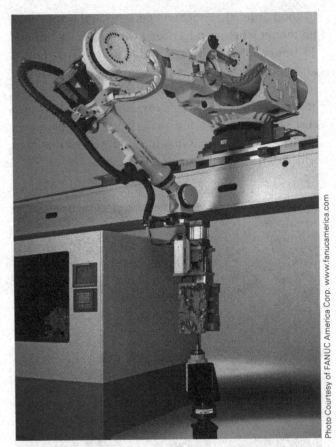

FIGURE 2-18 A FANUC robot bores out an engine body it pulled from the machine with a stand-alone tooling base. Often this type of tooling is tied to the robot controller and controlled in the robot program, along with all the positioning necessary.

REVIEW

You should now have a greater understanding of the events leading up to the modern robot as well as what an industrial robot is and why we use these kinds of robots. As you learn more about industrial robots in later chapters, you may find it beneficial to review bits and pieces of this chapter that were new to you. Case studies are another great place to learn about the uses of robots in industry, and you can often find them on robot manufacturers' websites or the RIA website. In this chapter, we covered the following topics:

- **What is an industrial robot?** We defined the industrial robot and began our exploration of industrial robotics.

- **A history of technology.** We explored some of the history outside of robotics that was specifically necessary to evolve the technology used by the modern robot.

- **The rise of the industrial robot.** We focused on events in the timeline of industrial robotics, including the first industrial robot.

- **Why use a robot?** We looked at some of the reasons behind robotic use in industry, including both human and technology factors.

KEY TERMS

3D printing	Dull	Mechanical relay	Relay
AND	Exoskeletons	NAND	Relay logic
Android	Four Ds	NO	Repeatability
Automata	Four Hs	Normally Closed (NC)	Robotic Industries
Automation	Fringe benefits	NOR	Association (RIA)
Character recognition	Gantry	NOT	Robotic integrator
Collaborative robots	Industrial automation	Numerically	Slide rule
Computer numerical	Industrial robot	controlled (NC)	Solid state
control (CNC)	International	OR	Stepped drum
Consistency	Organization for	Precision	Turnover rate
Dangerous	Standardization	Programmable logic	Voltaic pile
Difficult	(ISO)	controller (PLC)	Work cell
Dirty	Leyden jar	Punch card	XOR

REVIEW QUESTIONS

1. Where does the word *robot* come from, and who gave us this term?

2. What is an industrial robot as defined by RIA?

3. What was the Jacquard loom, who invented it, and when?

4. When was the first microchip introduced, and who invented it?

5. What are three things that Nikola Tesla is famous for?

6. Who created the first industrial robot, and when did that happen?

7. When did Kawasaki Robotics enter the robotics field?

8. What was the first commercially available micro-computer-controlled industrial robot?

9. When did KUKA start producing robots, and what was its first model?

10. Who created the first fully electric microprocessor-controlled robot and to which company was the robot shipped?

11. Who invented the first SCARA robot and when?

12. Which factors led to the increased robot sales and the robotics industry boom in the early 1980s?

13. What stopped the robotics boom of the early 1980s?

14. What are collaborative robots?

15. What happened in the early 1990s to boost robotic sales once more?

16. What is a robotic integrator?

17. Who introduced the first dual-arm robot?

18. Who released the first collaborative robot for industry, and in which year did it hit the market?

19. When did ABB's YuMi® first go on the market, and which award did it win?

20. What are FANUC's ZDT and FIELD programs?

21. List the four Ds of robotics and give an example of each as it relates to industry.

22. What are some of the other factors besides the four Ds or the four Hs that drive industry to use robotics?

References

1. Robotic Industries Association. (2017, March 13). About RIA. Retrieved from http://www.robotics.org/About-RIA

2. Britannica. (2017, March 18). *Alessandro Volta Italian Scientist*. Retrieved from Britannica. com: https://www.britannica.com/biography/Alessandro-Volta#ref1217529

3. Rehg, J. A. (2003). *Introduction to Robotics in CIM Systems*. New Jersey: Prentice Hall.

4. Vanian, J. (2017, March 26). *GM Is Using the Cloud to Connect Its Factory Robots*. Retrieved from Fortune tech: http://fortune.com/2016/01/30/gm-cloud-factory-robots-fanuc/

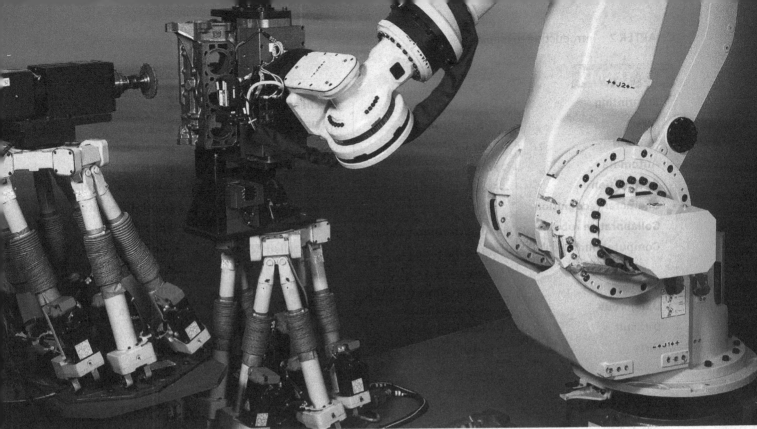

CHAPTER 3

Components of the Robot

WHAT YOU WILL LEARN

- How hydraulic, pneumatic, and electric power differ

- What the controller does for the robot

- How we use the teach pendant and what you will find on such a device

- How the manipulator of the robot relates to the human body

- The difference between major and minor axes

- How to number the axes of the robot

- How to mount a robot, including some of the options available

OVERVIEW

Robots have a wide range of uses, designs, and capabilities, but in the end a robot is a collection of subsystems that make up the whole. In this chapter, we look at each of these subsystems and their role in making the robot a functional unit. As a general rule, the more complex the robot, the more components and systems it contains. The better you understand the components and subsystems of the robot, the easier it will be for you to work with the robot regardless of whether you are operating, designing, repairing, or setting up the system for a new task. To help you understand the systems of the robot, this chapter covers the following topics:

- Power supply
- Controller/logic function
- Teach pendant/interface
- Manipulator, degrees of freedom, and axis numbering
- Base types

POWER SUPPLY

No matter the type, complexity, or function of the robot, it must have some source of energy to do work. The power supply used in a robot often depends on what the robot does, in which type of environment it is working, and what is readily available. Potentially, any power source we can harness could power a robot, but the main sources used for industrial robots are electric, hydraulic, and pneumatic power. As we look more closely at each power source, we will discuss the pros and cons of each.

Electric Power

Most of the robots used in factories today run on electricity, because this type of energy is readily available, easy to store, and fairly inexpensive. **Electricity** is the flow of electrons from a place of excess electrons to a place of electron deficit, and we route these electrons through the robot's components to do work. In an electrical system, **voltage** is a measurement of the potential difference or imbalance of electrons between two points and the force that will cause electrons to flow. We measure the flow of electrons in **amperes** or **amps**, where 1 A is equal to 6.25×10^{18} electrons passing a point in 1 second. It is this flow that does the work in the circuit. **Resistance** is the opposition to the flow of electrons in the circuit and the reason why electrical systems generate heat during normal operation.

Electrons can flow either in one direction, which we call **direct current (DC)**, or back and forth in a circuit, which we call **alternating current (AC)**. The power supplied by the electric companies to industry is AC, whereas the power from batteries, such as those that run forklifts or mobile robots, is DC. The type of robot and its functions will determine how much amperage it requires, which voltage level is needed to force enough current through the system, and whether the voltage is AC or DC. With mobile robots, we also have to take into account how long the batteries can run the system before they need to be recharged as well as how long the recharging process takes.

> ⚠️ **Safety Note**
>
> Always be cautious around electricity, as it takes very few amps passing through your body to cause severe damage. In fact, it takes 0.1 A passing through your heart for one-third of a second to cause ventricular fibrillation, which is a life-threatening situation.

In a DC system, the voltage stays at a constant level and all the electrons flow in one direction only. Because of this one-way flow, every component in a DC system will have a set **polarity**, or positive and negative terminals. With some components, such as switches or resistors, reversing the polarity of flow through the device has no effect. In other components, such as solid-state devices or motors, reversing the polarity can cause the component to run backward, block electron flow, and even cause permanent damage to the device. Another bad idea is hooking AC power to a DC system. Because the direction of flow changes every half-cycle of the sine wave with AC, everything the system is hooked to will get a reverse flow 60 times per second! On top of this, many DC systems are rated for incoming power of between 5 and 48 V, while the lowest voltage common AC source provides 120 V. Hooking 120 V AC into a 12 V DC control system can cause some truly catastrophic damage to everything the AC power encounters until finally it destroys something—in my own sad case, the copper trace on the card—to the point the voltage cannot jump across/through.

Many components will have the positive or negative terminal marked, making it easier to maintain polarity in the circuit. In addition, cable or card connections are often designed so they fit or connect in only one way, thereby preserving any polarity present. The positive side of each component in a circuit hooks to the negative

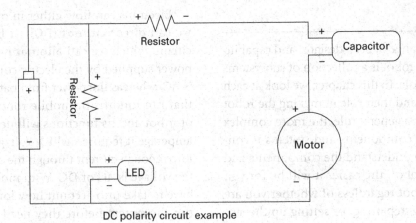

FIGURE 3-1 In this example, you can see the polarity of each component in the circuit. Notice that the polarity is component specific.

side of the component that precedes it and ultimately traces back to the positive side of the power supply. The negative side of a component hooks to the nearest component's positive side and ultimately traces back to the negative side of the power supply. If you look at the capacitor in Figure 3-1, you can see that the positive side connects to what would be the negative side of the resistor (resistors are not polarity sensitive) and ultimately the positive side of the battery. The negative side of the capacitor connects to the positive side of the motor and ultimately traces back to the negative side of the battery.

While a large number of worker-driven forklifts continue to be found in industry, some facilities are now using robotic solutions to move bulk or heavy items from one area to another. While any of the power systems utilized by forklifts could be used for material-handling robots, DC battery packs seem to be the power source of choice for these bots. When dealing with batteries, it is best to remember that a battery pack/system has only so many **amp-hours** (Ah; the number of amps deliverable over a length of time) in it. For example, a 10-Ah battery could deliver 1 A for 10 hours, 2 A for 5 hours, or 5 A for 2 hours. Regardless of the total potential or amp-hours of a battery, the more amps we pull from it, the sooner we will have to recharge or replace the battery.

We can increase the amp-hours of a system by adding batteries in parallel to share the load. In a proper connection, both of the positive leads of the batteries are hooked together and in line with the main supply's positive terminal; the negative terminals are connected in the same fashion (Figure 3-2). When setting up this kind of power supply, it is important to make sure that you use identical batteries; otherwise, the battery with the higher voltage and/or amperage will spend some of its energy trying to charge the battery with the lower voltage and amperage. At best, this is wasted energy; at worst, it could cause battery rupture. When we connect batteries this way, we

add the amp-hours of each battery to determine the total amp-hours. Example 1 illustrates this calculation.

Parallel battery connection

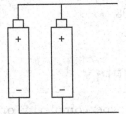

FIGURE 3-2 Two batteries in parallel, sharing the amperage requirements of the circuit and extending the amp-hours.

Example 1

If we connect three batteries in parallel, each of which has 15 Ah of energy, what would be the total amp-hours available?

Total amp-hours = sum of all the parallel connected batteries' amp-hours

Total amp-hours = number of batteries × amp-hours per battery in parallel (when batteries are identical)

Total amp-hours = battery 1 + battery 2 + battery 3

Total amp-hours = 15 + 15 + 15 = 45 Ah

Total amp-hours = 15 × 3 = 45 Ah

For industrial applications, we usually need a large amount of current to do the work necessary, which requires a large amount of voltage to push the electrons needed. Since many systems are built using 6 V or 12 V batteries, we have to boost the voltage somehow. Luckily, that just requires hooking the batteries in series with each other. To accomplish this, we hook the positive end of one battery to the negative end of another battery, continuing in this fashion until we have enough voltage for the system. The system is powered by connecting to the open ends of the series batteries (Figure 3-3). If you accidentally put one of the batteries in backward, such that either two positive terminals or two negative terminals are touching, the battery that is installed backward actually subtracts from the total voltage of the string—so be careful. Examples 2 and 3 illustrate the calculations for series connections.

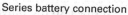

Series battery connection

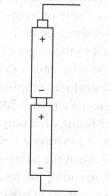

FIGURE 3-3 Two batteries in series.

Example 2

What would be the total voltage supplied by three 12 V batteries connected in series?

Total voltage = sum of the voltage of each battery in the series arrangement

Total voltage = number of batteries × voltage rating (when all batteries are equal in voltage)

Total voltage = 12 V + 12 V + 12 V = 36 V

Total voltage = 12 V × 3 = 36 V

Example 3

What would be the total voltage supplied by three 12 V batteries connected in series with the last battery placed in backward (i.e., negative to negative)?

Total voltage = sum of the voltage of each battery in the series arrangement minus the battery or batteries in reverse orientation

Total voltage = (number of batteries × voltage rating) − battery or batteries in reverse orientation (when all batteries are equal in voltage)

Recall that we subtract the last battery because it is placed in backward and subtracts from the total voltage of the series group.

Total voltage = 12 V + 12 V − 12 V = 12 V

Total voltage = (12 V × 2) − 12 V = −12 V

Because the last battery is connected incorrectly, the system is now running at 12 V instead of 36 V.

Most industrial systems use a bank of batteries that consists of parallel sets of batteries connected in series to meet the voltage requirements and extend the run time (Figure 3-4). The amp-hours of a set of series-connected batteries is the same as that of one individual battery, as each of the batteries is giving up electrons at the same time. In other words, if the batteries in Example 2 were 20-Ah batteries, the system would run at 36 V with 20 Ah of power. Even though each battery alone can provide 20 Ah, when hooked in series there is no gain of power—only a gain of voltage. This is why most of the industrial battery packs consist of a combination of cells with multiple series groups connected in parallel at the end points of the series chain. Example 4 provides a deeper look at the math involved.

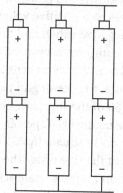

Series and parallel combination higher voltage and more amp-hours

FIGURE 3-4 Three sets of series batteries wired in parallel to increase the amp-hours while maintaining the proper voltage level.

Example 4

What would be the total voltage and amp-hours of a system that has five parallel groups or cells consisting of three 12-V batteries connected in series when each battery has 100 Ah of electricity?

Total voltage = number of batteries × voltage rating (when all batteries are equal in voltage)

Total amp-hours = number of batteries × amp-hours per battery in parallel (when batteries are identical)

Total voltage = 12 V × 3 = 36 V

Total amp-hours = 100 Ah × 5 = 500 Ah

Even though there are 15 total batteries in this system, there are no more than three in series at any given point, so we use only three batteries to figure the total voltage. The three batteries in series all give electrons at the same time to increase the voltage level and, therefore, act as a single amp-hour amount. Thus, we use the number of groups for figuring amp-hours, rather than the total number of batteries.

One weak or misconnected battery in this scenario could affect the whole system, which explains why we try to use the same type and quality of batteries when we create a power source of this nature. Proper maintenance of the batteries in this system is crucial for consistent operation, and any batteries that weaken or fail must be replaced quickly to maintain full functionality.

AC power does not have a set polarity because it is constantly changing the direction it flows through the circuit, which means we do not have to worry about the positive or negative polarity of AC components. Another difference between AC power and DC power is that the voltage or intensity is constantly changing in an AC system. A graph of DC power shows a straight line, whereas the graph of AC power is a sine wave (Figure 3-5). As shown in Figure 3-5, the AC power starts at zero, rises to a positive value, drops back to zero, falls to a negative value, and then rises to zero once more. One complete wave from zero to positive to zero to negative and back to zero is called a **cycle**. In the United States, the power grid uses 60-**hertz (Hz)** power, meaning 60 of the sine wave cycles per second.

Because of the nature of AC power, we measure it in terms of the **root mean square (RMS)**, which is a mathematical average of the sine wave. Therefore, when you measure 110 V at an outlet, you are actually measuring an average of the peaks and valleys of the sine wave. This is the effective amount of **electromotive force (EMF)**, which is technical term for the voltage of a system.

AC power is a good choice for any robot that will be stationary in an environment that has this power readily available, making it a favorite energy source within industry. Using supplied AC power negates the concerns about amp-hours and eliminates concerns about the space and weight requirements of batteries. With AC,

we do not have to worry about hooking batteries in parallel or series, but we do have to determine how much voltage is needed and whether it is single-phase or three-phase power. Let us take a moment to explore the difference between single-phase and three-phase AC.

Single-phase AC is AC power that has one sine wave provided to the system via a single hot wire and returned on a neutral wire. The power in a standard 110-V wall outlet is single-phase AC. You might wonder, "If there is only one sine wave delivered by one hot wire, why are there three prongs on the plugs I use?" While it is true that only one wire provides the power to the system, the other two wires are no less important. The second wire attached to the larger of the two prongs, known as the **neutral wire**, provides a return path for the electrons and allows for a complete circuit. Without this wire to complete the circuit, 120-V devices would not work. The third prong on a plug, the round one, is the **ground wire**; it provides a low-resistance path for electrons to flow when the wire insulation or some component fails. Grounding serves three important purposes. First and foremost, it prevents people from getting electrocuted. Second, it provides a safe path for electrons when they are present in amounts that will cause fuses or circuit breakers to open, killing power to the system. Third, it prevents damage to equipment and people due to fire, which often occurs when electrons freely flow outside of the normal system constraints.

Three-phase AC is AC power that has three sine waves located 120° apart electrically. This is the primary power source for most industrial facilities due to the great amount of work it can perform and the fact that it is very efficient. With single-phase AC, there are points during the cycle where no voltage flows in the system and thus no work is done. Granted, this happens for only a fraction of a second, 120 times per second, but all of those fractions add up and represent an inefficiency. Three-phase AC avoids this delay because one of the three phases is always supplying power, so no loss

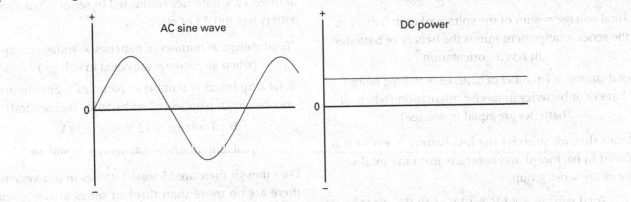

FIGURE 3-5 The left side shows the alterations of AC power; the right side shows the steady intensity of DC power.

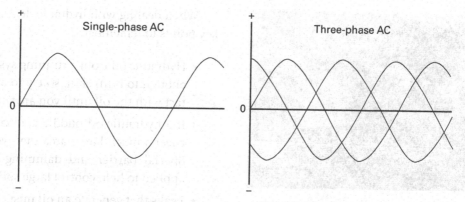

FIGURE 3-6 The three-phase AC (right side) always provides electricity to the system, whereas the single-phase AC (left side) has zero points—that is, times with no voltage flowing.

FOOD FOR THOUGHT

3-1 Single-Phase AC Versus Three-Phase AC

Whenever I am talking about the difference between three-phase AC and single-phase AC in my classes, I like to use the following example.

Imagine that your vehicle has run out of gas and you have to push it from the bottom of a hill (where you are currently located) to the gas station at the top of a small hill. As you push the vehicle up the hill, you take small breaks to catch your breath, which results in the vehicle stopping or slowing its forward motion. While you do not have to start over at the bottom of the hill, you do have to put extra energy into getting the vehicle moving once more. This is similar to single-phase AC.

Now suppose we have three people pushing the same vehicle up the hill. You start pushing the vehicle

up the hill, then a passerby stops to help, keeping the vehicle moving forward as you take a moment to catch your breath. About the time that your helper needs a break, a third person comes along and adds her muscle to the task. At times, only one person is really keeping the vehicle moving forward; at other times, two of the three are pushing the vehicle at least partially. Most important is the fact that the vehicle is always moving forward under force, so the forces of gravity and friction do not get the chance to fully stop or slow its forward movement. The vehicle ultimately gets to the gas station, but none of the three people is as tired as the single person completing the same task. This is similar to three-phase AC power.

If you would like to know more about electricity, consider doing your own research. You might also take a few courses in the field, as there is a wealth of information out there on this fascinating subject.

of momentum or inefficiency occurs due to times of no electron flow. Food for Thought 3-1 further clarifies the difference between single-phase and three-phase AC.

With three-phase AC, there are three hot wires, each of which supplies one of the three different sine waves. There is also a ground wire, just as in a single-phase AC circuit; it serves the same function as in the single-phase system. A neutral wire, however, may or may not be present. With three-phase power, because the hot wires are 120° apart electrically, one or two of the hot legs can act as a return or neutral wire, so that a dedicated neutral wire is unnecessary. A neutral wire provides additional safety in three-phase systems but is often omitted because it represents an added cost in terms of parts, wire, construction, and time. Figure 3-6 compares the sine waves for single-phase AC and three-phase AC.

Hydraulic Power

Hydraulic power involves the use of a non-compressible liquid given velocity and then piped somewhere to do work (Figure 3-7). We use hydraulic power when there is a need to generate great amounts of force with the precision control that is common with non-compressible fluids. This type of power allows robots to pick up car bodies and move them with ease, stopping precisely and holding position whenever and wherever needed. While some of these robots tend to work more slowly than their electric counterparts, they remain true workhorses in industry (Figure 3-8).

The tradeoff for this extra power is the addition of the hydraulic system to the robot and all the upkeep that goes with it. The oil needs to be filtered, monitored,

FIGURE 3-7 A hydraulic power supply. The pump is in the center, a pressure gauge and filter are on the left, and various points to access and return the hydraulic power are on the right.

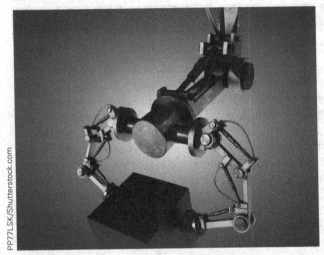

FIGURE 3-8 A hydraulically powered robot.

and usually changed out yearly. When changing out the oil, it is a good idea to open up the tank and remove dirt or other deposits. Filters also need to be monitored and changed, as failure to do so may result in damage to the system in the form of dirt and/or excessive pressure. While these are routine tasks for any hydraulic system, they do represent extra work running a hydraulic robot as compared to its electrical counterpart.

The other downside is that eventually most hydraulic systems leak. Leaks can range from a minor nuisance that is just a bit messy and easily fixed to a catastrophic failure that can result in gallons of hydraulic oil sprayed all over the equipment and anything in the immediate area. The most dangerous leak is the pinhole leak, which can turn the oil into a flammable mist and generate high pressure streams that can shear through metal and human tissue alike. (See Food for Thought 3-2 for more about a worst-case scenario.)

When dealing with hydraulic leaks, there are a few key points to remember:

- Hydraulic oil from a running system may be hot enough to burn skin, so try to avoid direct contact with the oil until you are sure it is safe.
- If a hydraulic oil puddle is uncontrolled, it can cover a much larger area than you might think. Special barrier and damming devices can be applied to help control large spills.
- Leaks that generate an oil mist can be a fire hazard. Most hydraulic oils are stable and take a lot of heat to ignite, except when dispersed as a mist in the air.
- Cleaning up hydraulic oil usually requires the use of some type of absorbing medium.
- You should not allow hydraulic oil to go down any drains connected to the sewer system, as it is a contaminant and will wreak havoc at water treatment facilities or in any natural waterways it might reach.
- When finished with the cleanup, make sure you dispose of any oil and oil-soaked materials properly. See the Safety Data Sheet (SDS) or a supervisor for more information on proper disposal. SDSs tell you about chemicals, their dangers, how to handle them properly, and how to dispose of them.

Pneumatic Power

Pneumatic power resembles hydraulic power but uses a compressible gas instead of a non-compressible liquid. One of the benefits of pneumatic power is the ability to vent the used air back to the atmosphere when it is no longer needed. With electricity, we have to return the electrons along some path to make it work; with hydraulics, we return the fluid to the tank. In contrast, with pneumatics, we simply vent, or release, the used air at a convenient point. When venting air, we need to be careful of two things. First, this process is often very loud. Second, small particles can become airborne due to the pressure of the escaping air. To combat both of these conditions, we use a muffler. The muffler slows the air as it passes through this device, reducing the sound and spreading the air that was flowing in a focused direction into a circular pattern (Figure 3-9). Even with the use of a muffler, you should use hearing and eye protection when working with pneumatic systems.

FOOD FOR THOUGHT

3-2 The Danger of a Pinhole Leak

During my time in the military, I worked on Huey helicopters. One of my sergeants at that time told me this story.

The Chinook helicopter is a workhorse for the military. With two main rotors, it is capable of lifting large and heavy cargo. This helicopter, like much of the equipment in the aviation field, depends heavily on its hydraulic system to run various things such as the cargo hatch in the rear, flight controls, and other systems in the aircraft. Because so many systems in the helicopter are dependent on hydraulics, any leaks are a matter of concern.

The Chinook in this story had a hydraulic leak. Personnel had noticed a reduced fluid level in the hydraulic reservoir, but no one could find any signs of the leak. After each flight the level in the tank would be lower, prompting the mechanics to look all over the helicopter for the tell-tale signs of a leak. A team of four looked for puddles, loose fittings, oily lines, and anything else that might show where the system was losing fluid, but they had no luck. This pattern continued for several flights, without the crew having any clue as to where the fluid was leaking from or where it was going to. The leak was not bad enough to ground the bird, but it was worrisome because eventually small leaks become big leaks.

Finally, the mechanic crew decided to start up the Chinook, and its hydraulic system, while the helicopter was sitting on the runway and see if perhaps they could find the leak that way. While making their inspections, one member was checking the outside of the helicopter for any signs of the leak and running his hand along the fuselage. Unfortunately, he was successful in finding it. It was a pinhole leak on the outside skin of the helicopter, which was venting hydraulic oil into the atmosphere. Because the leak was so small and the system pressure was so high, it cut right through the outer hull of the Chinook. The leak also cut four fingers from the soldier's hand as he ran it over the leak during the inspection. It happened so fast that the soldier did not even feel it at first! The last time my sergeant saw this soldier, he was on a medevac flight to a hospital in hopes of reattaching his fingers.

My sergeant told us this story so that we would realize how powerful and dangerous hydraulic systems are and to make sure we would pay attention to what we were doing. While this story is a bit graphic, it has stuck with me over the years and reminded me to be cautious when dealing with hydraulic systems.

FIGURE 3-9 On the left side are two mufflers: The gold cone is the muffler for the pressure regulation unit pictured, and the silver cylinder is a muffler for use where needed. To control the air pressure to the system, the yellow knob on top of the pressure gauge (center) is used. The brass connections on the right are points where air is supplied to the system.

Pneumatic power was the driving force for many of the early robots, as facilities had an abundance of air power, making it a cheap and easy power source to utilize. The early systems also used a simpler control system for positioning that consisted of a rotating drum with pegs on it, contacts or actuators for valves, and hard stops to limit the robot's travel. The pegs on the drum would rotate around and create various combinations of contacts or actuators to control how the robot moved, with the timing controlled by the speed the drum rotated and peg placement. The cylinders would extend until the robot hit a hard stop, preventing further motion. Because the cylinder was still under full pressure, this approach all but eliminated the issues encountered with mid-stroke positioning for air systems.

These early pneumatic robots were robust, fast, and easy to work on, but they also had a downside: They were often noisy due to the impact of the system with the hard stops. In addition, these sudden stops caused extra wear and tear on some of the robots' parts. These early robots picked up the nickname "bang-bang robots" due to the noise created as they hit one hard stop after another.

Today, due to refinements in the way we control and balance pneumatic power, some of the fastest robots in the world are air powered. These systems move so fast that their motion seems a blur. Parts or materials seem to be there one moment and then simply gone the next as the robot completes its tasks. Pneumatics is also a favored power source for end-of-arm tooling. Many grippers, suction cups, drills, dispensers, and other devices use air as the primary driver for their operation. One thing seems certain: Whether pneumatic power runs the whole robot or just various types of tooling, you can count on pneumatics being a part of the robotic world for the foreseeable future.

With each type of power having benefits and flaws, you may be asking yourself, "Which one is best?" or "How do designers pick the type to use?" To address these issues, we need to answer three questions:

- What do we want or need the robot to do?
- What do we have available?
- Where will the robot work or operate?

In terms of the appropriate tasks, pneumatic robots are good at light tasks that require speed, whereas hydraulic robots are mostly reserved for heavy work. Electric robots have found their way into numerous applications, and the electric servo motor has advanced to the point it can challenge hydraulic power for weight handling. Each of these options has both strengths and weaknesses. For instance, we would *not* want to use pneumatics for a robot that had to lift huge loads and maintain position if stopped at any point. In applications where strength and speed are needed, the electrical system often beats the hydraulic system.

When it comes to power availability, most industrial facilities have electricity and compressed air power sources available throughout the plants. Hydraulic power can be generated by adding a stand-alone system to the robot and powering the pump motor with electricity or some other means. Availability becomes a sorting factor mostly when one of the three sources is not present or should not be used for some reason. For example, a plant that is having trouble providing enough compressed air for its machines to run should avoid adding a pneumatic robotic system.

In terms of the area where the robot is operating, specific situations influence which type of power we should use. Hydraulic systems are often avoided when the area needs to be kept very clean or food processing is involved. In a highly explosive environment, electrical systems are less desirable due to the sparks they may produce. If the robot needs to roam around the plant, tying it to a compressed air system is a bad idea. Other factors, such as size of the work area, chemicals, heat, weight of parts, and people, must be taken into account as well.

Today, electrical power remains the go-to source for industrial robotics. This kind of power is cheap, plentiful, and easy to control. As improvements in servo motors have been made to point that they can now match the hydraulic power of old, the current electrical robots can achieve all the force with none of the hydraulic systems' cost, noise, or mess (Figure 3-10). Today's pneumatic robots rely heavily on electrical controls, making them more of a hybrid electrical system than a pneumatic robot. The only time you will likely find a robot that is purely hydraulic or pneumatic in nature is in applications in which electricity cannot or should not be used.

Photo Courtesy of FANUC America Corp. www.fanucamerica.com

FIGURE 3-10 A FANUC electrical robot moving a large assembly, a job that in years past was reserved for hydraulically powered robots.

CONTROLLER/LOGIC FUNCTION

Without some way to control the actions of the robot, the timing of those actions, and the sequence, the robot is essentially an expensive paperweight taking up space. The **controller** for the robot is the brains of the operation and the part of the robot responsible for executing actions in a specific order under specified conditions (Figures 3-11 and 3-12). It takes whatever sensor input is available for the robot, makes decisions based on a system of logic filters and commands called a **program**, and then activates various outputs as instructed by the program. (We will examine the process of creating and editing programs in Chapter 6.) When we want to modify the operation of a robot, doing so is often simply a matter of changing the program in the controller. Radical changes to the robot's operation, such as adding a new sensor or changing a gripper for a welding gun, often involve adding or changing wiring in the controller, possibly a firmware update, and a program change.

FIGURE 3-11 The controller for an ABB robot.

Image courtesy of ABB Inc.

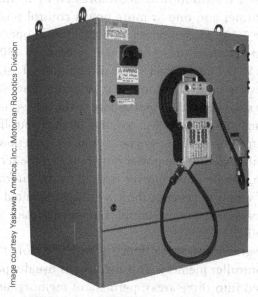

FIGURE 3-12 A Motoman controller for robots.

Image courtesy Yaskawa America, Inc. Motoman Robotics Division

Many of the early robots did not use feedback from the motors to monitor motion. In this kind of **open-loop control system**, commands are issued to various systems, such as stepper motors or control valves, and it is assumed that the robot did what it was told. The danger of this approach is obvious: If anything goes wrong, the robot has no idea. If an axis does not make it into position or a valve does not fire, there is nothing to warn the robot and in most cases the program continues. The crashes and damage that can result in these situations are pretty spectacular at times. Because of these risks, open-loop control is rarely used in modern industrial robots.

A **closed-loop control system** has some means of verifying proper operation, usually in the form of encoders or other position-sensing devices that confirm commands executed as directed. In today's robots, a microprocessor sends commands to a servo drive, which in turn controls the servo motor of a specific axis. As the motor moves, a feedback device (an encoder) sends positional information to the servo drive, which then sends information back to the microprocessor, which then verifies that this information matches the commands it issued in the first place. With this kind of system, the robot controller can detect when something is wrong and take some form of corrective action. In some cases, the corrective action is to stop and issue an alarm message for the operator. In collaborative systems, if the information received indicates there is a potential impact, the system usually initiates some type of damage control protocol, in which the motors are disengaged and the axis or axes involved soften to reduce any potential damage to people.

The robot controller has evolved as our understanding and technology have improved. The first controllers were open-loop systems that utilized drums with pegs in them or something similar to control the operation of the system, like the bang-bang robots. The second generation of controllers used relay logic to control motion but were still open-loop systems. These controllers required physical rewiring of the machine to change their operation. These early systems could perform tasks such as painting or spraying applications, but often they were used in material handling.

The third generation of controllers used computer control systems and could take advantage of system feedback. These controllers primarily used either a programmable logic controller (PLC) or computer numeric control (CNC) controller. With either type of controller, a change in operation was fairly simple to accomplish by changing either the ladder logic program in the PLC or the G-code program in the CNC. Because both of these systems can monitor inputs, the robot could actually monitor signals to verify correct operation.

The fourth generation of controllers utilized processors more like modern computers and took advantage of programming languages to get the job done. When they first emerged, these programming languages were designed for specific robots. This approach was soon abandoned, because people found it to learn the various languages, and they had no crossover from system to system. The next iteration utilized programming languages such as Basic and C+ to control the robot. Although this standardization was easier for programmers to work with, the languages themselves turned out to be less than great for the robot. Ultimately, a hybrid of the two approaches was developed that was easy for someone with a background in computer programming to work with but still did a good job of controlling the robot. Continued evolution of robotic programming languages led to those largely found in industry today. With the modern approach, the controller does most of the hard work in writing the code; you just have to set points, actions, and logic flow to create a program.

The fifth generation of controllers added in **artificial intelligence (AI)**, which is the ability of a computer program to make decisions when there is no clear-cut right answer or to learn from previous events. A standard system without AI often cannot make any decision where the choices do not provide some kind of mathematical advantage. In these instances, the system may fault out or lock up, waiting for a human to help it out. In addition, the system has no way to improve its process or programming without human intervention. With AI technology, the system can learn from data received, make certain modifications to the program, and make choices when no clear-cut best choice emerges from a group of options. As AI technology improves, we will see more of these systems utilized in industry.

The sixth generation of controllers now being used in industry comprises collaborative robot controllers. These systems utilize complex mathematics to calculate the amperage draw of various motors during motion and then carefully monitor the motors' operation. If the amperage draw is too high, the controller initiates a specialized fault routine to minimize any

damage to people or equipment. Collaborative robot controllers often monitor cameras or other systems for the presence of humans nearby; when a person is detected, they will slow down the robot to safe speeds. This and other advanced software functions allow the collaborative robot to work next to humans instead of behind a cage. Some AI functionality is present in these systems as well, typically in the ability to automatically offset programmed operation when parts are out of position and when other on-the-fly changes are needed to keep the robot performing its tasks. A side benefit is that most collaborative robots can be programmed by putting the system in teach mode, then grabbing the robot and moving it where you want it to go. This ability is a huge time saver over the traditional mode of driving the robot to positions using the teach pendant (discussed in the next section).

The robot controller is the main control for everything the robot does. It controls numerous aspects of the robot's operation:

- Power distribution
- Storing and executing program data
- Memory management
- Processing information from various sources
- Controlling servos
- Performing diagnostics
- Monitoring inputs and controlling outputs
- Interfacing with other machines and networks exterior to the robot

Power distribution is accomplished by using a transformer and one or more power control relays (Figure 3-13). Most modern industrial robots operate on standard three-phase AC power, typically at 480 V or 208 V. The transformer serves two purposes. First, it isolates the controller from the main power line. Second, it steps down the input voltage to the three-phase and single-phase AC power required by the controller. The three-phase AC power goes to one or more power relays that control the power distributed to the servo system. Single-phase power is used by an internal power supply to generate the low voltages required by the processor and other control components inside the controller.

Programs and other data are typically stored in the controller memory. This memory is usually partitioned into three areas: permanent memory, user memory, and working memory. Permanent memory is

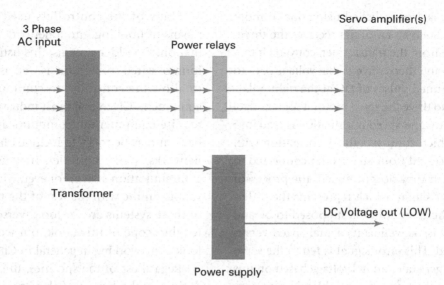

FIGURE 3-13 Power flows through the robot controller.

frequently stored on a series of electrical programmable read-only memory (EPROM) chips and comprises data that the user cannot modify. This data generally consists of a core operating software package, an application package, and a robot data file. The core operating package is developed by the robot manufacturer for the specific controller type, much like the operating system on a personal computer. The application package is developed by the robot manufacturer to enable its robots to perform a specific function, such as material handling, palletizing, or arc welding, similar to a word processor or media applications used on a personal computer. Some robot manufacturers load multiple application packages into a controller and allow customers access to other applications with the use of a password or code key. The robot data file, sometimes referred to as a robot library, contains information about the particular robot with which the controller is being used. It contains information such as robot arm length, payload ratings, servo characteristics, and other information specific to the type or model of robot. If the controller has the capacity to control several models of robots, it will either contain the data files for these robots or provide some way for the user to install them as needed.

The robot user sets up user memory, which stores all the programs written by the robot user. This memory also contains the robot input/output configuration, various communication parameters, and robot-specific data such as axis limits and calibration data, frequently stored in static **random-access memory (RAM)** chips. RAM allows information to be entered, changed, and

deleted, unlike **read-only memory (ROM)**. The user RAM memory usually has an on-board battery that allows the memory to retain the data if power is lost to the controller. This battery must be replaced periodically, usually every year or two, to prevent the loss of all the user data. If you do not replace the battery when it is running low, you may come in one morning and find everything entered previously by the user has disappeared, so that hours of time need to be spent getting the robot ready to go once more.

The controller processes information using one or more microprocessors. Most modern controllers use a dual-processor architecture. One processor is a dedicated motion processor, which handles all servo system processes. It monitors and controls robot position and speed, performs servo system diagnostics, controls communications with the servo amplifiers, and communicates with the other processor. The second processor handles all non-motion processes. It monitors and controls all input/output signals, monitors external safety circuits, communicates with PLCs and other devices, performs general system diagnostics, and communicates with the motion processor. Both processors communicate with each other via a common data bus, which is an internal communication path in the robot.

Some controllers may utilize more than two processors, in which case pieces of the tasks would be the responsibility of one processor, while another processor picks up the remaining tasks. The broad divisions would remain the same, except that more processors would be working on the tasks and, therefore, the entire process would be faster.

Servo control is accomplished using one or more servo amplifiers. Such an amplifier receives the three-phase AC voltage from the transformer, converts it to a high-voltage DC, and then converts that voltage back to AC via carefully timed pulses of DC at the right polarity and intensity to drive the servo motor. Program data containing position and speed information is read into the processor, which compares this information with feedback data received from an encoder connected to the servo motor. If they do not match, the processor produces an error signal, which represents the difference between where the robot is supposed to be and where it actually is, as well as its actual speed versus its program speed. This error signal is fed to the servo amplifier, which generates an AC voltage based on this data that is fed to the servo motor, which in turn controls the direction and speed of the servo motor. The processor continuously monitors feedback data from the encoder and adjusts the error signal accordingly.

In modern servo controllers, a series of diagnostics is built into the software that monitor both the hardware and the software of the controller. If a malfunction occurs, a fault message is sent to the operator via the teach pendant or some other type of interface. This message includes the fault type, fault number, a descriptive message, and possibly axis or motor identification. The fault type is an alphabetic or numeric code that tells the operator or maintenance technician where the specific fault has occurred. The fault number is simply an identifying number attached to the fault code by the programmer. The descriptive message gives a brief description of the type of fault. The axis or motor identification may appear if the fault can be isolated to a specific axis or motor. These messages are intended to aid the operator or maintenance person in troubleshooting and repairing the controller.

The controller, through the software, monitors and controls the various inputs and outputs used by the controller. It can check the status of an input, turn an output on or off, and perform diagnostics on various inputs and outputs. Diagnostic messages are generally limited to configuration errors and other software malfunctions; they generally do not indicate whether a malfunction has occurred with a specific input/output device. With some creative programming, users can create fault routines that give information about missing inputs, monitor for inputs or outputs that are changing too rapidly, or trap other helpful information that goes beyond the data produced by the factory diagnostics.

Many of the controllers used today have some means of hooking into a network and sharing information. On older systems, this usually requires some form of wired connection as well as an adaptor or two to allow the controller to communicate with other equipment. Given that most industrial networks today require equipment to communicate on a plant-wide basis, it may be tricky to include older systems in such networks. Many controllers now utilize standardized communication systems or even wireless devices to get robots online with the rest of the plant. The specifics of these systems are far too diverse and complicated for the scope of this book, but we will take a closer look at networking in general in Chapter 9.

Regardless of the specifics, the controller is most definitely the brains of the operation. Without the controller, a robot is just a high-priced piece of industrial equipment taking up space. Several robotics companies' claim to fame is based on their advancements in the controller field. For example, Yaskawa Motoman is well known for pushing the bounds of the controller by increasing the number of axes controlled by a single controller. When you get out in the field, you would do well to learn all the specifics of the controllers used by the robots you are working with.

TEACH PENDANT/INTERFACE

To accompany the controller, we need some way to communicate our desired changes to the robot's programming or operations. In most industrial robots, we use the **teach pendant** for this purpose. It allows the operator to view alarms, make manual movements, stop the robot, change/write programs, start new programs, and perform any of the other day-to-day tasks required to run robots (Figure 3-14). In industrial settings, anyone who is in the robot's danger zone *must* have the teach pendant present to help ensure safety. We call the teach pendant a human interface device because we use it to control or interface with the robot's operation. Often, we use the teach pendant to write new programs without the need for using any other software or engaging in other types of computer interaction with the robot, saving time and aggravation in the programming process.

Teach pendants vary in style, operation, size, and complexity from manufacturer to manufacturer and sometimes from model to model of the same brand of robot. Even though they come in a wide variety

FIGURE 3-14 The teach pendant for an ABB robot. Notice the red E-stop on the upper-right corner.

FIGURE 3-15 The two yellow switches pictured here are the dead man's switches for a Panasonic robot's teach pendant. It does not matter which one is depressed, but at least one has to be depressed at all times to allow manual movement of the robot.

Safety Note

OSHA requires operators to take the teach pendant with them whenever they are in the danger zone, so they always have access to an E-stop. The last thing you want is for the robot to start up when you are nearby and have no way to stop the system.

of configurations, some standard features are found on most pendants, such as an E-stop and dead man's switch. The E-stop is an emergency stop that shuts down most of the systems of a machine. The **dead man's switch**, which is usually a trigger- or bumper-type switch on the back of the pendant, is required to move the robot manually (Figure 3-15). If you release the switch or press down too hard, the robot will stop moving. If something goes wrong—say, the robot hits you on the head—your natural reaction is to either let go or grip the switch tighter, so both of these actions stop all movement when a dead man's switch is present.

Other items you can expect to find on a teach pendant include a display to give you information, some way to move the robot, and a method to record the robot's position for programming purposes (Figure 3-16). What you push to move the robot, the other options that may be available, and the data you can find on the pendant are up to the manufacturer.

Generally speaking, there is a good chance that you can control and modify most, if not all, of the desired functions of the robot from the teach pendant. Figures 3-16 and 3-17 give you some idea of the various teach pendant configurations you may encounter.

Some of the new industrial systems, like the Baxter robot, are moving away from the traditional teach pendant in favor of coupling special modes and the sensors of the robot with AI-enhanced software to learn new tasks. (In the case of Baxter, the tablet head is the closest thing it has to a teach pendant.) Many of the collaborative robots with the newer controllers allow

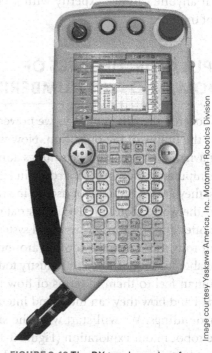

FIGURE 3-16 The DX teach pendant for use with Yaskawa Motoman systems.

FIGURE 3-17 The teach pendant for the FANUC system my students use in class.

FIGURE 3-18 A robot arm produced by Motoman.

direct manipulation of the robot for programming, negating the need to use the teach pendant to drive the robot to each point.

As industrial robots continue to evolve, human interface devices like the teach pendant will change and evolve as well. The ultimate goal is to create a robot that anyone can use expertly with a minimal amount of training.

MANIPULATOR, DEGREES OF FREEDOM, AND AXIS NUMBERING

We have covered the different ways we power robots as well as the brains of the operation. Now we focus on the brawn of the system that gets things done—that is, the manipulator. **Manipulators** come in all shapes and sizes; they are what the robot uses to interact with and affect the world around it. For industrial systems, these are often robotic arms or overhead systems with a series of rods that move the tooling around. Many of the collaborative robots used in industry today have a very human feel to them in terms of how they are constructed and how they can move and interact with their surroundings. We will start with the standard arm-style robot in our exploration (Figure 3-18), and then expand our discussion to some of the other types found in industry toward the end of this section.

Manipulators consist of various parts such as motors, linkages, housing, cables, and other parts needed to complete the various tasks of the robot. Each part of the robot that has controlled movement is called an axis. Each **axis** of the robot gives the robot a **degree of freedom (DOF)**, or one more way that the robot can move. The more DOFs a robot has, the more complex and organic looking the movements. For instance, many gantry-style robots in industry have only three axes. They can move up and down (their z-axis), they can move back and forth (their y-axis), and they can rotate the end-of-arm tooling (their a-axis). As you can imagine, the motions of this system are very square and limited. Most industrial robot arms have five or six axes, with six axes being the more common. A six-axis robotic arm is capable of mimicking most human motions, giving the system a great deal of flexibility with six DOF (Figure 3-19).

To identify the axis of an arm-style or similar robot, it is common practice to start at the base and number outward. The **base** of a robot is where a nonmobile robot is mounted to or bolted on a solid surface for stability, or where a mobile robot's platform for a manipulator is mounted. From the base, we travel toward the end-of-arm tooling and number each axis as we go, starting with 1. Figure 3-20 shows the axis numbering for an ABB robot. When the manipulator is placed on a mobile base, we commonly assign this axis's number last, as an additional or optional axis to

Image Courtesy of Miller Welding Automation

FIGURE 3-19 Panasonic has mounted one of its welding robot arms to a gantry base, combining the large area of the gantry style with the flexibility of a six-axis arm. This creates a "best of both worlds" kind of system.

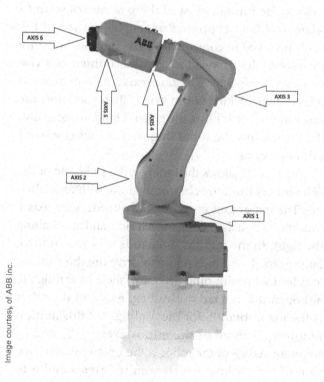

Image courtesy of ABB Inc.

FIGURE 3-20 Numbering the axis of an arm-style robot.

general area of use, while the **minor axes** are responsible for the orientation and positioning of the tooling. If we compare the robot to the human body, the major axes (1–3) would be like your torso, axis 1; your shoulder, axis 2; and your elbow, axis 3. The minor axes (4–6) would be like your wrist, with your hand representing the tooling. We often define the minor axes as **pitch** (axis 4), which is the up-and-down orientation of the wrist; **yaw** (axis 5), which is the side-to-side orientation of the wrist; and **roll** (axis 6), which is the rotation of the wrist. When these two systems are combined in robotics, the resulting six-axis robot uses the major axes to copy our body motions and the minor axes to copy our wrist motions while performing tasks (Figure 3-21). In fact, the robot beats the human body in terms of mobility: Many of its rotary joints can turn nearly 360°, with axis 6 often capable of performing several complete revolutions before any damage to the system, usually in the form of airlines or cables under too much tension.

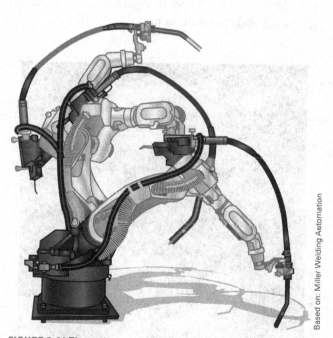

Based on: Miller Welding Automation

FIGURE 3-21 The major axes of an arm-type robot in action.

External axes are the axes of motion that often move parts, position tooling for quick changes, or in some other way help with the tasks of the robot. They are not part of either the manipulator or the major or minor axes of the robot. These axes are controlled by the robot controller and often assist the robot in its tasks but are considered separate from the manipulator. When eternal axes are utilized, they often come as part of a packaged robotic system where they are

the system. If the robot in our example of axes numbering were mounted on a mobile base, we would refer to the base as axis 7. Axes numbering is of primary importance when troubleshooting robot malfunctions. The alarm will often tell you which axis is at fault, but this information is relatively useless if you are unable to determine which physical axis of the robot faulted out.

The axes of the robot are broken into two main groups, the **major axes** and the minor axes. The major axes are responsible for getting the tooling into the already integrated and ready for use once the initial installation of the robotic system is complete. Since external axes are not part of the manipulator, we do not count them in our DOF, but they do present added motion options and so increase the overall flexibility of the system.

While the robotic arm is the most common configuration for industrial robots, other configurations are also possible. The **delta robot** or **parallel robot** is designed for either an overhead mount or a side mount; it consists of three rotary major axes at the top with any minor axis motion being driven by shafts that connect to the tooling and rotate. Figure 3-22 shows the linkages for the major axes at the top attached to, and coming out of, the big yellow base. If you follow the rods in the center down to the tooling, you can see three relatively straight shafts that attach to the middle of the yellow base. These shafts drive the minor axes for the tooling. Many delta- or parallel-style robots have only three or four axes. By comparison, the FANUC model shown in Figure 3-22 has six axes, giving it a wide range of motion.

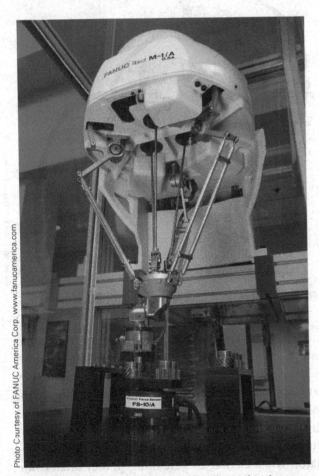

FIGURE 3-22 One of FANUC's six-axis delta-style robots.

Photo Courtesy of FANUC America Corp. www.fanucamerica.com

Numbering the axes of a delta- or parallel-style robot is a bit difficult, as the configurations vary by manufacturer for these types of robots. Because of the circular nature of the base of these robots, the starting point is not nearly as universal as that found in the standard robot arm, so the first thing you need to do is identify axis 1. My recommendation is to use the teach pendant to manually move the robot in joint mode, which moves only one motion axis of the robot at a time; select axis 1; and see which set of linkages moves. Once you have identified axis 1, the industry standard seems to be to number the other axes, in order, in a counterclockwise direction. Nevertheless, you should be aware that several patents and papers on delta-style motion show the axis numbering in a clockwise format. Most of these resources seem to relate to either 3D printers or general analysis of the math involved in coordinating axis motion. To verify your robot does indeed follow the counterclockwise trend, you simply have to move axis 2 in joint mode. If it is counterclockwise from axis 1, then you know the axis numbering follows the general guidelines; if not, then you know the robot in question uses clockwise axis numbering.

Figure 3-23 shows the inner workings of one of the delta robots from my classroom along with axis labeling. The major axes are along the outside, with axis 1 near the back, axis 2 along the left side, and axis 3 along the right. In the middle, three rods stick out of three large gears. These rods rotate to drive the three minor axes for tool positioning. They are labeled 4 through 6, although this is a tad misleading. Axis 4 of the robot is the main rotation for the tooling, but this motion requires all three of the minor axes (4, 5, and 6) to rotate. Axis 5 of the robot is the up-and-down rotation of the tooling, which requires axes 5 and 6 to rotate. Axis 6 of the robot is the rotation of the tool specifically; only axis 6 rotates when this is in motion. Because of this setup, a problem with axis 4 could involve one, two, or three different motors and/or their linkages. This is another reason why numbering the axes of delta-style robots is a bit difficult.

Figure 3-24 shows the part of the minor axes that moves when you move the robot in joint mode for each axis number. Four-axis delta robots will have only one minor axis rod, making it simpler to identify the axes. If your delta robot has five or six axes, I would recommend moving each axis separately, watching what the robot does and how it moves, and then recording this information for reference.

FIGURE 3-23 One of the delta-style robots used in my classroom, minus the top cover so you can see the axes better. The three major axes are positioned around the outside ring, and the three minor axes on the inside are driven by the large gears attached to the rods sticking up.

FIGURE 3-24 To clarify how the robot is actually moving, the axis 4, 5, and 6 labels point to the specific rotation they cause instead of specific drives (as shown in Figure 3-23).

Another industrial robot in which it can be tricky to number axes is the dual-arm robot. You might think that this style would use the same axis numbering as the standard arm-style robot, but the addition of another full set of axes complicates things a bit. First, you need to determine if there are any rotary axes that move the torso of the robot and, therefore, both sets of arms. If any are present, likely these are labeled first. Next, you have to determine how the system designates

the right and left arms of the robot. For example, Yaskawa Motoman uses R1 for what would correspond to a human's right arm and R2 for what would correspond to a human's left arm on the DA20 dual-arm robot. Once you have the arms identified, you must then identify the various axes in each arm. On the DA20, instead of axes 1–6, the designations L, U, R, B, and T are used for the various axes of arm motion. Other robots, such as ABB's YuMi®, use standard numbers for each axis, but you will still need to figure out how they designate the right and left arms.

If you encounter one of the newer robots in your workplace or if you forget the commonly used axis numbering pattern for arm-style robots, remember that you can simply put the robot in manual mode and use joint motion to identify each specific axis. I recommend that you do this *before* you need to troubleshoot your robot: It may be too late to figure out the axis numbering after the robot alarms. When working with robots, it is always a good idea to create some "cheat sheets" with information such as the axis numbering, any commands specific to that robot or robot model, and any other information you might need to find out but rarely use. As the field of robotics continues to evolve, it is difficult to predict which configurations or labeling systems you will encounter over the course of your career. You will need to learn all you can about the new systems as you work with and around them in industry.

BASE TYPES

Industrial robots are often bolted to the floor or a solid steel structure and can work on only things within a very finite area of reach. This model works great for many applications of the robot, but what happens when you need a really large work envelope or need the robot to work with multiple machines in an area? In such a case, you must figure out how to make the robot mobile. In this section, we look at both the solid mount and mobile options in greater detail so you can have an idea of what to expect in industry.

Solid Mount Bases

A **solid mount base** involves mounting the robot firmly to the floor or to other structures using bolts and fastening systems. Solid mounting of the robot allows the system to maintain a very specific coordinate base to work from, which in turn allows industry as a whole to make full use of the precision of robotic systems. Solid bases may be anchored to concrete floors, secured to building walls, mounted on overhead structures, or even secured inside of machine systems, depending on the requirements of the job (Figure 3-25).

FIGURE 3-25 Two ABB robots. The one on the right is mounted parallel to the floor, and the one on the left is mounted to the side of the pedestal or a wall-type mount.

No matter how the base is mounted, there are few key points to remember:

- Make sure whatever holds the robot in place is robust enough to bear the forces and weight of the system. Bolts, nuts, mounting plates, and so on all have limits regarding the amount of force they can withstand.

- Make sure what you mount the robot on/to can handle the weight of the system *and* whatever load it will be maneuvering.

- Check the security/tightness of any mounting hardware periodically, paying special attention to any noted wear.

- Crashes are conditions in which the robot endures unexpected forces, so check the base when you are inspecting the system (especially for wall or overhead mounts).

If you follow these simple guidelines, you should be able to avoid most of the worst-case scenarios. As each setup will have its own unique circumstances and requirements, you may find the need to add to these basic rules on a case-by-case basis.

Mobile Bases

As the name implies, **mobile bases** use systems to move the manipulator to various locations so it can perform its functions. Some of these bases are restricted to liner rails that allow the robot to move back and forth over a finite area, whereas others give the system a great range of mobility and freedom. The type of mobile base used generally depends on what the robot does as well as the environment in which it works. For instance, a wheeled base would not be the best fit for a robot designed to feed parts precisely into machines from an overhead position. Just like many other aspects of robotics, the base should fit the task.

Linear bases with a finite reach are often referred to as a **gantry base**. This name comes from their similarity to the gantry robot as far as the movement is concerned. A gantry base usually consists of a rigid track along which the robot moves back and forth. These tracks can be mounted on the floor, mounted over machines on frames, or attached to walls as needed (Figure 3-26). The size of the gantry base depends on the type of parts or tasks the robot is expected to work with—it may range from 10 to 15 feet in length, all the way up to systems stretching well over 100 feet in length. Sometimes a single robot runs up and down the base; other systems have two or more robots sharing this base. In the latter case, users must pay special attention to the accuracy and strength of the track's

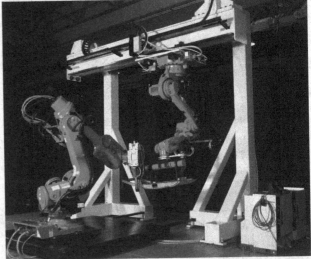

FIGURE 3-26 A standard solid mount and a small overhead gantry mount working together to get the job done.

mounting as well as how the robots share the length of said base. Misalignment of the track can cause the robot to alarm out or miss its designated positions. Improper programming or robot management can cause the robots to impact one another, which often leads to a spectacular mess of damaged robots and busted parts.

Another base used on the industrial floor is the **wheeled base**. This base consists of wheels, a drive system, and some form of navigation system so the robot can find designated points and avoid obstacles along the way. This type of base is very popular in the material-handling and -moving industry and has been the basis of automating parts retrieval and storage. In some completely automated warehouses, wheeled robots perform the tasks once accomplished by humans and forklifts. Such systems are also used to take pallets of parts between departments in industry, in which they are equipped with sensors and other guarding devices so that any people who happen into the robot's path are not in danger.

Industry is warming to the idea of using mobile bases to perform tasks besides material handling. For example, the partnership between Yaskawa Motoman, which supplies the robot arms, and Clearpath, whose OTTO division provides the mobile base, has yielded a mobile robotic arm that can tend machines or other such tasks over a much larger area than the traditional gantry base. If industry buys into this configuration for mobile robots, you can expect other companies to form partnerships of this nature or develop their own mobile bases for the robots they sell.

Other bases may also be encountered in industry, as some unique needs may sometimes call for unique solutions. The majority of industrial robots will fall into one of the categories mentioned here, but even in these cases unique situations and requirements arise from time to time. The major point to remember is that, regardless of base type, the robot must be mounted in such a way it can perform the required tasks accurately and safely. If you notice excessive wear, mounting hardware working loose, instability of position, or anything else amiss that traces back to the base mount of the robot, you may have to change the mounting style or make some drastic changes to the system to get things secure, stable, and safe once more.

REVIEW

By now, you should be familiar with the basic systems of the modern robot and have an understanding of the role of each system. We will continue to build on this system as we explore the world of industrial robotics. If you have access to an industrial robot, I encourage you to take the time to learn the specifics of the system, as this information will prove very beneficial for whatever you do with the robot. Over the course of this chapter, we covered the following topics:

- **Power supply.** This section focused on the common forces used to run robots and provided information about how each of these power sources works.

- **Controller/logic function.** We learned about the brain of the robot and its importance.

- **Teach pendant/interface.** This section explored how we communicate with the robot, direct its actions, or make changes to its operation.

- **Manipulator, degrees of freedom, and axis numbering.** We learned how to number axes, how we move the robot around, and what degrees of freedom are.

- **Base types.** We saw how to mount the robot and the benefits of each base type.

KEY TERMS

Alternating current (AC)	Delta robot	Mobile base	Resistance
Amperes	Direct current (DC)	Muffler	Roll
Amp-hours (Ah)	Electricity	Neutral wire	Root mean square
Artificial intelligence	Electromotive force	Open-loop control	(RMS)
(AI)	(EMF)	system	Single-phase AC
Axis	External axis	Parallel robot	Solid mount base
Base	Gantry base	Pitch	Teach pendant
Closed-loop control	Ground wire	Pneumatic power	Three-phase AC
system	Hertz (Hz)	Polarity	Voltage
Controller	Hydraulic	Program	Wheeled base
Cycle	power	Random-access	Yaw
Dead man's switch	Major axes	memory (RAM)	
Degree of freedom	Manipulator	Read-only memory	
(DOF)	Minor axes	(ROM)	

REVIEW QUESTIONS

1. What is the difference between AC and DC electricity?

2. One amp is equal to _____ electrons flowing past a measured point in 1 second.

3. What could happen if you reverse the polarity of DC components?

4. How would you connect a group of batteries to increase both the voltage and the amp-hours?

5. Describe what happens to AC power through one complete cycle.

6. What is the difference between single-phase AC and three-phase AC?

7. List at least three things to remember when dealing with hydraulic leaks.

8. What do we have to be careful of when venting pneumatic power and how do we avoid these dangers?

9. Describe the operation of a pneumatic robot controlled by a drum.

10. What is the function of the robot controller?

11. What is the difference between an open-loop control system and a closed-loop control system?

12. What is artificial intelligence in the context of robotics?

13. List four things the robot controller is responsible for.

14. What are some of the things we can do with the teach pendant?

15. Why does the dead man's switch kill the robot's manual actions when released or pressed too hard?

16. What is the benefit of having more degrees of freedom in a robot?

17. How do we number the axes of an arm-style robot?

18. What are the two main groupings of axes, and what is the function of each?

19. What do we call axes 4, 5, and 6 in most arm-style robots?

20. Do we include external axes in our DOF count? Why?

21. What is the best way to identify axis 1 in a delta-style robot?

22. What is the recommended method to identify the axis of the robot when you encounter one you are not familiar with or you forget the conventional numbering methods?

23. What are some of the key points to remember when solid-mounting robots?

24. What are gantry bases?

Image courtesy of ABB Inc.

CHAPTER 4

Classification of Robots

WHAT YOU WILL LEARN

- How to classify robots by their power source

- How to classify robots by their work envelope and the kind of reach they have

- How to classify robots by their drive system

- How the International Standards Organization (ISO) classifies robots

OVERVIEW

As a part of human nature, we compare things that we perceive to have equivalency as part of our decision-making process. We prefer to first put all the things that are similar together, and then whittle down our choices by needs or desires. In the robotic world, this decision-making process can be quite tough, as there are many choices—that is, many types of systems that will do the job, and many options within robotic subsets to choose from. In this chapter, we look at some of the classifications for robots in an effort to help with the initial steps in grouping robots to find what you need for an application. As we look at the various ways to classify robots, we will cover the following topics:

- How are robots classified?
- Power source
- Geometry of the work envelope
- Drive systems: classification and operation
- ISO classification

HOW ARE ROBOTS CLASSIFIED?

When we group industrial robots, we have to first define a unifying characteristic. We can classify them by which type of power source they use, how we program them, what they do, how their internal systems are organized, how they move, or any other criteria that we deem important. The most commonly used groupings for industrial robots are based on power source, drive type, work envelope, and application, but these are by no means the only ways robots can be classified. Some agencies, such as the **International Standards Organization (ISO)**, which specializes in creating uniform standards of operation, have created specific classifications for robots, including industrial and nonindustrial types. In general, we classify robots by some primary feature, function, or operation that loosely describes a finite set of industrial robotic systems.

Once we have decided on a unifying characteristic of a class, we can then begin to find all of the various robots that fit the criterion. Often, a large number of systems may match the criterion. For example, if we specify "electrically powered robots" as our criterion and then find all the various systems that fit this classification, we would have thousands of robots to choose from. In such a case, we would likely need to narrow things down further by identifying an additional classification and possibly some specific parameters, such as **payload** (i.e., how much weight the robot can move). If the specifics of the job are more important than how the robot is powered or how it moves, then a classification based on application would be best. While robots are flexible machines that can perform a wide range of tasks, many industrial robotics companies specialize in systems that are optimized for industrial tasks such as painting, welding, material handling, and so on. Again, the *how* of classification often depends on the *why* or *what* we are looking for.

As we explore the common classifications for industrial robots, keep in mind that these are just some of the ways in which we can group robots. In 2012, Baxter was the primary collaborative robot on the market; only five years later, there were at least 18 different systems billed as "collaborative." This is a great example of how the field of robotics continues to grow and change—and as it does, so will our classifications. While new classifications, such as collaborative robots, may emerge in the field, the groupings covered in this chapter have stood the test of time and will serve you well as a basis for classifying and understanding the operation of the various robots in industry.

POWER SOURCE

Sorting robots by their power source is an easy and obvious starting point when classifying robotic systems. Under this rubric, the groups consist of robots that use the same type of power to perform their functions. This classification method works well for robots and their tooling, as everything that performs work needs some kind of power to make that happen. In this section, we will take a quick look at the three main types to make sure everyone is on the same page.

Electric Power

Electric robots use electricity to run the motors and everything else, with the possible exception of tooling, on the robot. Two types of electricity may be used: alternating current (AC) or direct current (DC). (If you have forgotten the difference, refer back to Chapter 3.)

DC provides a greater amount of torque, but if the motors use brushes, they will require more maintenance. **DC brushes** are made of carbon and transfer the electricity from the power wires going into the motor to the rotating portions of the motor. Because they are in contact with the rotating portion of the motor, they tend to wear down over time and generate sparks that could become an ignition hazard under the right conditions, such as when a robot is painting parts with a flammable paint. DC brushless motors avoid the spark and mess problem but cost more. DC is also the common power choice for mobile systems, as batteries provide direct current.

AC is the most common power choice for industrial systems, as it is readily available, has low maintenance needs, and drives both stepper and servo motors (Figure 4-1). **Stepper motors** move a set portion of the rotation with each application of power: The more steps per rotation, the finer the position control. Stepper motors were often used in early robots and are still favored for many applications where there are other means of verifying position. **Servo motors** are a continuous rotation-type motor with built-in feedback devices called **encoders**, which provide feedback about the motor's rotational position (Figure 4-2). High-end encoders also provide information about the speed and direction of rotation as well as the total number of degrees the motor has moved. The robot controller uses feedback from encoders to determine when to stop the motor for proper positioning and to verify that the system is working correctly.

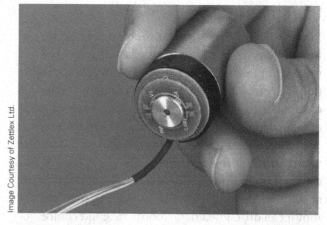

FIGURE 4-2 The cover has been removed to expose the encoder on this small precision motor. These types of encoders allow for high levels of precision in small robotic systems.

Image Courtesy of Zettlex Ltd.

the robot and allows it to generate great amounts of force. Improvements in AC servo motors have eroded the numbers of hydraulic robots in use, but these powerhouses of the robotic world are still found in industry today.

The great power of the hydraulic robot does come at the following costs:

- Hydraulic leaks
- Cost of oil
- Fire hazard
- Increased maintenance
- Increased noise

Unless a rigorous preventive maintenance program is in place, most hydraulic systems will leak sooner or later. Leaks cause big messes, present fire hazards under certain conditions, and can damage other components as the fluid interacts with them. When it comes to maintenance costs for hydraulic robots, the dollars will add up as mechanics replace parts worn out from normal operation, test the oil to ensure nothing is going wrong, and change out the oil every six months to one year, on average, depending on the number of running hours and the type of oil used. In addition, the hydraulic pump and any cooling system add to the noise level in the work environment.

FIGURE 4-1 This FANUC robot is stacking bags of chocolate on a pallet. The black objects with the red cap are the servomotors, and the large gray square under the robot that has the red cables by it and the slits is the AC power supply for the robot.

Photo Courtesy of FANUC America Corp. www.fanucamerica.com

Hydraulic Power

Hydraulically powered robots use a non-compressible fluid given force through velocity to drive the moving parts of the robot. The name assigned to this category can be somewhat misleading, as some electrical power is also required, primarily for the hydraulic pump, valve control, and system controls, plus any additional sensors. Nevertheless, the hydraulic power moves

⚠ Safety Note

Pinhole leaks are the most dangerous type for hydraulic systems. They can turn the oil into a flammable mist and generate forces great enough to cut through metal.

Pneumatic Power

Pneumatic systems work in much the same way as hydraulic systems but with one key difference: They use a compressible gas instead of a non-compressible liquid. One of the main problems with pneumatic systems relates to positioning of the robot. Because gas is compressible, the only way to hold the robot's position is at the extremes of travel or by keeping the system under constant force against some form of stop. You can stall out a pneumatic motor and not damage the system, but as soon as you stop applying pressure, the motor can drift. Pneumatic robots are good choices for areas that have physical stops the robot can work with but are less adept at holding position without the aid of additional equipment.

Pneumatic systems are excessively noisy, especially at points where we vent the air. They are generally inexpensive to run, as many plants already have a supply of compressed air plumbed to most, if not all, areas of the facility. Many of the early industrial robots used pneumatics for the prime moving power, but this power source has since given way to the AC servo motor.

One area where pneumatics has really taken root is in tooling. Pneumatics is the go-to power source for grippers, drills, sprayers, and vacuum systems in industry (Figure 4-3). Thus, many electrical robots use pneumatic tooling.

Pneumatic systems do require some extra maintenance but not as much as their hydraulic counterparts. The main concerns are protection of air lines from damage, the compressibility of gas, and the noise the system generates. Since most of the lines for a pneumatic system are manufactured from thin-walled rubber or plastic, it does not take a lot of force to damage them. Any holes or leaks will lower system pressure and can send chips flying. If the line is cut into two, it can whip around under pressure, damaging people and equipment alike. Compressed air is not returned to the storage tank, but the used air has to go somewhere. At some point in the system, air is returned to the atmosphere, usually under pressure, which creates a lot of noise—even with a muffler in place to diffuse it.

While power source is one of the first classifications used for robots, it is often a poor way to select a robot for industry today. For most applications, we end up going with an electric robot—and that category includes a huge number of robots. As a consequence, sorting by how the robot moves and where it can reach (the topic covered in the next section) is often a more useful method of identifying robots with similar capabilities.

FIGURE 4-3 The two blue lines running to the gripper on this robot provide the force for gripper motion.

GEOMETRY OF THE WORK ENVELOPE

Another popular way to classify robots is by the area they can reach—that is, by the **work envelope**. This basis has long been one of industry's favorite ways to classify robots, as it gives the user a good idea of how the robot will move and how it can interact with the world around it. When analyzing the work envelope of a robot, you have a chance to utilize what you know about geometry and quite possibly may pick up some new insights as well.

While at first glance it might appear that the number of possible combinations of robot arm geometry is unlimited, the types of robots in a modern industrial plant can actually be categorized into two basic types: serial link or parallel/delta. In the serial link robot, all the arms are connected to each other in series, one after the other, much the same way the human arm is formed by starting at the shoulder, moving to the elbow joint, moving down to the wrist, and then out to the individual finger joints. In the parallel or delta configuration, the robot arms are mounted to a common base and set up parallel to each other. All of the arms are connected to a common platform at the end, where the tooling or work piece is mounted. Figure 4-4 shows two parallel- or delta-style robots in a work cell with a serial link or articulated arm robot. One of the parallel arm robots has a deburring tool attached to the faceplate; the other has a holding fixture for a small engine block. The large serial link robot is used to load the engine block onto the holding fixture.

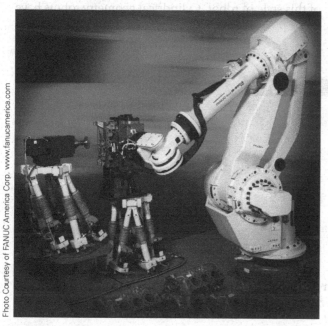

FIGURE 4-4 Two parallel- or delta-style robots work with an engine block, while the large serial or articulated arm-type robot does the heaving lifting involved with loading and unloading.

The serial link robot is by far the most popular and most commonly seen robot in industrial manufacturing facilities. Five basic types of serial link robots are used: Cartesian, cylindrical, spherical, articulated, and selective compliant articulated robot arm (SCARA). These types of robots have been around for many years, although the articulated robot is the type most commonly seen in modern industrial facilities. The configuration of the axis and the type of motion they generate create the work envelope for the robot. In turn, the similarity of this envelope to geometric figures is how many of the divisions were named. In this section, we take a closer look at each kind so you can have a better idea of how these various robots move as well as some of their strengths.

Cartesian Geometry

Cartesian robots have a cubic or rectangular work envelope. Many of the gantry-type robots fall into this category. They tend to move in a linear or straight-line fashion. These robots often have two or three major axes to move in (x, y, and z), with x being front to back, y side to side, and z up or down. When there are only two major axes, x is often the axis that is omitted.

Cartesian robots are popular for loading and unloading parts as well as for moving materials over large distances. The system may be mounted over the equipment it serves, which saves floor space. This style has been revitalized by mounting an articulated-type robot on a Cartesian/gantry-style base, resulting in a huge work envelope with the flexibility of an articulated robot (Figure 4-5).

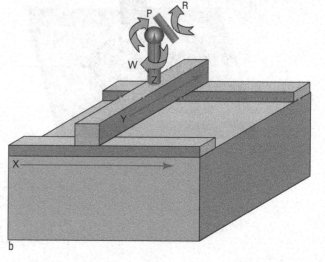

FIGURE 4-5 (a) A gantry-type robot welding a large tank. (b) The rectangular work envelope and axis motion of a Cartesian system.

The Cartesian geometry model offers several advantages. As mentioned, such robots can cover large areas and can be mounted over equipment, saving floor space. The overhead mounting also offers easier control over the position of the robot wrist.

The Cartesian geometry model also has some disadvantages. The large size of the work envelope creates difficulty in accessing and working on the robot due to other machines, electrical conduits, guarding, fixtures, and other such obstructions getting in the way. Also, depending on the location of the operator station, visibility may be less than ideal. One of the most important safety considerations when working with industrial robots is the ability to see the robot, its location, and the work piece that it is manipulating.

Cartesian geometry robots have found applications in a variety of manufacturing systems. They are ideally suited for material handling, particularly of large components, as the overhead mounting makes it easy to move large objects without worrying about obstructions or other equipment on the manufacturing floor. They are also used in assembly of components, particularly automotive components, and especially heavy-duty components.

Cylindrical Geometry

Cylindrical geometry robots, as their name implies, have a cylindrical work envelope and in many ways are the "rotary cousin" of the Cartesian geometry robot. The common structure for this type of robot has axis 1 rotary, axes 2 and 3 linear, with the usual two or three rotary orientation axes for the tooling (Figure 4-6). In essence, the *y*-axis of the Cartesian geometry robot has been replaced with the rotary axis of the cylindrical geometry robot.

These systems are good choices for reaching deep into machines, save on floor space, and tend to have the rigid structure needed for large payloads. On the downside, with large payloads, a counterweight may be required, which increases the load on the motors. Also, the loss of the *y*-axis travel limits the work area of this style of robot. Cylindrical geometry robots have

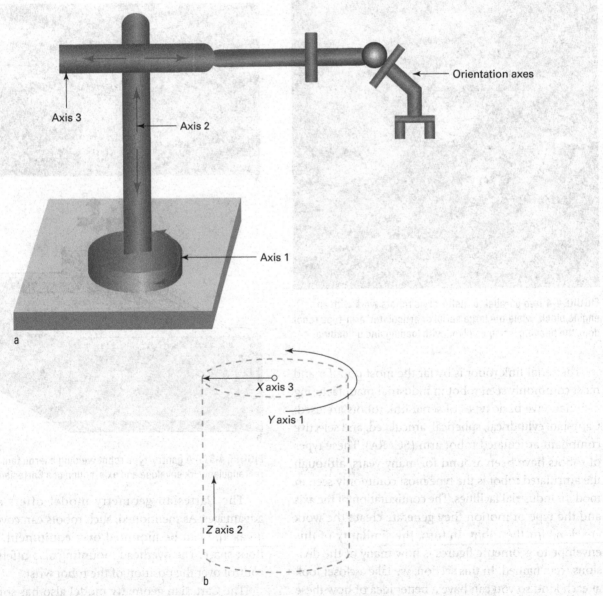

FIGURE 4-6 (a) This drawing lets you get a feel for how a cylindrical robot moves. (b) A cylindrical work envelope, showing how the *X*, *Y*, and *Z* components relate to the motions and axis.

been mounted on linear bases to increase the length of the work area and regain some of this flexibility. These robots find applications in material handling, palletizing, packaging, and the assembly of components.

Spherical Geometry

With spherical geometry, you can imagine the robot being in the middle of a ball instead of a cylinder. Spherical geometry, also known as polar geometry, gives the user a wide range of options for robot positioning, although the full range of the sphere is not available due to the physical constraints of the robot's construction and the surface on which it is mounted. This geometry essentially involves taking the cylindrical robot and replacing the linear z-motion of axis 2 with a rotational axis (Figure 4-7). The robot still has the same sweep and reach as with the cylindrical geometry, but angular options for the major motion of the arm are now available. In fact, some models might be considered "double-jointed" due to the arm's ability to

fold back over the top of the robot without having to turn around. Spherical geometry was more of an evolutionary step in robotics and is not often seen in use today, but some of these robots can still be found in industry moving parts or assembling products.

Articulated Geometry

Articulated geometry robots have a spherical-type envelope that is constrained by the construction of the robot. The articulated robot leaves the linear motions behind for rotational motion at all of the various axes. It is also known as jointed arm, revolute, and even **anthropomorphic** because the motions look very organic and lifelike in many cases. Articulated geometry is the most common option in today's industrial robots due to the flexibility of the design and the fact it can replicate a wide range of human motions. It has a chunked-up spherical work envelope due to the limitations of rotary motion caused by the robot's shape, design, and construction constraints such as wiring

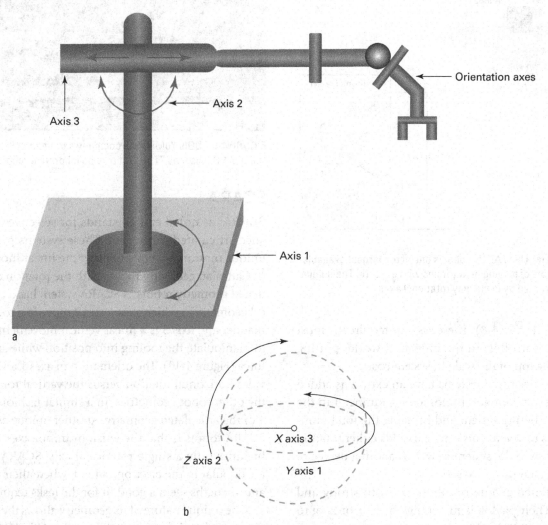

FIGURE 4-7 (a) The modified cylindrical robot that creates the spherical geometry. (b) The X, Y, and Z components for a spherical robot with the related axes. Notice how the bottom is cut off by where the robot mounts.

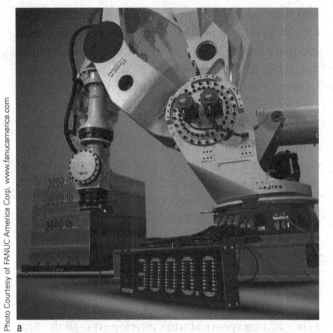

a

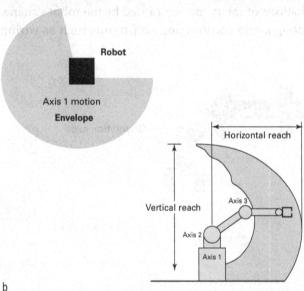

b

FIGURE 4-8 (a) This FANUC robot is one of the largest available and is capable of moving heavy items with ease. (b) The unique envelope created by using only rotational axes.

and hoses (Figure 4-8). These systems require the most complex controllers in the industrial world, putting them at the top of the scale in system cost.

Sometimes these systems have an extra axis added between the major and minor axes—number 4 in the robot numbering system and bringing the total number of axes to seven. This extra axis adds either rotation or extension to the system as well as another degree of freedom (DOF).

Articulated geometry robots are both strong and flexible. Their payloads may range from 1 or 2 kg to more than 1000 kg. You can find this style of robot

performing all aspects of industrial applications, including painting, handling parts, assembly, welding, inspection, and any other repetitive types of tasks that people might perform.

A new addition to the articulated family is the dual-arm robot that has a human-like torso, which may or may not rotate, with two articulated arms mounted to what we would consider the shoulder area. With this type of robot, the articulated work envelope is roughly doubled: Each arm on each side has its own work envelope, and these two envelopes overlap in the middle. Some of the collaborative robots utilize this style, as it closely mimics what humans can do, but offers the advantage of a greater range of rotary motion and positioning that humans can only dream of bending into (Figure 4-9).

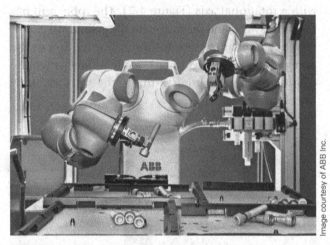

FIGURE 4-9 ABB's YuMi is the company's premier collaborative robot. ABB bills it as "The first true collaborative robot."

SCARA

SCARA, as noted earlier, stands for selective compliance articulated robot arm. These systems provide a unique motion type by blending the linear movement of Cartesian geometry robots with the rotation of articulated geometry robots. A SCARA system has a cylindrical geometry in which axes 1 and 2 move in a rotational manner and axis 3 is a linear vertical movement meant to manipulate the tooling into position while applying force (Figure 4-10). The orientation of axes 1 and 2 provides horizontal rotation versus the vertical rotation of the other robot geometries, in a similar fashion to axis 1 of the articulated geometry. Another unique aspect of SCARA robots is that the wrist, or minor axes, usually includes only a single rotational axis. SCARA robots are popular in the electronics field, where their motion and strengths seem a good fit for the tasks required.

The unique nature of its geometry allows the SCARA robot to mimic some of the reach of the Cartesian

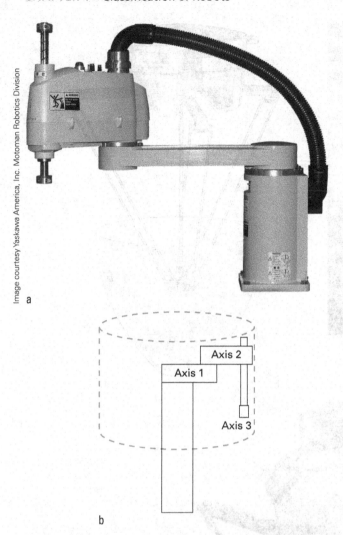

a

b

FIGURE 4-10 (a) A SCARA robot with a cylindrical work envelope. (b) While cylindrical in nature, the depth of the cylinder of the SCARA work envelope is limited by the depth of travel of axis 3 and the reach of the tooling.

geometry robot, but with a much smaller footprint. SCARA systems are faster and cleaner than their Cartesian cousins, but the payload of this system is usually 20 kg or less; thus, this is not the system of choice for toting heavy parts. These systems are also costly and tend to be more difficult to program for linear motion, due to the orientation of the rotational axes. If a large number of linear moves are needed for the task, it may be better to use a different style of robot.

Horizontally Base-Jointed Arm

The horizontally base-jointed arm geometry is an adaptation of the SCARA system in which the linear axis is axis 2 instead of axis 3. Instead of the tooling rising up and down, as with the standard SCARA robot, this system moves the whole arm up and down. Such robots also tend to have a minor axis complement of two or

three axes versus the single rotational axis of the traditional SCARA types. This configuration provides the power of the SCARA robot in the vertical direction plus flexibility in tooling orientation that rivals any of the other systems considered in this section (Figure 4-11).

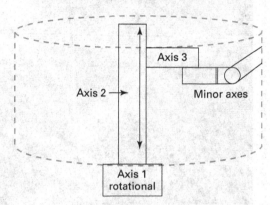

FIGURE 4-11 While the geometry is basically the same, the horizontally base-jointed arm geometry gives the robot a much larger work area as well as options in how the tooling is oriented or twisted.

Delta Geometry

Delta robots, also known as parallel robots, have become popular in industry and 3D printing over the last few years due to their speed and unique design. With that unique design comes a unique geometry. Delta robots are mounted over the workspace, like the Cartesian geometry systems, but that is where the similarities end. In Figure 4-12a, you can see that the system consists of three vertical arms coming to a pyramid-type point at the tooling below. The three major axes still drive the tooling around and the three minor axes orient the tooling, but the work envelope does not resemble those we have discussed to this point. Instead, this arrangement, due to the sweeping motion of the three major axes, produces a cone-shaped work envelope similar to an acorn or the nose cone of a rocket. The majority of the work envelope is closer to the base of the robot, but then the envelope narrows as the tooling moves farther away from the overhead unit (Figure 4-12b). In essence, these systems sacrifice a large portion of their work envelope for speed and the benefit of mounting over the work area.

Delta systems have become a popular choice for part sorting when coupled with machine vision. The mechanical speed of the system allows the controller more time to process images and determine any necessary offsets without sacrificing cycle time or increasing

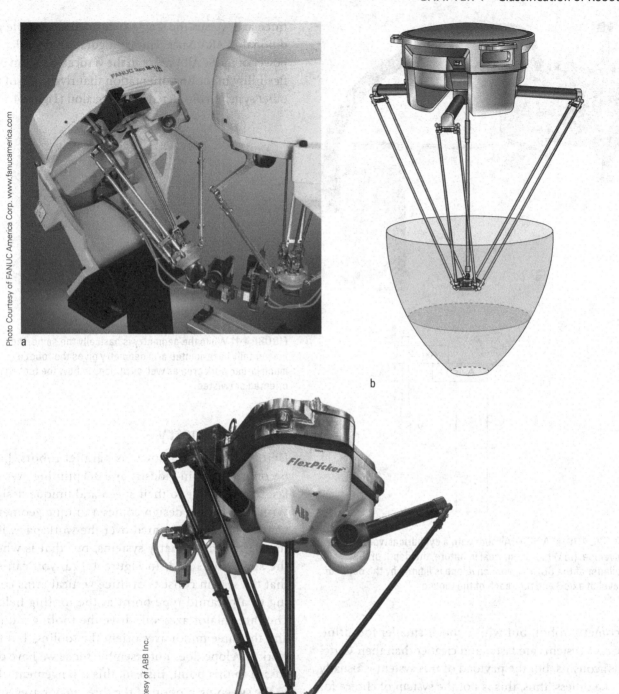

a

b

c

FIGURE 4-12 (a) Two delta-style robots working together, complete with vision systems attached to the tooling. Notice the unique configuration of the arms used to create the unique motions of this style of robot. (b) The unique shape of the delta robot configuration. (c) ABB's large delta-style robot along with the pneumatics running to the tooling attachment point.

the amount of time it takes the robot to complete its programmed actions.

Most of the latest developments in the realm of geometry have consisted of combinations of the systems examined in this section. The new dual-arm articulated robots are a great example of the new and

exciting ways to utilize the robot geometries that we already know and love. The mounting of articulated robots on Cartesian-style gantry bases is another marriage of technologies that offers the best of both worlds. As described in Chapter 2, the combination of Yaskawa Motoman's MG12 robot mounted on top of

Clearpath's OTTO 1500 self-driving vehicle takes the articulated work envelope wherever it is needed in the facility. At the moment mobile bases are not considered to be a separate geometry, but that could change down the road if they gain enough popularity.

DRIVE SYSTEMS: CLASSIFICATION AND OPERATION

Another way of sorting robots is by the method they use to connect the motors to the moving parts, known as the **drive system**. Drive systems represent another broad classification that, like using the power source as the basis of classification, will lead to many different robots being grouped together. For this reason, the drive system is more often used as a criterion in specifying a robotic system rather than as the sole sorting method. We explore the drive system side of things briefly in this section. If you would like to know more about the parts of the drive systems, you should consider taking a class or two on electric motors and mechanical power transmission.

Direct Drive

In **direct drive** systems, the rotating shaft of the motor is connected directly to the part of the robot it moves. This design was first patented in 1984 by Takeo Kanade and Haruhiko Asada. Because this kind of system greatly increases the speed and accuracy of the robot, it has become a favorite among industrial robots. The direct drive design couples the robot directly to the output shaft of the motor, so there is a one-to-one ratio of movement. In other words, a full rotation of the motor shaft creates a full rotation of the robot joint.

One consideration with this method is the lack of mechanical amplification of force; thus, the **torque**, or rotational force generated by the motor, is the limiting factor when determining the robot's payload. Luckily, we have developed electric motors capable of generating large amounts of torque to help circumvent this limitation, making direct drive systems suitable for handling loads of up to 250 pounds.

When calculating the capacity of direct drive systems, we have to deduct the weight of the robot from the payload of the motor. To do this, we first determine the payload from the motor carrying the bulk of the load, then verify the payload with the subsequent motors of the system to ensure no part of the system is overloaded.

Example 1

In this example, we will figure out the payload of a robot with the following motors set up for direct drive:

Axis 1: 250 ft-lb of torque

Axis 2: 175 ft-lb of torque

Axis 3: 100 ft-lb of torque

Axes 4 and 5: 75 ft-lb of torque

Axis 6: 25 ft-lb of torque

For the sake of our example, we will ignore the complexity of figuring accelerated forces into the mix and instead work with just the pure weight of the system. The tooling weighs 10 lb, the portion of the robot from axis 5 to axis 6 weighs 15 lb, the portion of the robot from axis 4 to axis 5 weighs 20 lb, the portion of the robot from axis 3 to axis 4 weighs 50 lb, the portion of the robot from axis 2 to axis 3 weighs 70 lb, and the portion of the robot from axis 1 to axis 2 weighs 70 lb. All the weights in this example include the weight of the robot parts as well as any motors and other equipment installed on the portions of the robot.

First, we have to see if there are any places where the torque is insufficient for the weight. For axis 1, we add all the moving weights together to determine the total:

10 lb + 15 lb + 20 lb + 50 lb + 70 lb + 70 lb = 235 lb

Looking at the torque for the axis 1 motor, we are good to go, with 15 lb (250 lb of torque − 235 lb of weight = 15 lb of force) to spare.

Next is axis 2. At this point, we could add all the weights up again, leaving out the weight for the portion between axis 1 and axis 2, or we could just take our answer for the axis 1 motor and subtract the weight between the two axes from it:

235 lb − 70 lb = 165 lb

Again, we are good to go with 10 lb to spare (175 lb of torque − 165 lb of weight = 10 lb of force).

Now we move to axis 3 and do the same thing:

165 lb − 70 lb = 95 lb

We are under the limit again, but only by 5 lb (100 lb of torque − 95 lb of weight = 5 lb of force).

Next is axis 4:

95 lb − 50 lb = 45 lb

Here we have 30 lb to spare and no issues (75 lb of torque − 45 lb of weight = 30 lb of force).

Next is axis 5:

$$45 \text{ lb} - 20 \text{ lb} = 25 \text{ lb}$$

Here we have 50 lb to spare and plenty of power (75 lb of toque − 25 lb of weight = 50 lb of force).

Last but not least is axis 6:

25 lb of torque − 15 lb of weight = 10 lb of force or just the weight of the tooling at this point

This leaves us with 15 lb of force free to use. So that would make our available payload after this set of tooling 15 lb (25 lb − 10 lb = 15 lb), right?

Wrong. Look back to the amount of force we had to spare with axis 3. We had only 5 lb left over after accounting for the weight of the robot with the 10 lb of tooling in place.

If this robot were used to pick and place parts, then, we would be fine if those parts weighed 5 lb or less, right? The answer is not actually that cut and dried. Recall that we ignored acceleration in the problem statement: Now we need to consider it. Force equals mass times acceleration ($F = M \times A$). Stating this idea in other words, if we move the part quickly, it could generate more than 5 lb of force due to the acceleration, which might cause problems when the robot starts to slow/stop a movement or change direction. Some concerns also arise regarding where the weight of the part is centered—though we will save that discussion until we look at robot tooling more closely.

In this example, we have only 5 lb of force left, which may not be ideal for moving parts. Even so, we should be good to go if we are just using the designated tooling, as it weighs 10 lb, correct? For the most part, yes. We could still run into issues with excessive force due to acceleration because axis 3 is relatively under-powered and a good portion of the robot is dependent on this axis. Certain movements or changes in movement may focus the force on this axis and exceed our 5 lb of force safety margin.

Luckily for us, the companies that build and supply robots take these kinds of issues into consideration when building and testing their robots. Thus, if you purchase a robot that has a payload of 25 kg or 50 kg, it should be safe to consume all of that payload in tooling weight or tooling plus part weight combined. My advice is to always select a robot that has a larger payload than the application requires to avoid the kind of issues described in this example. For those of you who would like to dig deeper into this kind of math, taking a physics course would be a good place to start.

Reduction Drive

Reduction drive systems take the output of the motor shaft and alter it via mechanical means. As a rule, these drives slow the rotational speed of the system so as to increase the torque or force out, but they can also change the direction of rotation or turn rotational motion into linear motion. These systems often require more maintenance because they contain additional moving parts: The more complex the system, the greater the chance something will require repair. At the very least, reduction drive systems will require more preventive maintenance than their direct drive counterparts. Given the variety of ways we can reduce the motor output, several subgroups of this drive type can be distinguished.

Belt Drive

Belt drives utilize a flexible continuous loop known as a belt to transfer power from the motor to the output of the system. By changing the size ratio between the **drive pulley**, which is connected to the motor or force, and the **driven pulley**, which is connected to the output of the system, we can alter the torque of the system. To increase torque, we must sacrifice speed, which means a larger driven pulley. To increase speed, we must sacrifice torque, which means a larger drive pulley.

Because of the precision required with most industrial robots, the belt of choice is a **synchronous belt**, which has teeth at set intervals along its length (Figure 4-13). The upside of this design is that it resists **slippage**, or loss of output power due to the belt slipping on the pulley. The downside is that if the robot stops unexpectedly, often teeth are ripped from the belt.

Over time, all belts used in a drive system wear and expand. In severe cases this can leave enough slack in the system that the belt can jump a tooth, skipping ahead or behind by one or more teeth, so that it does not fully transmit the motion of the drive motor. This results in the robot physically missing its designated location, which can cause all kinds of mischief, if not outright disaster.

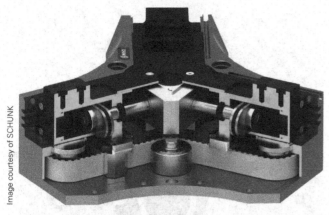

Image courtesy of SCHUNK

FIGURE 4-13 A synchronous belt used to drive a gripper system for a robot. Notice the teeth on the white belt at the bottom of the cutaway image. This is similar to the setup used for robotic axis drives.

Chain Drive

For the most part, chain-driven systems work in the same way as belt-driven systems, with a few exceptions. These systems use **sprockets**, which have teeth designed to fit into the links of the chain instead of pulleys, and a chain, usually made of metal, which connects the drive sprocket to the driven sprocket (Figure 4-14). Sprockets resemble gears, albeit with teeth that are widely spaced apart and longer than standard gear teeth.

Like synchronous belts, chains do not slip, but they do wear out. Over time, a chain's links will become longer due to the stress on the chain and will wear down from the contact with the sprockets. This equates to a change in the length of the chain, which affects the relationship between the drive and driven sprockets. When the difference becomes severe enough, the chain may jump a tooth on the drive or driven sprocket.

Another major difference between a belt-driven system and a chain-driven system is the maintenance. We *do not* oil belts, as this would increase the chance of slippage and could cause the belt to break down; in contrast, most chains require lubrication to work properly.

Chain drives combine the rigidity and force transmission of a gear system with the flexibility and forgiveness of a belt system. Gear systems are discussed next.

Gear Drive

While it is difficult to determine when humans first started to use gears, history shows that we have been

APNP/Shutterstock.com

FIGURE 4-14 While this system is not from a robot, it does show the components of a chain system as well as the slack that develops over time.

using gears to transfer force for thousands of years. Gears come in many shapes, sizes, and varieties, but they all have cogs or teeth, which are projections that match up (**mesh**) with similar projections on other gears to transmit force. The gear tied to the power supply is called the drive gear, and the gear tied to the output is called the driven gear—similar to the nomenclature used with pulleys in a belt system. Two or more gears connected together are called a **gear train** or **transmission**, as they connect the output(s) to the prime driving force.

Because of how gears connect and the way they transmit force, the driven gear rotates in the opposite direction to the drive gear. If this is a problem, we can use an **idler gear**, which consists of one or more extra gears added to a system to change the direction of rotation or to cover large gaps between the drive and driven gears. An idler is placed on a dedicated shaft that is not an output shaft.

If there are an even number of gears in a drive system, the last gear will rotate opposite to the input gear (Figure 4-15). If there are an odd number of gears in a system, the last gear will turn the same direction as

Aha-Soft/Shutterstock.com

FIGURE 4-15 Here you can see the rotation of two gears and the fact the second gear rotates in the opposite direction of the first. With any even number of gears, the last gear rotates opposite to the first gear.

Aha-Soft/Shutterstock.com

FIGURE 4-16 By adding in a third gear, the last gear rotates in the same direction as the first gear. This is true of any gear train containing an odd number of gears.

the input gear (Figure 4-16). The same system works for gears in the middle of a gear train as well, as long as you count the number of gears from the drive gear to the gear in question, including the gear for which you are trying to determine the direction of rotation. For instance, if the system includes a transmission with seven gears, and you want to know how the fourth gear in the system rotates, you would count from the first gear to the fourth gear, 4, which is an even number. That means gear 4 is rotating opposite to the drive gear.

We often combine multiple gears to create complex gear trains that have compound gears and multiple output points with their own torque and speed. **Compound gears** are two or more gears on the same shaft, often made from one solid piece of material. When dealing with compound gears, usually one of the gears will be a driven gear while the other will be a drive gear, with multiples of each possible when the compound arrangement includes three or more gears. One or more gears in the compound arrangement serve as the power source for a completely new gear train, thus making it the drive gear for that transmission system.

When trying to figure out the rotation of gear systems using compound gears, there are a couple of things to remember. First, multiple gears on the same

shaft must rotate in the same direction. Second, when counting gears for rotation determination, always start at the drive gear. I recommend determining the rotation of the driven gear of the compound set first, as the drive gear will have to rotate in the same direction since it is on the same shaft. Treat each drive gear as the start of a new gear train for determining rotation.

Several styles of gear shapes are shown in Figure 4-17. **Spur gears**, for example, are made by cutting teeth into the edge of a cylindrical object.

A mechanical power course can teach you about gearing in general, so we will finish our discussion of gear drives by focusing on harmonic drives, which are

fotofocus/Shutterstock.com

FIGURE 4-17 Part of a gear transmission system, which includes compound gears.

commonly used to drive the tooling of robots and are found in the last axis of many robotic arms. **Harmonic drives** are a type of specialized gear system that uses an elliptical wave generator to mesh a flex spline with a circular spline that has gear teeth fixed along the interior (Figure 4-18). The circular spline is typically the driven portion of the system, whereas the flex spline contacts the circular spine only at two points, located 180° apart. The wave generator is inside the flex spine but usually separated from the flex spine by ball bearings. The flex spline has fewer teeth than the circular spine.

FIGURE 4-19 Some of the various sizes of ball screws that you might encounter in robotic applications.

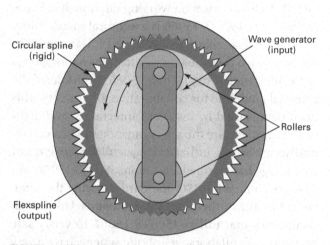

FIGURE 4-18 The various parts of a harmonic drive system, which is the gearing used for axis 6 of many robots.

A harmonic drive system can generate torque ratios up to 320:1 and has no backlash. **Backlash** is the distance from the back of the drive gear tooth to the front of the driven gear tooth; it represents loss of motion in the system. Most gear systems need a certain amount of backlash to function properly, so harmonic drives are unique in eliminating this loss.

Ball Screw

Another type of drive that is frequently encountered in robotics is the ball screw (Figure 4-19). A **ball screw** consists of a large shaft with at least one continuous tooth carved along the outer edge, along with a nut or block that moves up and down the length of the shaft. The prime mover connects to the shaft either directly via a coupler or through a belt, chain, or gear drive system. The block or nut that moves up and down the ball screw usually rides along the tooth via ball bearings and attaches to whatever is being moved.

Ball screws are highly efficient and precise to one-thousandth of an inch per foot or finer, making them a favorite for precision motion. These drives do suffer from backlash, like gear systems, but only when they change directions. A small portion of the direction change takes up the backlash and from that point on backlash is no longer a concern. In cases where the block has to carry heavy loads, the system may also include a guide rod to ensure that the block travels straight without binding.

Given the broad nature of the drive system categories, it is rare that this is the only search or grouping criterion used when finding a robot for an application. More often, the drive type is added to one of the other criteria for grouping robots mentioned in this chapter. In many applications, the drive type is of little importance and, therefore, something ignored all together. Just remember that the reduction drive-type robots will require more maintenance than the direct drive-type robots due to the additional systems and their requirements. This type of categorization does work well for robot tooling, so keep that point in mind.

ISO CLASSIFICATION

The International Standards Organization (ISO) develops, updates, and maintains sets of standards for use by the industries of the world. An ISO certification guarantees that a company is complying with a defined set of specifications on quality, safety, and reliability when manufacturing its products, giving its customers greater peace of mind when purchasing from that company. A company has to go through a great deal of preparation to receive ISO certification, and it then faces periodic inspections to ensure it is still meeting the requirements of any certifications it holds. ISO certification is not a free service, but the money spent on this endeavor is often well worth it: ISO certification can be a great marketing tool to attract new clients and maintain current ones.

ISO breaks robots into three broad classifications: industrial robots, service robots, and medical robots. The industrial robotics classification is the one that ISO has worked with the longest, with the other two being newer designations.

ISO defines an industrial robot as "actuated mechanism programmable in two or more axes with a degree of autonomy, moving within its environment, to perform intended tasks" (Harper, 2012). This is an updated version of the first robot definition created by the ISO, as initially industrial robots were the organization's only focus. Since 2004, ISO has classified industrial robots by their mechanical structure:

- Linear robots
- SCARA robots
- Articulated robots
- Parallel robots
- Cylindrical robots
- Others

Notice the similarities between the ISO robot classification list and this chapter's section on robot classification by geometry of the work envelope. ISO includes Cartesian robots and gantry robots in the linear group. SCARA robots include the variants described earlier in this chapter, and delta and parallel robots are the same types of robot. ISO defines the articulated robot category in a way that would include the spherical geometry robots described in this chapter. The "others" category includes new styles of industrial robots that do not fit into any of the other categories (such as the OTTO mobile base), giving them some place to be categorized until a unique category can be created.

ISO 8373:2012 provides the ISO classifications for industrial robotics, along with the information needed to work safely with these systems. Another standard, ISO/TS 15066, focuses on working safely with collaborative robots. ISO/TS 15066 is a technical specification, rather than a standard, which means it was a work in progress when this book was written. Even though it is not a full standard yet, ISO/TS 15066 is considered the essential guideline for collaborative robot safety. This document is used by installers, integrators, and manufacturers to ensure the safety of workers when collaborative robots are utilized. Once collaborative robot technology has matured a bit more, ISO/TS 15066 will likely become a full ISO standard, carrying the same weight as any of the other ISO standards. Until then, companies that utilize ISO/TS 15066 to verify safe operation of collaborative robotic systems have a very good legal leg to stand on should anything unfortunate happen.

REVIEW

Robots can be classified in many ways, such as by application or payload. The classification methods described in this chapter have been used in the robotics industry for years. Of course, when it comes time to pick a robot for a task, the specifics involved in that task will play a large role in determining the precise selection criteria and, therefore, the classification system used. Moreover, because robotics is an ever-changing and expanding field, new robotic systems are likely to emerge that do not easily fit into the categories mentioned here.

Over the course of this chapter, we covered the following topics:

- **How are robots classified?** This section discussed why we classify robots and some of the broad categories we might use.

- **Power source.** Classification based on power source was the first option we examined, and we looked at some of the strengths and weaknesses of each category.

- **Geometry of the work envelope.** We can group robots based on their work envelope as well as by considering their axes of movement.

- **Drive systems: classification an operation.** We discovered the difference between the direct drive and various indirect drive systems used to move robots.

- **ISO classification.** We explored how ISO classifies robots—namely, based on how they work in a mechanical sense. This led to nearly the same groupings as we found in the geometry section.

KEY TERMS

Anthropomorphic	Driven pulley	Mesh	Stepper motors
Backlash	Encoders	Payload	Synchronous belt
Ball screw	Gear train	Reduction drive	Torque
Compound gears	Harmonic drives	SCARA	Transmission
DC brushes	Idler gear	Servo motors	Work envelope
Direct drive	International	Slippage	
Drive pulley	Standards	Sprockets	
Drive system	Organization (ISO)	Spur gears	

REVIEW QUESTIONS

1. Name at least three of the common ways we group industrial robots.

2. Why is payload a characteristic we might use to further refine robots within a grouping?

3. What are the three main types of power sources used for industrial robotics?

4. What is a DC brush, and what is the danger associated with this component?

5. Describe the operation of a stepper motor.

6. What are some of the costs associated with using a hydraulic robot?

7. What is the primary difference between hydraulics and pneumatics?

8. Where is pneumatic power used extensively in the robotics world?

9. What are the benefits of a Cartesian work envelope?

10. What are the pros and cons of a cylindrical geometry robot?

11. What is the work envelope of an articulated geometry robot?

12. Which geometry is the most common for robots and why?

13. What is the difference between a SCARA robot and a horizontally base-jointed arm?

14. What is the tradeoff inherent in the work envelope of the delta robot?

15. Describe the operation of a direct drive robot.

16. What is the difference between a chain-driven and belt-driven system?

17. Define the motion of the last gear in a transmission that has an even number of gears. Define the motion of the last gear in a transmission that has an odd number of gears.

18. How is it possible for compound gear systems to have both drive and driven gears on the same shaft?

19. What are the benefits of harmonic drives?

20. What is the ISO definition of an industrial robot?

21. What are the categories used by ISO for industrial robot classification?

22. What is the focus of ISO/TS 15066?

Reference

1. Harper, Chris. Current activities in international robotics standardisation. 2012. http://europeanrobotics12.eu/media/15090/2_Harper_Current_Activities_in_International_Robotics_Standardisation.pdf. Accessed July 14, 2012.

Image Courtesy of SCHUNK

CHAPTER 5

End-of-Arm Tooling

WHAT YOU WILL LEARN

- What robot tooling is and how we use it

- The varieties of grippers commonly used in industry

- The math required to figure out the amount of force a gripper needs to generate

- Other types of tooling found in industry

- How robots use multiple tooling

- The ways tooling can flex to align with out-of-position parts

OVERVIEW

Before we get into programming in the next chapter, it is wise to take a bit of time to examine the tooling that transforms robots from very expensive toys to highly functional equipment in industry. While robotic tooling is so diverse there is no way we can examine all the options, we will cover the common options you are likely to encounter in the field. While exploring the different tooling types, we will review some of the optional equipment and peripherals utilized with tooling to provide a broad overview of this area. During our exploration of tooling, we will hit on the following topics:

- What is EOAT?
- Types of tooling available
  - Grippers
  - Gripper force
  - Payload
  - Other grippers
  - Welding tooling
  - Other types of EOAT
- Multiple tooling
- Positioning of EOAT

WHAT IS EOAT?

End-of-arm tooling (EOAT) consists of the devices, tools, equipment, grippers, and other tooling at the end of a manipulator that a robot uses to interact with and affect the world around it. Think of EOAT as the "doing" portion of the robot, which works in conjunction with the positioning aspects of the system. EOATs attach to the end of the robot's wrist or minor axes. In many cases, they are a separate purchase from the robot. It is common to purchase a robot and then spend several hundred to several thousand dollars more, depending on the device, to get the tooling necessary to perform the desired task(s). A number of companies, such as SCHUNK and SAS, specialize in providing tooling for industrial robots. Some of these companies produce robots as well, but often tooling companies focus primarily on creating tooling that work with most, if not all, of the robotic systems in industry. Robotic tooling is a booming field, with new types and configurations of tooling constantly under development to meet the ever-changing needs of industry.

When it comes to tooling for industrial robots, one can find suction cups, welding guns of various configurations, sprayers, grippers in a wide array of configurations, and about any other tooling the robot could need to mimic or surpass human action. Sometimes more than one kind of tooling is needed so that the robot can perform its tasks. When it comes to tooling, the key is to remember that the robot's job in industry is to perform useful work; thus, any tooling that people or machines have used to do that work in the past has a high probability of being adapted for use by robots. Robot tooling is driven by end-user desire/need, which is why the field continues to advance as we continue to find new and unique ways to utilize robots. In the next section, we discuss some of the common types of tooling, though by no means do we cover them all.

TYPES OF TOOLING AVAILABLE

When robotics first came onto the scene, they were referred to as "machines looking for a purpose." When industry took note of robotics and began to buy into its use to offset the rising costs of labor and materials, one could say that robotics had finally found its purpose. Because industrial robots often perform tasks that were previously done by human hands, they need some way to grip and manipulate parts—which is why grippers are by far the most diverse realm of tooling. Given the high probability that you will work with this type of tooling, we will begin our exploration of the tooling types with grippers.

Grippers

Grippers are a type of end-of-arm tooling that applies force to secure objects for maneuvering. Such a simple definition for this complex group of tooling, which is a common workhorse in all fields of robotics! In industry, grippers range from simple mechanisms with two moving parts to complex arrangements that resemble human hands (Figure 5-1). Some of the grippers found in industry mimic human motions. Others create motion types and ways of securing parts that no human hand could ever mimic—which gives the robot an advantage in certain applications. The materials, power source, and shape of a gripper are defined by the tasks it performs in industry, which explains why so many different options are available. Nevertheless, grippers typically have one or more of the following characteristics:

- They are capable of gripping, lifting, and releasing parts.
- They may be capable of sensing the presence of the part.
- The weight of the gripper itself must be kept to a minimum.

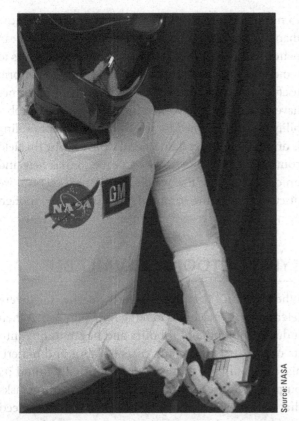

Source: NASA

FIGURE 5-1 The Robonaut, created through a partnership between NASA and GM, utilizes an advanced hand-style gripper that allows the robot to perform tasks in a very human manner.

- Ideally, it must retain the part even if power is lost.
- The design should be as simple as possible.

With actuating grippers, two motion types are commonly used: *parallel* and *angular*. **Parallel grippers** have fingers that move in straight lines toward the center or outside of the part to close and grip or open and release the part, respectively (Figure 5-2). **Angular grippers** have fingers that hinge or pivot on a point to move the tips outward to release parts or inward to grip parts (Figure 5-3). Angular grippers are good choices for parts or items that are of consistent size. By comparison, parallel grippers have the ability to grip a wider range of parts, depending on how far the fingers can travel.

Fingers, sometimes called **jaws**, are the part of the gripper that moves to hold parts (Figure 5-4). The **tooling base** is the part that attaches to the robot and holds the mechanisms to move the fingers.

We power these motions with the standard power sources used in robotics: electricity, hydraulics, or pneumatics. The task we use the gripper for, the readily available power supply, and the environment in which we use the system all influence the choice of power source for gripping action. Hydraulic and pneumatic

Image Courtesy of SCHUNK

FIGURE 5-2 A set of parallel grippers.

FIGURE 5-3 A simple angular set of grippers that create a scissors-type motion.

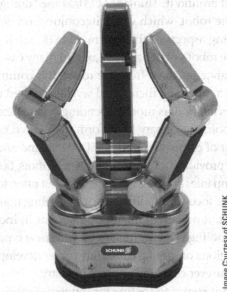

Image Courtesy of SCHUNK

FIGURE 5-4 A set of angular grippers that have an extra joint on each finger to make them more like human fingers.

grippers are great for wet, dusty, or explosive environments, with hydraulic power generally reserved for manipulating heavy parts. Electrically driven grippers have grown in popularity due to improvements in torque and size reduction.

Grippers can be powered in both directions, or they can be powered in one direction (open or close), with the other passive motion being controlled by another means, such as spring pressure or mechanical tension. With passive methods, as soon as you remove the power source, spring tension or some other force will cause the gripper to return to its unpowered state. When working with grippers, regardless of the type, you should take special care to ensure that no damage occurs to parts and/or people during shutdowns or unexpected power loss. The last thing we want is for a robot to start throwing 100-pound parts at people!

Many of the grippers sold for industrial applications have generic fingers that are a solid piece of metal. The assumption with this approach is that the company buying the gripper will machine the fingers as needed to match the shape of its parts. Tooling suppliers will machine the fingers to customer specifications, but this generally adds a fair amount of cost to the gripper system—which is why many companies prefer generic fingers. Companies that have internal machine shop operations sometimes order just the gripper base with no fingers and then create their own. Whether modifying the manufacturer-provided fingers or creating their own, end users have to ensure they have enough material left in the fingers to support the part and withstand the forces involved with moving that part while providing enough friction to prevent the part from slipping out. We will dig deeper into these factors shortly.

The number of fingers a gripper has provides information about how it positions a part when grasped. Two-finger grippers give side-to-side centering, but nothing else. With these systems, some other means is necessary to provide centering along the length of long or odd-shaped parts. Three- and four-finger grippers are great for circular or other standard geometry parts, as they will center the part in two directions consistently (Figures 5-5 and 5-6). Some four-finger grippers and those having five or more fingers are great for working with **odd-shaped parts** or parts with unique shapes and proportions. These grippers are often highly specialized and, therefore, more expensive than the more common three- and four-finger varieties.

Grippers that resemble human hands are less about how the gripper centers the part and more about manipulating parts as a person would (Figure 5-7).

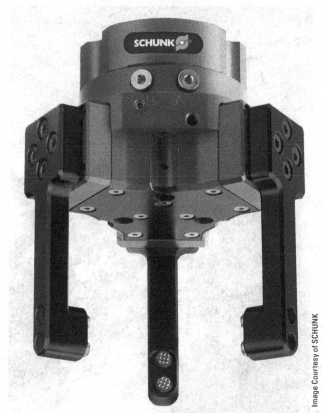

FIGURE 5-5 A three-finger parallel gripper.

Hand-shaped grippers tend to be versatile, but often come with a high price tag, low payload capacity, and complexity of control and operation. Hand-style grippers are typically used for intricate tasks such as part manipulation for assembly or manipulating tools and devices designed for people, as opposed to raw-strength tasks or generic part lifting.

At this point, you may be asking yourself, "How do I know which gripper to use?" The answer depends on what you want the gripper to do and which system you are using. The following guidelines are intended to help you with the selection process:

1. The tooling must be capable of holding, centering, and/or manipulating the range of parts the robot works with.

2. There needs to be some way of sensing when a gripper has closed on the part or if it is empty. This sensing can be internal to the gripper or it can be provided externally, depending on the system. Newer grippers can determine the amount of force exerted on the part as well, which is very beneficial when working with delicate materials (Figure 5-8).

3. The weight of the gripper should be kept as light as possible, as this weight is deducted from the robot's total payload.

FIGURE 5-6 A four-finger system centering a tire rim.

FIGURE 5-7 Human hand–style grippers.

FIGURE 5-8 Too much force would turn this light bulb into a pile of broken glass. This demonstrates some of the force control monitoring and control options available.

4. The gripper must have proper safety features built in for the environment in which it works and the parts with which it works (e.g., the ability to hold parts even when power is lost, or fingers made from non-sparking materials for flammable environments).

The cheapest gripper that meets these four criteria is usually the one chosen. A fact that is often overlooked is that EOAT is a growing and evolving field, so the solution chosen today may not be the best option in as little as three to six months in the future. With this caveat in mind, from time to time you may want to reevaluate the tooling in use, especially if you start having problems, such as not enough payload left after you add the tooling, or if the gripper is wearing out and needs to be replaced.

Gripper Force

No matter how specialized or advanced a gripper is, if it cannot create enough force to hold the parts during normal operation, it will be a source of frustration, downtime, and cost. In this section, we look at some of the math involved in selecting robot tooling. For those readers who are going into engineering, consider this discussion to be just the tip of the iceberg: People who design tooling for robots must also account for power source torque, force vectors, lever multiplication of force, and a multitude of other considerations during the process. Yes, there truly are real-world applications for the math you have spent years learning (or in some cases, avoiding).

While working to determine the force required from the gripper, we need to look at several factors: part size and shape, direction in which the part is being moved, friction, size of the gripper, and any safety factor we want to build in. Part size and shape are fairly straightforward. Whenever the center of gravity is outside of the gripping area, however, the part will act like a lever and multiply any forces created by movement. The **center of gravity** of a part is where we consider the mass to be centered; if we support the part at this point, we consider all the forces to be in balance or equilibrium. As we move the part, if the center of gravity is outside or out of line with the gripper, it will cause excessive force issues that we need to take into account.

Friction is the force resisting the relative motion of two materials sliding against each other; it acts as a resistance to slippage of the part. The greater the resistance or friction, the more force it takes for the part to slide in the gripper. Once the part starts to move, the friction actually decreases, thus making any slipping of the part very dangerous. The larger the gripper, the larger the area to which the force is applied and the more surface to interact with the part on a friction basis.

The **safety factor** is the margin of error we build into the process. In other words, the safety factor is how we account for variance in part weight, length, errors in math, and other unaccounted forces. The larger the safety factor, the more room for error in the calculations. Unfortunately, larger safety factors often equate to a higher cost for tooling. In the case of fragile parts, too much force can be just as bad as, if not worse than, not enough force.

Let us look at a couple of examples to see how this all fits together.

Example 1

We will use Figure 5-9 for this example with the following information:

- The part weighs 1 pound.
- The gripper's jaws are parallel.
- The part is gripped 2.5 inches from the center of gravity.
- The gripping surface is 0.75 inch long.
- The part is 0.375 inch wide where it is being gripped.
- The part is being lifted with a maximum acceleration of 2.5 Gs, including normal gravitational force.
- The coefficient of friction between the gripper and the pen is 0.80.
- The engineer in charge wants a safety factor of 2 to be included.

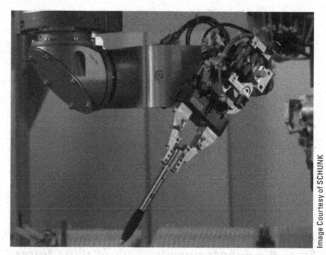

FIGURE 5-9 A two-finger gripper working with a part that is fairly long in length and has a high probability of having a misaligned center of gravity.

Image Courtesy of SCHUNK

Because the part is being lifted/moved sideways, we must take the torque of this action into account, based on the differences in the center of gravity. The overall torque (T) on the jaws equals weight (W) times distance (d). The weight equals mass (m) times gravity (G), or in this case 1 pound. The distance equals 2.5 inches. The width of the fingers' grip (b) is 0.75 inch. The width of the part (p) in the jaws is 0.375 inch.

$$T = m \times G \times d$$
$$T = \text{torque}$$
$$m = \text{mass}$$
$$G = \text{gravity: } 9.8 \text{ m/s}^2 \text{ or } 32.2 \text{ ft/s}^2$$
$$d = \text{distance}$$
$$W = m \times G$$
$$W = \text{weight}$$
$$m = \text{mass}$$
$$G = \text{gravity: } 9.8 \text{ m/s}^2 \text{ or } 32.2 \text{ ft/s}^2$$

This means we can substitute the weight of the part for the $m \times G$ portion of the formula as needed.

Because the gripper has a center of force, two forces are involved in the torque: the force of the part below the center of force and the force exerted by the part above the center of force. This changes the torque equation:

$$T = F_2(b/2) + F_1(b/2)$$
$$T = (b/2)(F_1 + F_2)$$

Via algebraic substitution for torque, we get the following expression:

$$F_1 + F_2 = 2(m \times G \times d)/b$$
$$F_1 = \text{force below the gripper}$$

F_2 = force above the gripper

m = mass

G = gravity: 9.8 m/s² or 32.2 ft/s²

d = distance

b = width of the gripping contact

When we add in the width of the part in the fingers (p), we get the following expression due to force vectoring:

$$F_1 + F_2 = \frac{2(m \times G \times d)}{\sqrt{(b^2 \times p^2)}}$$

p = part width in the gripper

By Newton's first law or the law of inertia, $F_1 = F_2$. Since the gripper provides both of those forces, $F = F_1 + F_2$. Thus,

$$F = \frac{2(m \times G \times d)}{\sqrt{(b^2 \times p^2)}}$$

$$F = \frac{2(1 \times 2.5)}{\sqrt{(0.75^2 \times 0.375^2)}}$$

$$F = 5.96285 \text{ pounds of force}$$

Next, we have to factor in the acceleration. We do so by multiplying the force times 2.5 Gs of acceleration for 14.907125 pounds of force.

We then factor in friction by dividing the new force answer of 14.907125 by 0.80, the coefficient of friction, giving us 18.63391 pounds of force.

Last, but not least, we need to factor in the safety factor we specified earlier. We multiply the force with friction factored in by 2, giving us a total need for 37.26781 pounds of force for our gripper. If we just look at the pen in Figure 5-9, we might never dream that the specification for the gripper for this system would need to be nearly 40 pounds of force exerted by *each* of the jaws!

Example 2

For the second example, we will use the same pen setup as in Example 1, but this time we will move only in the vertical direction, with no side swing. We will use the same basic set of data:

- The part weighs 1 pound.
- The gripper's jaws are parallel.
- The part is gripped 2.5 inches from the center of gravity.
- The gripping surface is 0.75 inch long.
- The part is 0.375 inch wide where it is being gripped.
- The part is being lifted with a maximum acceleration of 2.5 Gs, including normal gravitational force.

- The coefficient of friction between the gripper and the pen is 0.80.
- The engineer in charge wants a safety factor of 2 to be included.

Because this will be a vertical move, we do not need to worry about the part's center of gravity—it will be in line with the motion. We also do not need to know how much of the gripper is in contact with the part, or the width of the part, so we can ignore the part's center of gravity distance of 2.5 inches, the gripping surface of 0.75 inch, and the part thickness of 0.375 inch.

First, we will determine the force needed for the weight of the part under acceleration:

$$F_r = W \times G$$

F_r = required force

W = weight of part

G = Gravitational force, including acceleration

$$F_r = 1 \text{ lb} \times 2.5 \text{ Gs}$$

$$F_r = 2.5 \text{ pounds of force}$$

Normal force is the force that an object pushes back with when acted on by a force. If the object cannot generate enough normal force, it will be damaged in some manner as the force acting on it overcomes the structure of the item. Friction, or the resistance of two items to slide past or against each other, also figures into the normal force:

$$F = \mu N \text{ or } N = F/\mu$$

F = friction force

μ = coefficient of friction

N = normal force

Previously we figured that the part exerts 2.5 pounds of force during the move. Because two jaws hold the part, we divide this force by 2 to get the requirement for each jaw: 2.5/2 = 1.25 pounds for each jaw. We use this to figure out the normal force required by each jaw: $N = F/\mu = 1.2/0.80 = 1.5$ pounds of force exerted by each jaw on the pen.

Of course, we cannot forget our old friend, the safety factor. When we multiply the 1.5 pounds of force by the safety factor of 2, we end up with 3 pounds of force needed by each jaw for the vertical move.

In Example 1, we figured that we needed 37-plus pounds, or more than 12 times the amount of force needed for a vertical move! Together, Examples 1 and 2 illustrate the importance of knowing your part, the movements involved in the process, and the forces involved with nonlinear movements. Imagine what would happen if we had a gripper that could generate only 3 or 4 pounds of force, which is fine for vertical moves, but we expected the tooling to grip the pen during a swinging motion. Based on our calculations, we can almost guarantee that at some point in the swinging motion the pen will slip from the grippers and become a projectile! This is why it is important to do the math *before* buying or specifying a gripper for the robot.

Payload

While we are on the subject of math, this is a good time to talk about payload and the effect that tooling has on it. **Payload** is a specification of a robotic system that informs the user how much weight the robot can safely move. Often payload is specified in kilograms (the metric system is the standard measurement system in all but a few countries, with the United States being one of the holdouts). If the robot has a payload capacity of 25 kg, then the maximum weight it can safely move, in addition to the weight of the robot, is 25 kg. Anything we add to the robot, such as end-of-arm tooling, reduces the amount of force we have left to move parts. Therefore, the equation for payload looks like this:

$$Ap = P - Wt$$

Ap = available payload

P = specified payload capacity for robot

Wt = weight of the EOAT and any added peripheral systems or equipment attached to the robot besides the core system

Example 3

For this example, we will work with the following information:

- The payload for the robot is 25 kg.
- The weight of the end-of-arm tooling is 5 kg.
- An additional sensor for the end-of-arm tooling is required and weighs 1.3 kg.

First, we need to add up the total weight added to the robot:

$$5 + 1.3 = 6.3 \text{ kg}$$

Next, we plug the information into our base formula:

$Ap = P - Wt = 25 - 6.3 = 18.7$ kg of available payload

Of course, this does not take into account any calculations for acceleration, but as a rule, the manufacturer of the robot has made those calculations in advance. From the previous examples, you know that how we pick up the part and how we move it affect the system. Therefore, while you may be able to pick up and slowly move a part that is heavier than the stated payload, any swinging-type or high-speed motions will have a high probability of causing problems with the system. At best, you will get an alarm that one of the axes is overloaded; at worst, you will damage the system or throw the part. In addition, working a robot close to or at its rated payload will wear out the system more rapidly than working it at 75% or less of the rated payload. This might be a way to justify the purchase of a larger robot, as the investment could save the company in the long run.

Other Grippers

In the industrial robotics world, some other types of tooling are lumped in with the gripper family that you may not expect, as these tools bear no resemblance in shape or movement to the human hand or fingered grippers. Because industry works with materials that are not always easy to grasp with a closing motion, some grippers use electromagnets to attract ferrous metal parts, balloon types inflate or deflate to grip a part, and suction devices use less than atmospheric pressure (a **vacuum**) to secure parts. In any case, if the primary design is intended to pick up and move items, we generally refer to the EOAT as a gripper.

We use magnetic grippers in applications that work with magnetic metals. These EOATs come in a wide range of sizes and shapes, though the basic operation of all types is the same. When a coil of copper wire has current running through it, it generates a magnetic field. If we coil this wire around or inside a metallic frame and pass current through the wire, we create an electromagnet that attracts **ferrous metals** (i.e., metals that contain iron). Once the electricity stops flowing, the magnetic force dissipates and the gripper releases the ferrous metals it attracted. Magnetic grippers are widely used in steel mills, junk yards, sheet metal manufacturing facilities, and other places where iron-based metals need to be moved from one point to another.

A suction cup or **vacuum gripper** works by creating a pressure that is less than atmospheric pressure around it (Figure 5-10). Because of this lower pressure, the pressure of surrounding atmosphere will exert an upward force on the object and hold items against the suction cup. The process is similar to what happens with the wing of an aircraft, as the low pressure over the wing creates lift under the wing and keeps the plane in the air. Suction grippers are great choices for moving heavy and light objects alike. They are widely used in the glass, food, and beverage industries because there is no moving mechanism and, therefore, no crushing motion that might damage the items moved. Vacuum grippers also work well for moving large, flat items like sheets of metal or glass that would be difficult to pick up by other means. When used in this way, the tooling will often use multiple vacuum grippers to create a lifting grid (Figure 5-11). Many of these types of grids offer individual control over the vacuum grippers, allowing the programmer to turn on only those grippers needed, saving on both operating cost and maintenance.

FIGURE 5-10 This vacuum gripper can be used singularly or as part of a grid.

FIGURE 5-11 A vacuum gripper at work picking up packages and putting them in boxes.

To create the vacuum needed for these grippers to work, we have several options, but for the sake of clarity we will look at the two most common methods. First, we can hook the vacuum cup to an external vacuum generator that basically works like an air compressor in reverse, pulling air from the location of the suction cup and venting it elsewhere, usually the atmosphere. This type of system requires additional equipment, produces increased noise, and consumes electrical power, which increases the overall cost of operation. On the upside, such a system can create a high vacuum; that is, it evacuates nearly all of the air from the system, providing a good seal between the suction cup(s) and the object. The greater the pressure differential between the outside atmosphere and the inside of the vacuum cup, the greater the lifting force.

The second, and more common, option when designing vacuum grippers is the Venturi effect, which uses a Venturi valve to create a vacuum or suction (Figure 5-12). The Italian physicist Giovanni Venturi first discovered that when air is forced through a conical nozzle, its velocity increases and the pressure decreases. In these vacuum gripper systems, we pass compressed air through a conical Venturi orifice. As the air passes from the restricted area to the larger area, the pressure falls and the velocity increases (Figure 5-13). Because of the large difference between the conical restriction and the line after it, an intense reaction is created that sucks the stationary air out of the area of the vacuum cup or wherever the air opening is attached into the main line after the Venturi valve. The air continues on and is often vented to the atmosphere, thus requiring nothing more than an air muffler in most cases to complete the system. The incoming stream of air can come from any readily available source, such as the facility's central compressed air system.

Venturi vacuum generators have many advantages:

- They do not have any vibration, heat generation, or moving parts.
- The vacuum turns on and off immediately with the air supply.
- Multiple Venturi valves in series can increase the vacuum generated.
- This approach works well in harsh environments.
- The low-cost systems are much less expensive than other mechanical systems.
- The Venturi valves are easy to repair, replace, or change out.

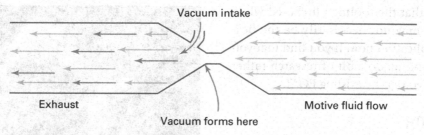

FIGURE 5-12 The working principle of the Venturi vacuum generator.

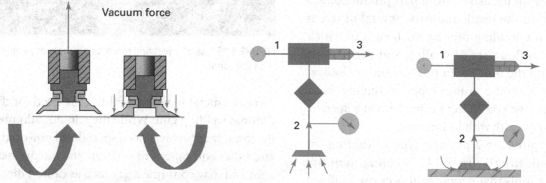

FIGURE 5-13 These images illustrate what happens when the Venturi valve has air flowing through it, the parts of the system, and the force vectors that lift the part.

- The Venturi valve is often located in the vacuum cup, reducing air consumption and improving response time.
- Venturi vacuum systems are lightweight and mobile, and do not require an electrical connection.

Balloon-type grippers take a different approach: They work with pneumatic power and rely on inflation or deflation to grip the part. Originally these grippers worked by placing the deflated balloon or flexible bladder inside some portion of the part and then inflating it with air. The balloon would take the shape of the part and exert force, allowing the system to pick up the part for manipulation. An advancement in this technology has been to fill the balloon or flexible bladder with coffee grounds or a similar substance that can easily conform to the shape of the part. The air-filled balloon then allows the material inside free movement as the system presses the balloon against the part. While holding the balloon against the part, the system vacuums the air out, locking the loose media together in a tight matrix. The result is a gripper that holds parts securely in much the same manner as a human hand would in terms of rigidity and flexibility.

While balloon-type systems are effective for gripping rigid and flexible items as well as delicate items,

they are not widely used in industry today. As this technology matures, you may well encounter these systems more often in the field.

Pin and mandrel grippers work like the balloon-type grippers. They inflate a bladder or bellows with air to create pressure against the part. The pin gripper fits around a projection on the part and then inflates the bladder to create pressure against the part. The mandrel gripper works in the same fashion, but fits on the inside of the part and presses outward against the walls of the part to create the gripping force. With the advancements in gripper technology, these grippers seem to be fading out. They are limited in the type of parts they can work with and have a limited range of movement for part variation.

These are by no means all of the gripping systems available, but the majority of the systems in use today do fall into one of the classes mentioned in this section. When you get into industry and start to learn more about the systems utilized at your facility, be sure to take the time to learn all the ins and outs of your tooling as well. Many of the maintenance calls I responded to in industry related to the tooling in some manner. I became intimately familiar with the EOAT operation, as I often had to repair and rebuild these systems.

Also remember that the tooling purchased when the robot was new may not necessarily be the best tooling for your application now. If you find that you need to buy new tooling, do a bit of research rather than automatically reorder the existing EOAT.

Welding Tooling

Welding operations are another common application for industrial robotics. These operations are utilized in many areas of industry—from part production, to automation, to the textile industry. Several different versions of the welding process exist, each of which has its own unique set of tooling and peripherals that help with the process. In this section, we look at the big picture on the various types of welding, but I highly encourage you to dig deeper into any specific field you might work with in industry.

Welding guns are a popular type of tooling for robots in industry (Figure 5-14). **Welding guns** are the tube-like tools that people or robots use to direct the electrically charged wire used to fuse metal during welding operations. This tooling requires its own power supply, which is in addition to or separate from the power requirements of the robot, and uses either MIG or laser technology.

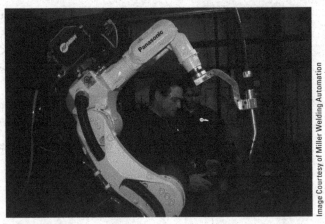

Image Courtesy of Miller Welding Automation

FIGURE 5-15 Use of a welding torch as a couple of people work with the robot.

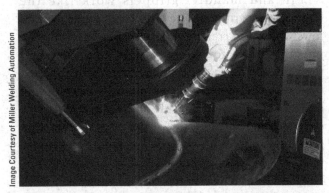

Image Courtesy of Miller Welding Automation

FIGURE 5-14 A welding robot in action. Notice how bright the welding process is and the smooth path of metal left behind. You should always wear proper darkened eye protection when watching or working around welders.

MIG welders use electrically charged metal wire, fed through the welding gun, to fuse metal together; the wire is shielded from oxygen during this process by an inert gas such as carbon dioxide (CO_2). The current causes high levels of heat at the point of contact that melts the wire and both pieces of metal to create a new, solid metal connection (Figure 5-15).

Laser welders use intense beams of light to create the high temperatures needed to melt and fuse the two pieces of metal together, without the need for the traditional welding gun. While they do not add metal to the joint, laser welders need specialized power supplies and other equipment to generate the intense beam of light and may consume a gas such as carbon dioxide in the process. Key benefits of these systems are that they are consistently repeatable, fast, reduce the required raw materials, and can get to places that are difficult to access with the human hand.

A welding torch may include a clutch or other collision-sensing device to prevent damage to the torch in the event of a collision. These devices send a signal to the robot indicating that there has been a collision, and the robot stops moving to prevent further damage to the torch. Newer robot software may have collision detection or force detection provisions built into the software, eliminating the need for a clutch or other sensing devices. The software monitors the motor current and, in the event of a sudden spike or increase in motor current, it stops robot motion to prevent damage to the torch. In these cases, a solid mount can be used with the torch instead of breakaway or movable mounts that allow the welding gun to move out of position to prevent damage. With the movable mounts, an impact often requires the operator to reset the torch and perform some kind of calibration sequence.

Many manufacturers of robotic welding torches have designed automatic cleaning equipment to periodically clean the torch during the welding process. The stations are designed to clean out the torch, remove any excess slag, and add more antispatter compound to reduce the probability of slag adhering to the welding torch (Figure 5-16). While it may add to the costs of a robotic welding station, this equipment eliminates the need for manual cleaning of the torch and prolongs the tooling's life. Simple programs can be incorporated into the robot to periodically move it to the torch cleaning station, where the torch can be cleaned, dressed, and antispatter applied before continuing with the welding operation.

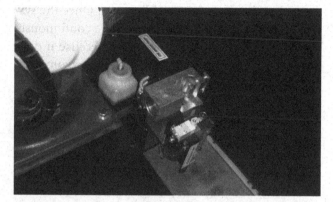

FIGURE 5-16 This tooling cleans the spatter out of the welding gun for a Panasonic 102S welding robot. Once the torch tip is clean, it applies an antispatter spray that makes the tip easier to clean the next time.

In metal inert gas (MIG) welders, the wire feeder feeds wire from a spool out to the torch tip where the wire is used in the welding process. Two types of wire feeders are available: the push type and the pull type. The push type pushes the wire from the back of the robot through a feed tube up to the welding process; it works best with hard metal wire. Recent advancements in welding technology have now made it possible to robotically weld aluminum and other soft metals, in which case a pull-type wire feeder is necessary. The pull-type wire feeder is located at the front of the robot arm near the torch. It pulls the wire through the wire feed tube and then pushes it on through the torch, which prevents soft metal wire, such as aluminum wire, from bunching up inside the torch feeding tube.

The weld controller controls the welding process via a schedule created and stored in the weld controller. The weld schedule contains all the parameters necessary to complete the weld process, such as weld voltage, weld current, and wire feed speed. The weld controller is typically connected via an electrical interface to the robot controller. The robot controller tells the weld controller which weld schedule to use, when to start welding, and when to stop welding, and provides any other special instructions such as weaving from side to side or using edge detection to find where to start the weld.

Many automotive assembly plants and other plants now use robots as part of their spot-welding process, replacing manual welding equipment and reducing human involvement in this phase of the manufacturing process (Figure 5-17). The result is consistent, precision welds that prevent wear and tear on human workers and improve quality. **Spot-welding** uses a device called a gun, which resembles a large copper clamp. It fuses metal together by passing current from one tip to the other, through the material fused (Figure 5-18). The robot can be set either to move the spot-welding gun or to position the parts inside a stationary gun, depending on the size of the part and the locations of the spot welds.

FIGURE 5-17 A spot-welding process welding the frame of a vehicle together.

FIGURE 5-18 A close up of a spot-welding gun that you might find in use by industry.

Spot-welding gun tips require periodic maintenance and replacement, as they are slowly eroded by the welding process. This process is also controlled by a weld control system, but the parameters used differ as there is no wire feed to such a system. The weld controller provides a weld schedule that includes weld voltage, weld current, pressure, and **gun time**, which is the amount of time that current flows through the spot-welding gun tips.

There are two primary types of spot-welding guns. The "C"-type gun consists of a fixed tip and a movable weld tip resembling a large C clamp (Figure 5-18). When setting up a tool frame for a C-type gun, the tool's center point is set at the fixed tip of the welding gun. (Tool frames are covered in detail in Chapter 6.) In the "X"-type gun, the jaws form the letter "X" and act like a scissors (Figure 5-19). The tool's center point for an X-type welding gun is set at the point where the tips of the weld gun come together.

Rainer Plendl/Shutterstock.com

FIGURE 5-19 The X-type spot-welding gun. The primary difference between it and the C style is that both jaws move on the X type.

Pneumatically operated welding guns use pneumatic pressure to control the force of the welding gun, which is measured as the amount of pressure exerted at the point where the tips of the gun come together. One problem with pneumatically operated guns is potential inconsistency in force due to variations in air pressure, leaks, or worn-out pneumatic cylinders on the gun. Hydraulically powered welding guns often use pneumatic pressure to control a hydraulic servo valve, which in turn controls the amount of weld force applied at the gun tips. This type of welding gun is being discontinued in most robotic spot-welding applications due to the additional dress-out required for the hydraulic system as well as the problems associated with the force and operation of this gun.

Recent advances in controller and servo technology have made the servomotor–driven welding gun a practical alternative to both the pneumatic and hydraulic

welding guns. A servomotor is used to bring the tips of the welding gun together, with the weld force then being controlled by the amount of torque generated by this motor. The weld controller monitors the amount of current drawn by the servomotor and, in turn, the amount of torque generated by the servomotor. A pressure gauge is used to calibrate the amount of force exerted at the tips of the welding gun. Once the system is properly calibrated, it provides a continuously variable force at the gun tips. Also, because it does not rely on air or hydraulic pressure, the servomotor–driven welding gun provides a much more consistent weld force. As they become more cost-effective, these guns are increasing popularity and replacing their fluid power–driven predecessors.

Ultrasonic welding is another type of welding used to join textiles, plastics, and some metals. This welding technology uses high-frequency vibrations, above the realm of human hearing, to generate heat via friction, which in turn fuses materials together. This process is utilized in the following applications, among others:

- General-purpose sewing
- Lacing and quilting
- Filters
- Medical disposables
- Cuffs and sleeves
- Ribbons and trims
- Ballistic vests and body armor

Ultrasonic welding offers several advantages. It does not require adhesives or other consumables, which eliminates the need for solders, threads, glues, and so on. Weld bonds are produced fairly quickly, in a matter of 2–4 seconds, with no heating of the surrounding area. Ultrasonic welding also produces accurate repeatable and reliable welds and bonds at an economical price. In addition, it is environmentally friendly, as the welding process does not generate any slag, blowout, or other by-products.

An ultrasonic welding system differs in the way the weld is produced. Thus, we find a different set of equipment involved in the process:

- A power supply
- A transducer
- A booster
- A horn
- An anvil or nest

The power supply converts the incoming power into the high frequency and high voltage needed by the transducer. The transducer contains a number of piezoelectric crystals, which changes the high-frequency signal generated by the power supply into a mechanical vibration. The booster transmits the vibrating mechanical energy and increases its amplitude. The horn delivers the mechanical vibration energy to the weld joint. The anvil or nest supports the parts that are being welded or fused together.

The process of an ultrasonic weld begins when the materials to be fused are placed in a fixture known as the nest. The horn is then brought into contact with the materials, and pressure is applied to clamp the parts together between the horn and fixture. At this point the welding process is engaged and the horn vibrates at whatever frequency for which the system is set (Figure 5-20). Although the horn travels only thousandths of an inch or microns of distance, it does this thousands of times per second, generating frictional heat in the clamped materials. As the materials heat up, they fuse or weld. Once this process is complete, the hold time begins; during this time, the horn applies pressure, but no vibration, so the parts can solidify once more. Finally, the horn retracts. If all has gone well, a new welded joint is left behind.

Other Types of EOAT

Sprayers are another type of robot tooling commonly encountered in industry. Originally these systems applied liquid paint in the auto industry, but the technology has broadened considerably since the early years. Today's spraying tools apply both liquid and powder-based paints, adhesives, lacquers, and any other sprayable substance desired for use in industry (Figure 5-21). As a result of the robot's precision, these systems save thousands of dollars each year in wasted raw materials while providing the high-quality bonds and coatings that customers demand (Figure 5-22). Application of these materials is a task that can be both dull and dangerous in many cases, making it a perfect fit for robotic systems.

Add a motor to the robot's EOAT, and you now have a drill, threading tool, milling setup, polishing station, or other rotary system ready to create, modify, or finish parts. In essence, this type of tooling turns the robot into a very flexible industrial machine. In some facilities I have seen, this type of setup has been used to create control panels for boats, allowing for quick changes of programming and a smaller footprint than most industrial computer numerically controlled (CNC) machines. A similar application is deburring or deflashing of metal and plastic parts, which is the

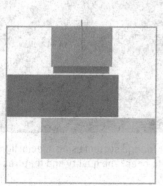 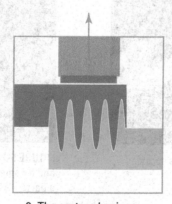 

1. The process begins with the materials in the fixture or nest and the horn makes contact to apply pressure to hold everything in place.

2. The system begins to vibrate the horn at frequencies of 20,000 cycles or more per second, generating heat through friction and melting the materials.

3. As the materials heat they blend or weld together and the horn stops vibrating, but stays in place for a period of time called hold time to allow the materials to solidify once more while ensuring a strong weld.

FIGURE 5-20 Ultrasonic welding.

Photo Courtesy of FANUC America Corp. www.fanucamerica.com

FIGURE 5-21 A paint-spraying robot working on bumpers.

Photo Courtesy of FANUC America Corp. www.fanucamerica.com

FIGURE 5-22 There is very little wasted paint in this spraying application.

Courtesy ATI Industrial Automation

FIGURE 5-23 A few examples of deburring or deflashing tools that a robot can utilize.

Photo Courtesy of FANUC America Corp. www.fanucamerica.com

FIGURE 5-24 A robotic system using special lighting and a camera to check the quality and features of a part.

process of removing metal burrs or excess material from casting and stamping operations (Figure 5-23). With the proper tooling setup, a robot can unload parts, load in any raw materials necessary, clean up the newly produced part, and send it on its way to the next stage in the production process.

Inspection is another field in which robots are widely used. End-of-arm tooling in these applications may include specialized sensors, cameras, infrared systems, or any other sensing device (Figure 5-24).

The idea behind this type of tooling is to use the robot for either quality control or to help ensure the safety of humans. At one of my former workplaces, a small robotic arm with a specialized gas sensor was used to check the welds of the product by detecting leaks of the pressurized nitrogen gas inside. If the robot detected the gas, it would kick the part off to the side for an operator to inspect more closely. This is but one example of the thousands of uses for inspection tooling.

As you explore the world of robotics, you will encounter variations and adaptations of the tooling discussed in this section of the chapter as well as completely new ways of manipulating the robot's world. When this happens, I encourage you to take a few moments to figure out how the tooling works, what its limitations are, and how it could help you in your chosen field of robotics. The more you understand about tooling and the deeper your knowledge of the existing tooling, the better your chances of picking the right tooling for whatever job your robot will perform.

MULTIPLE TOOLING

We have examined some of the various types of tooling available and discussed how the flexibility of the robot is one of its key features, but what should we do when we need to use more than one type of tooling on a robot? The simple answer is that we change out the EOAT or use multiple tools at once. As we drill down into this issue, you will see that several options for these choices exist as well. To determine which option is best, you first must determine how many different tools you need and how often you need to change the tooling. As we discuss the various methods for using multiple tooling, we will look at the pros and cons of these options.

A low-tech method to use multiple tooling involves having the operator or another appropriate party physically take off the current tooling and put on the new tooling. The prerequisite here is that the new tooling must align with the mounting plate of the robot and work with the robot's systems. If this is the first time using the tooling, some modifications to the tooling, the robot, or both may be required. This method is suitable for systems that are changed only rarely or when adapting the robot to a new task it will perform for a long period of time. It is also common to use this method to replace tooling damaged by normal wear or unprogrammed contact (a crash). The downsides to this approach are that the robot will likely be out of production for a fair amount of time and that someone needs to perform all the work of changing the tooling, plus making sure everything is set up properly. In some cases, this can take hours or—in severe cases—even days of work.

Another low-tech option is to mount multiple tools to the robot simultaneously. This approach avoids the need to change tooling and provides for increased functionality, but comes at a high price.

First, it reduces the payload of the robot by the weight of two or more tools instead of just one tool, which reduces the force remaining to move parts and other materials. Second, this method often requires specialized tooling bases and/or tooling systems to allow for the multiple functionality. Anytime the tooling is specialized, the price increases dramatically. In situations where the robot does not need to move parts and it performs the same operations repeatedly, this option can be worth the cost—which is why you will often encounter these setups in industry.

The next step up is to use quick-change adaptors like the one shown in Figure 5-25. The alignment pins provide consistent placement for the tooling, and the coupling system provides for positive locking while making it easy to detach the current tooling and swap in new tooling. These setups can transfer various power sources as well as make communication connections for any sensors the tooling may have. The system works by attaching one plate to the robot, usually at the end of the wrist; the other plate goes on the base of the tooling. When it is time to change the tooling, a release mechanism allows the two plates to separate. As long as the new tooling has a proper adaptor plate, all you have to do is attach the new tool and enter any offsets needed to get the system running once more. Quick-change units are great choices for systems that require frequent changes, such as daily or even hourly. The adaptors are also a crucial part of the automatic systems we will discuss next. The downsides to this approach are the cost of the connection plates and the initial setup required to attach the base plate to the robot and the quick-change plate to the new tooling.

Automatic systems, like the one shown in Figure 5-26, are used when the robot requires multiple

FIGURE 5-25 Some of the various tooling adaptors that allow for quick change of tooling either by hand or other mechanical means.

Image Courtesy of SCHUNK

FIGURE 5-26 This tool changer system removes and replaces the robot's tooling as needed.

Image Courtesy of SCHUNK

FIGURE 5-27 In this type of quick-change tooling, the base connects around the specially machined base of the tooling. This type of tooling would be perfect for a system where the robot does all the moving and the tooling remains static.

tooling to perform each cycle of its operation. In this type of operation, the robot may pick up parts from a conveyor or bin, load them into a machine, take finished parts out, deburr or thread holes, and then place the parts on another conveyor. If the robot needs only two or three different tooling options, the EOATs may be mounted on the robot continuously—although this cuts into the maximum payload of the system. More often, the robot will use quick-change adaptors (Figure 5-26) and simply change them out as needed, as in the system pictured in Figure 5-27. In Figure 5-27, the tooling is stored on a rotary wheel and rotated into position as needed; an empty spot takes the current tool after it is detached, the new tooling needed is rotated, and that tooling is held in position while a positive lock with the robot is completed. Other systems may store the tools in a stationary location and let the robot do all the positional work of dropping off the current tooling and then picking up the new tooling. The only downside to this approach is the amount of time lost to changing tools during each cycle. Compared to the flexibility it gives the system, however, this is usually a small price to pay. Misalignments of these systems can lead to crashes where part of the robot impacts something solid or other alarm conditions, so it is wise to add some type of sensor system to ensure proper tooling position before detachment and attachment.

As you can see, multiple options can be used when multiple tools are needed. Often more than one of the methods mentioned in this section can perform the tooling change, so the user must determine which one offers the most benefit for the least cost. When rapid change-outs of tools are needed during the production cycle, some form of automatic tool change is essential, because shutting the machine down every few minutes for a person to change the tool is just not economical. In other types of scenarios, you should weigh the pros, cons, and cost of each method to determine which one best fits your specific need.

POSITIONING OF EOAT

The best tooling in the world becomes useless weight if you cannot get it into position to perform its tasks. Sometimes we run into trouble because we cannot move the robot into the position needed; other times the problem relates to inconsistency in the parts. Whatever the case, we need to make sure that the robot can do its job without the operator making adjustments every few minutes—or we might as well get rid of the robot altogether. This section focuses on some of the ways we get around the problem of positioning tooling.

The common industrial robot has six or seven degrees of freedom (DOF), but have you wondered why? The reason is that the three major axes and three minor axes of movement in a six-axis robot allow for a wide range of positioning options and are often all we need for industry. The three minor axes orient the tooling in the proper position for the task it must perform. Problems with getting the tooling into position, from the robot's perspective, typically occur when we have only one or two minor axes or parts that are not consistent in position/dimension. When the problem lies in the parts, we have several options to resolve this situation. By comparison, if the robot simply does not have the DOF needed to reach the desired position, we have only two options: adapt the process to the robot or change the robot. Here, we consider those situations where the parts are inconsistent or out of position, as we have a tooling solution to help with this case.

Remote center compliance (RCC) is a simple way for tooling to respond to parts that are not always in the same position. RCC devices allow the tooling to shift a small distance from the center position without causing the robot to alarm out or exerting excessive force on the tooling. By using springs or materials that can flex, as shown in Figures 5-28 and 5-29, systems achieve passive RCC. Such passive RCCs often have an added shear plate that breaks loose in the case of excessive side force to preserve tooling and prevent damage. RCCs are good choices for reaming, taping, or other operations in which a small amount of flexibility is required. In these instances, the tooling comes in, makes contact with the part, and flexes in the required direction to complete the programmed task.

Many of the modern RCC devices are active RCCs, in which sensors inside the unit can detect how much side-shift is occurring (Figure 5-30). While these RCCs are more complex, they do alleviate the need for a shear plate inside the setup. If the tooling is being forced too far out of position, the sensor detects this condition and stops the robot, which in turn sends an error to the teach pendant screen. Some of these systems can also detect the amount of torque the tooling is using to perform its tasks, giving valuable feedback to the operator. With proper programming, the robot can alert the operator to dull tools, holes that have

FIGURE 5-28 Notice the springs between the tooling base and where the base attaches into the tooling holder on the robot. This arrangement allows for some side-to-side motion of the tooling without incident.

Image Courtesy of SCHUNK

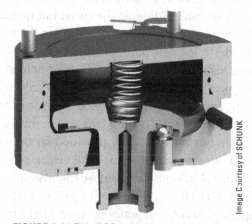

FIGURE 5-29 This RCC unit allows the tooling room to compress and move while engaging the part, but it centers up once the pressure is released via the pin shown on the right-side interior.

Image Courtesy of SCHUNK

not been threaded properly, or other situations that require operator intervention. Some of the better systems even allow the robot to make small offsets to the programmed position so it can complete the assigned task without damaging the tooling or part.

Another (and better) solution to the problem of part alignment is to use vision systems (Figure 5-31). With vision systems, the robot uses a camera to take a picture of the part and then compares this picture

FIGURE 5-30 This active compliance device is mounted between the tooling and the robot to detect issues. Inside there are strain gauges or other such sensors that detect when something is amiss.

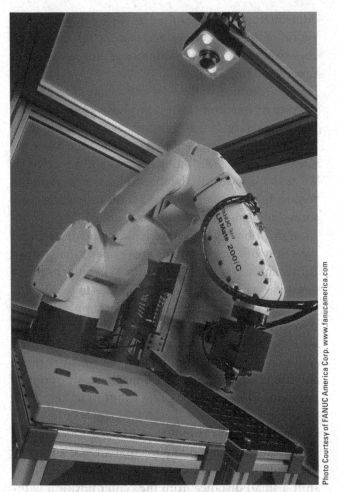

Photo Courtesy of FANUC America Corp. www.fanucamerica.com

FIGURE 5-31 This vision system identifies parts and their locations, so that the robot can then pick up those parts with a vacuum gripper and place them in the tray next to the robot.

to sorting criteria predefined by the programmer or stored by the software during the teaching process. Once the system has filtered the visual information, it runs a subroutine that allows it to offset the tooling in relation to the new part position. Over the last few decades, vision technology has advanced dramatically, to the point that it is becoming the standard for dealing with part positioning variance. The use of vision systems increases the overall cost and complexity of the robot, but the versatility provided far outweighs the cost. Not only can we use the vision system to adjust the robot's position, but we can also sort items by color, pick specific pieces out a pile of parts, carry out quality checks, and perform many other functions that require visual information.

Just as we did not cover every type of tooling out there in our earlier discussion, we have merely scratched the surface of ways to adjust for changes in

part position or misalignments in this section. While vision systems and advanced programming are popular solutions within the modern robotic world, many mechanical systems can perform the same functions with consistently reliable results. There is a good chance that you may find yourself working with systems in the field that use some combination of mechanical offsets, sensors, and vision systems to meet the specific challenges of the tasks that the robots must perform.

REVIEW

By now, you should have a better understanding of the diversity of robotic tooling and have some idea where to start in learning more. While new systems and processes for robotic tooling are frequently introduced,

the reality is that there are a lot of old, tried-and-true systems in use today that will be around for many years to come. I encourage you to take a bit of time and research some of the advances in robotic tooling so you can stay current in the field, especially if you are planning to order new tooling for a robot or need to replace some worn-out tooling in the field.

Here is a quick recap of what we covered in this chapter:

- **What is EOAT?** In this part of the chapter, we discussed what tooling is and why it is important to the world of robotics.

- **Types of tooling available.** This section covered grippers, welders, spraying guns, suction cups, and

other types of tooling for robots, and looked at some of the math involved with tooling.

- **Multiple tooling.** This section described the options for attaching tooling and identified the situations that these options best serve.

- **Positioning of EOAT.** We wrapped up our coverage of tooling by discussing how to deal with misalignments of the part and ways we can shift the tooling to fit the task at hand.

FORMULAS

Payload:

$Ap = P - Wt$

Ap = available payload

P = specified payload capacity for robot

Wt = weight of the EOAT and any added peripheral systems

Vertical grasp formula:

$F_r = W \times G$

F_r = required force

W = weight of part

G = gravitational force, including acceleration

Normal force:

$F = \mu N$ or $N = F/\mu$

F = friction force

μ = coefficient of friction

N = normal force

Torque:

$T = W \times d$

T = torque

W = weight

d = distance at which weight is applied

Weight:

$W = m \times G$

W = weight

m = mass of item

G = gravity

Required force for horizontal grasp:

$F = 2(m \times G \times d)/\sqrt{(b^2 + p^2)}$

F = force required

m = mass

G = gravity ($m \times G = W$)

d = distance from center of gravity, CG, to center of force, CF

b = width of jaws contact

p = thickness of part in gripper

KEY TERMS

Angular grippers	Grippers	Parallel grippers	Ultrasonic welding
Center of gravity	Gun time	Payload	Vacuum
End-of-arm tooling (EOAT)	Jaws	Remote center compliance (RCC)	Vacuum gripper
Ferrous metals	Laser welders	Safety factor	Welding guns
Fingers	MIG welders	Spot-welding	
Friction	Odd-shaped parts	Tooling base	

REVIEW QUESTIONS

1. Where do we attach the end-of-arm tooling?

2. What is the difference between angular grippers and parallel grippers?

3. If you need a gripper for a wide range of part sizes, which would be better: parallel or angular grippers? Why?

4. Which types of environments are hydraulic or pneumatic grippers well suited to?

5. When we power grippers in only one direction, how is the other motion achieved?

6. Which precaution should one take regardless of the type of gripper or power source for the gripper?

7. To center a part in two directions at once, what is the recommended number of fingers on a gripper?

8. For odd-shaped parts, which type of gripper will you need?

9. What are the four rules for determining which gripper to use?

10. When determining the force required from a gripper, which factors do we consider?

11. How does an electromagnet work?

12. How does an Venturi valve work?

13. What is the difference between MIG welding and laser welding?

14. Explain how collision software is utilized to protect robot tooling.

15. What are the two types of wire feed devices for welding, and which type of metal does each work with?

16. What are some of the features of welding controlled by a weld schedule?

17. Describe the process of spot-welding.

18. What are the two configurations of spot-welding guns, and what are the power source options for the closing action?

19. Describe the process of ultrasonic welding.

20. What are the two drawbacks to mounting multiple tools on the robot at the same time?

21. What are the benefits of quick-change adaptors?

22. Describe the operation of an automatic tool-change system.

23. What is the difference between active and passive RCC units?

24. How do vision systems help with part misalignment or variance in parts?

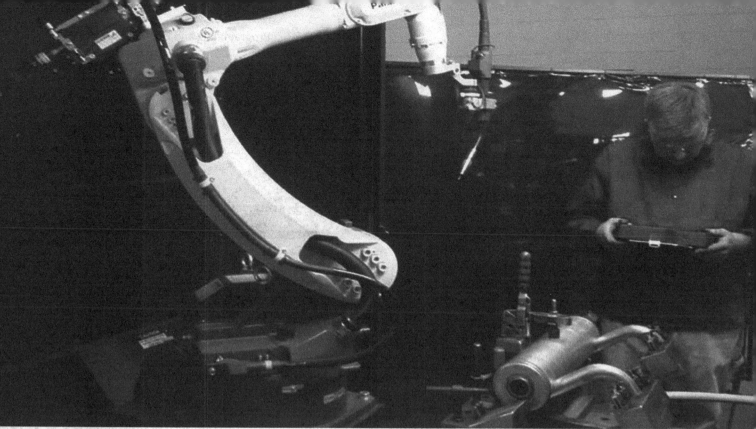

CHAPTER 6

Programming and File Management

WHAT YOU WILL LEARN

- The basic evolution of the robot programming language

- The five different levels of programming languages

- The key points in planning out your program and what each step entails

- What singularity is

- What subroutines are and how they help the programming process

- The difference between local and global data

- The basics of writing a program and the main motion types available

- How several common logic sorting commands work

- How to test a program that you have written, including what to look for

- What to do when the robot is ready for normal operation, including some potential sources of problems

- How to manage the data of the robot to ensure proper operation and save yourself time and effort down the road

OVERVIEW

Programs are user-defined sets of instructions that control the *what*, *when*, and *how* of robot action. A program is only as good as the person or persons who create it, and those individuals' programming skills are crucial in getting the full versatility out of a robot. The programming side of robotics is an ever-evolving field as we look for new ways to make the robot easier to program while expanding the control options. The downside to programming is that each robot manufacturer uses a slightly different method to program everything, which makes it a bit challenging when you have multiple types of system in one facility. The upside is that, once you learn the basics of programming and how to apply them to a specific robot, it is much easier to learn the specifics of programming a new robot.

With those points in mind, this chapter covers the basics of writing a program and considers some industrial programming examples to get you ready for the field. During our exploration, we will hit on the following topics:

- Programming language evolution
- Planning
- Subroutines
- Writing the program
- Testing and verifying
- Normal operation
- File maintenance

PROGRAMMING LANGUAGE EVOLUTION

Before we begin our exploration of how programming languages evolved, let us take a moment to define what a robot program is. A **robot program** is the list of commands that run within the software of the robot controller and dictate the actions of the system based on the logic sorting routine created by the user/programmer (Figure 6-1). As mentioned in the introduction to this chapter, the performance of a robot is only as good as the person(s) who create(s) the program. If we tell the robot to move through an object in the way, the robot will try to do so: At best, it will error out; at worst, it will crash into something. If we tell the robot to add in several stop points that do nothing productive, the robot will waste time. Whatever we put in the program is what the robot will do or attempt to do. Luckily, we can continue to modify and change the program until

FIGURE 6-1 A Panasonic welding program on its teach pendant.

we have the best program for the application. Moreover, with proper maintenance of the robot and its programs, once we have the ideal program the robot can continue to perform its tasks in an ideal way for days, weeks, months, and even years. **Programming languages** are the rules governing how we enter the program so the robot controller can understand the commands.

Before the digital age, we controlled or programmed robots by using devices that opened and closed contacts in specific sequences. Punch cards, peg drums, relay logic, and other methods were used to control these systems. Thus, these setups required creating punch cards with a different set of holes in them, moving the pegs on the drum, rewiring the relays, or some other physical alteration of the control structure to change the operation of the system. As you can imagine, making such a change in operation often took a long time and the process was less than user friendly. Today it is rare to see this type of system, but you may occasionally run across one of these older systems in the industrial world. If you do, there is a good chance it is one major crash or breakdown away from replacement, as parts for such systems are likely either hard or impossible to find.

In the early 1970s, computer technology changed the way we program robots, but there were some growing pains along the way. The first approach to computer programming of the robot involved using a language designed around the needs of the robot. The result was a programming language or technique that was great for controlling the robot, but difficult for users to work with. Programming these systems was unlike any other programming of the day, and the proprietary languages were not well suited to data processing. Several companies produced systems with this type of programming, such as Cincinnati Milacron's T3 language, but the complaints from users made this a short-lived method of programming.

Since the customer is always right, robotics companies took the users' complaints to heart and tried another approach to robot programming. This time the manufacturers started with a known computer language, such as BASIC or Fortran, and added commands to control the robot. The result was a language that computer programmers of the day understood, and that offered some data processing capability. From the programmer's side of things this method was great, but it was not the optimal way to control the robot. Since these languages were primarily intended for computers, rather than robots, there were problems with the motion commands, which translated into inefficiency of motion, as well as other commands that were not well suited to robot control. Unimation was one of the companies that tried this approach to robot programming: It developed the VAL language for its PUMA robots, with this proprietary language being based on the general-purpose computer language BASIC.

Over time, manufacturers found a way to give users the best of both worlds: They created programming languages that combined the efficiency of those languages specifically designed for the robot with the established programming flow provided by common computer languages. The resulting languages were easy to understand for those users with a programming background, yet avoided the inefficiency of using a computer language only. These hybrid languages continue to evolve, to the point that some systems no longer require users to have a computer programming background. In fact, many users can successfully program modern systems with no more than 40 hours of hands-on training. Obviously, it takes more training and time to reach the expert level, but a week of specialized training or its equivalent in on the job experience allows most workers to edit existing programs or write new, effective programs for the system. These hybrid languages are responsible for the differences in programming methods both between manufacturers and sometimes between various models from the same manufacturer.

The evolution of robot programming can be broken into five different levels:

- Level 1: No processor
- Level 2: Direct position control
- Level 3: Simple point-to-point
- Level 4: Advanced point-to-point
- Level 5: Point-to-point with artificial intelligence (AI)

As we explore these five levels of robot programming, I encourage readers who have experience with robots to classify the systems you have worked with. With the emergence of collaborative robots and the widespread integration of vision systems, we are getting closer to level 6 languages with advanced AI. Currently, though, we need to complete a bit more development before we suggest the new languages constitute a full-blown new level. Keep an eye out for developments on this front, as you may well work with a system that has advanced AI functions someday soon.

Level 1: No Processor

As the category name implies, level 1 systems lack computer or processor control. These systems work with relay logic or mechanical drums with pegs to control the timing of actions and include any system that works without the need for a computer or digital processing chip. To change the operation of these systems, the user must physically change something in the system—change the pegs on the rotating drum, create a new punch card to control which contacts are made and when, rewire the relay system, or some other physical manipulation of the system. These systems are mostly a part of industry's past, but a few of the simpler systems may still be in use somewhere.

Level 2: Direct Position Control

As the category name implies, the level 2 system of control requires the programmer to enter the positional data for each axis as well as all the motion, processing, and data-gathering commands to create a program. Level 2 programming is the most basic level of processor control and the most labor intensive for the programmer, requiring knowledge of the position of each axis when the robot is in the desired location. If there is no way to gather the information directly from the robot, the programmer has to do the complex math necessary to find the position, and any errors in the math or the system equate to missing the desired position. As you can imagine, this tedious style of programming has a high risk of error and is not a favorite among programmers. Few, if any, industrial systems still use this form of programming.

Level 3: Simple Point-to-Point

Simple point-to-point programming was a common way to control robots in the mid- to late 1970s and throughout the 1980s, and many older industrial systems continue to use this method of programming. The programmer must enter the motion type, data gathering, and other aspects of the program, but does not manually enter the positions. When programming level 3 systems, it is common practice to write the basic program offline, with each position having

```
10 'knm
20 '4-16-08
30 'box template
100 MOV P_SAFE
110 POFFZ.Z = 10
120 MOV P1 + POFFZ
130 MVS P1
140 MVS P3
150 MVS P3 + POFFZ
160 MOV P5 + POFFZ
170 MVS P5
180 MVS P2
190 MVS P4
200 MVS P4 + POFFZ
210 MOV P8 + POFFZ
220 MVS P8
230 MVS P6
240 MVR P6, P7, P8
250 MVR P8, P9, P10
260 MVS P10 + POFFZ
270 MOV P11 + POFFZ
280 MVS P11
290 MVS P11 + POFFZ
300 MOV P12 + POFFZ
310 MVS P12
320 MVS P13
330 MVS P13 + POFFZ
340 MOV P14 + POFFZ
350 MVS P14
360 MVS P15
370 MVS P15 + POFFZ
380 MOV P16 + POFFZ
390 MVS P16
400 MVS P17
410 MVS P17 + POFFZ
420 MOV P18 + POFFZ
430 MVS P18
435 MVS P22
440 MVS P19
450 MVR P19, P20, P21
460 MVS P23
470 MVS P23 + POFFZ
480 MOV P_SAFE
P1 =(160.710,-89.430,502.100,111.960,110.010,0.000)(6,0)
P2 =(221.920,-89.430,506.690,111.960,110.010,0.000)(6,0)
P3 =(286.430,-97.050,511.610,111.960,110.010,0.000)(6,0)
P4 =(288.360,-42.670,511.840,111.960,110.010,0.000)(6,0)
P5 =(170.600,-34.020,501.570,111.960,110.010,0.000)(6,0)
P6 =(268.230,29.240,511.020,111.960,110.010,0.000)(6,0)
P7 =(256.280,19.570,509.430,111.960,110.010,0.000)(6,0)
P8 =(268.450,4.780,511.020,111.960,110.010,0.000)(6,0)
P9 =(283.020,1.360,511.740,111.960,110.010,0.000)(6,0)
P10 =(288.140,25.940,511.760,111.960,110.010,0.000)(6,0)
P11 =(247.520,72.240,509.920,111.960,110.010,0.000)(6,0)
P12 =(261.970,70.880,511.830,111.960,110.000,0.000)(6,0)
P13 =(289.730,65.980,514.110,111.960,110.000,0.000)(6,0)
P14 =(252.980,110.240,511.950,111.960,110.000,0.000)(6,0)
P15 =(292.690,101.820,515.120,111.960,110.000,0.000)(6,0)
P16 =(269.820,95.340,512.500,111.960,110.000,0.000)(6,0)
P17 =(269.820,119.110,513.100,111.960,110.000,0.000)(6,0)
P18 =(189.260,156.660,508.670,111.960,110.000,0.000)(6,0)
P19 =(258.110,139.580,513.790,111.960,110.000,0.000)(6,0)
P20 =(249.680,160.070,512.990,111.960,110.000,0.000)(6,0)
P21 =(257.540,171.670,514.020,111.960,110.000,0.000)(6,0)
P22 =(293.830,127.760,515.400,111.960,110.000,0.000)(6,0)
P23 =(291.210,166.100,515.970,111.960,110.000,0.000)(6,0)
POFFZ =(0.000,0.000,0.000,0.000,0.000,0.000,0.000,0.000)
```

FIGURE 6-2(A) A level 3 program with the positional data entered at the end. This was written on a Mitsubishi five-axis robot.

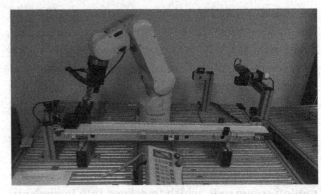

FIGURE 6-2(B) The robot for which the program in (a) was written.

a label, but no coordinate data. Once the programmer is happy with the program, he or she uploads it to the robot and then physically moves the robot to each point and records the positional data. Sometimes this requires moving through the program one line at a time; at other times it requires moving the robot to the desired position and then saving the data under the proper position label. The exact method depends on the model of robot, controller, teach pendant, and manufacturer of the robot (Figure 6-2).

Some systems allow the programmer to create the program directly from the teach pendant using this method. Having programmed systems using this method both by creating the program in advance and by creating it on the fly from the teach pendant, I can honestly say that I prefer to create the program offline and upload it. It is easier to type on a keyboard than on the teach pendant, because the latter works like the older cellphones, where each key has three letters on it. To type out a word, you must press a key one to three times for each letter, which takes more time and effort than typing on a keyboard. For editing a line or two of code, the teach pendant works well; to write a long program, a computer is a much better option.

Another drawback of programming directly on the teach pendant is that the display screen usually shows only a few lines of code at any given time, making it hard to keep the flow of your program clearly in mind as you go along. If you do plan to enter the program from a teach pendant, I recommend having a paper copy of your program in advance so you can enter it in without having to worry as much about the logical flow.

Level 4: Advanced Point-to-Point

Advanced point-to-point programming is an advancement of simple point-to-point programming that makes the process of writing a program much simpler. With level 3 languages, the user bore the brunt of inputting all the action commands and direction commands. With level 4 languages, programming is as simple as creating a new program, entering a string of points with the proper motion label to reach those points, and testing out the program. There is no need to memorize the large number of movement and logic commands that are required by simpler programming methods. The programmer has to determine the key points of the program, how the robot moves between those key points, and any logic filters that might be necessary. The software of the robot handles the rest. (We will look more closely at movement types and logic filters later in this chapter.)

Level 4 programming languages truly changed the industrial robotics world. Suddenly, operators who had little or no experience with the system were able to create simple programs (Figure 6-3). Instead of needing a strong background in programming earned over years of classes or experience, a week's worth of training at the manufacturer's site was enough for most users to begin writing working programs. Manufacturer training usually lasts 30 to 40 hours and is presented over the course of one week, usually at the manufacturer' straining facility, with a general cost of approximately $2000, give or take a few hundred dollars. Some manufacturers waive the training cost if a company buys a certain number of their robots, helping to sweeten the sales deal. Advanced programs still require the programmer to have a good understanding of programming as well as some form of advanced training. Many robot manufacturers offer this training as well, but it is another week or two of training beyond the initial 30−40 hours, and only those familiar with programming the robot should attend.

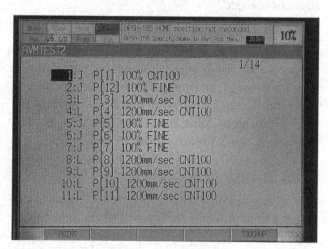

FIGURE 6-3 A level 4 program on the FANUC controller written in my robotics class. In this program, linear and joint movements are used between the various points.

The lion's share of industrial systems fall into this category, as level 4 languages make it simple and easy to modify or write new programs. Another benefit is that the simplicity of the programming language makes it easy to write a program directly from the teach pendant. In fact, the teach pendant is the most common way programs are entered for these systems, as the variables and logic for programs are options selected from menus. Offline programming is still an option, though: It involves the use of advanced software that can simulate the motions of the robot with accurate models of the work envelope, allowing the programmer to create viable programs with little need for adjustment. There is a high probability that you will work with this programming style if you spend any amount of time with industrial robots.

Unfortunately, there is no universal level 4 language, so you still have to learn the specifics of the language developed for the particular brand of robot you are working with. On the plus side, those programmers who learn a level 4 language tend to pick up other manufacturers' level 4 programming languages quickly and easily.

Level 5: Point-to-Point with AI

The newest advancement in computer programming is the addition of artificial intelligence to advanced point-to-point programming. The inclusion of AI takes the convenience of level 4 languages and adds the ability to correct errors as well as advanced teaching methods. While it is easy to set points in a level 4 language, the only leeway for error is provided by tooling compliance. Level 5 systems use vision, active compliance, or some other advanced sensing method to determine the difference between the desired position for the system and the taught position; they then make offsets to correct this difference. This ability adds a completely new level of flexibility to the industrial system and allows the robot to work in conditions that were infeasible for robotics in the past.

Another development in the level 5 programming styles relates to the way we teach points. The standard way to teach a point is to use the teach pendant in manual mode to move the robot into that position via various motion types and a combination of movements. Some level 5 systems allow users to physically grab the robot and move it into whatever position they desire. This turns teaching from an exercise resembling trying to get a stuffed animal out of a claw machine into something as simple as taking someone's hand and showing that person how to do a job. This type of teaching is very organically natural and makes it easier to position the robot as desired. The Baxter robot is a great example of this type of system, as it offers advanced safety features, camera vision to make offsets on the fly, and manual positioning teach mode in which the user can move the robot's arm instead of using a teach pendant.

Many of the currently available collaborative systems use this kind of teaching protocol and functionality, adding to their already attractive package of features. If your robot lacks this option, there are some third-party options that can add this feature to the robot, such as Robotiq's Kinetiq Teaching system. Robotiq's system is designed to work with multiple manufacturers' robotic welding systems. It gives users the ability to program by grabbing the tooling and moving it into position for welding applications, making it easier for welders to program welding robots. While third-party options for this type of programming are not as plentiful as they could be, they do exist.

PLANNING

Now that you know a bit about how programming evolved, it is time to learn the basic process of programming—and this begins with planning. Just as most things in life require a certain level of planning to achieve success, so you need to have a game plan in place before you start writing a program. The level of planning depends on the complexity of the task you have in mind for the robot. If you want the robot to move from point A to point B in a straight line, the plan is fairly straight forward: Make sure the robot can reach both points and there are no objects in the straight-line path for the robot to hit. If you want the robot to hit multiple points while welding two pieces of metal together and to avoid impact with any of the fixture parts, then you will need a more complex plan. Breaking the planning process down into steps will help to make sure you do not overlook anything (Figure 6-4).

Step 1: Goal Setting

What do you want the robot to do? Before you can start to plan, you have to know what the goal is. Remember, a program is a series of logical steps designed to control the actions of a robot. A robot cannot wing it or just make it happen; it can perform only within the confines and rules set by the program you create. If you create a program with no clear-cut idea of what you desire out of the robot, do not be surprised when

FIGURE 6-4 Not all robot tasks are simple, with easy avenues of approach. Many programs take a large amount of planning to avoid positioning issues, hitting objects such as clamps or fixtures, and other things that might affect the quality of work.

the robot fails to meet your expectations or do anything of value (Figure 6-5). When you add in the fact that robots are often faster and stronger than humans, a poorly planned program could literally be dangerous to the people around the system.

FIGURE 6-5 This part has multiple welds laid on top of one another to create the strength required. Without proper planning, these welds could look fine but actually have imperfections and other flaws that would reduce their structural strength.

Take the time to figure out what you want the robot to do from start to finish. I recommend writing down the tasks you have in mind or sketching out what you want the robot to do. During this step, it is important to make sure the tasks you have in mind are something the robot is capable of doing, rather than something beyond the system's reach, strength, speed, or tooling type.

Step 2: Task Mapping

Now that you know what you want the robot to do, take the time to figure out how it will do it. Ask yourself these questions:

- Which kind of tooling does the robot need?
- How should the robot move between points?

- Does the robot need to avoid any obstacles?
- What is the robot doing at each point?
- What is the robot doing between each pair of points?
- Do any conditions or other factors in the process need to be addressed?
- Is the process logical?

We will look at each of these questions in detail, as answering them can save you a lot of time, trouble, and frustration down the road when you start writing the program.

Which Kind of Tooling Does the Robot Need?

Before you begin mapping out any of the tasks, it is important to make sure the robot has the necessary end-of-arm tooling (EOAT) installed or that you have it available. If not, then you will need to order the proper tooling or come up with a different plan. For instance, a pick and place system with a gripper installed that was not designed for welding would need new tooling, an extra power supply, and proper software to become a welding robot (Figure 6-6). Sometimes the tooling change needed may be as simple as modifying the fingers on a gripper to hold the geometry of a new part. If the process requires more than one type of tooling, you will need to either allow for the swapping of tooling or install multiple tooling on the robot. Regardless of the need, I recommend addressing this concern right out of the gate as having the wrong or improper tooling is a fast way to insure failure.

FIGURE 6-6 Some pick and place operations require something beyond the generic gripper.

Photo Courtesy of FANUC America Corp. www.fanucamerica.com

How Should the Robot Move Between Points?

When conducting a motion analysis, one of the major concerns is whether the process is path dependent or position dependent. A path-dependent process requires careful analysis of the path(s) involved in process execution. For example, in an arc welding process, the paths must be carefully laid out to avoid damage to the work piece from excessive heat buildup. For painting operations, the paths should be laid out to ensure adequate coverage with minimum overlap. A position-dependent process, such as palletizing, requires planning of all positions to ensure the most efficient use of robot motion. As the end point is the main focus instead of the path, the only concern here is to make sure the path between points is free of obstacles.

Robot motion is controlled by robot motion instructions. When determining robot motion, careful consideration should be given to each of the characteristics of a motion instruction. A motion instruction contains up to five parts:

- Type of motion
- Position
- Speed
- Termination
- Options

Each part is essential to controlling robot motion accurately and must be understood to properly control robot motion. Here is an example of a motion instruction you would find in a FANUC system:

JOINT P[X] 100% FINE

This instruction starts with the motion command. There are four primary motion types in robotics: joint, linear, circular, and weave.

Joint motion is point-to-point motion in which all the axes involved usually move at the speed of the slowest axis, with no correlation between the separate axes involved. Because all the axes are moving independently, there is a good chance the motion between the points will not be a straight line. The tooling may curve, dip, or move in odd ways due to the amount of distance each axis moves. The robot should move between the two points in the same way each time—so if it did not hit anything the first time, you should be safe unless something changes in the work envelope. I have seen several programs where a joint movement caused a crash with objects in the work envelope due to this unexpected motion. I recommend using joint motion only when the robot is away from fixed objects and there is no real chance for impact. On the plus side, singularity is typically not an issue with joint movement. (We will look more closely at the dreaded singularity soon.)

In **linear motion**, the controller moves all the axes involved at set speeds to create straight-line motion between two points. Take a yardstick or other straight object and place it between your two points, and you will see the path of the robot engaged in linear motion. These motions do include angularity in case of a difference in height between the two points, so keep that factor in mind as well.

Be careful when following an object that must navigate uneven elevation changes during its linear motion path. For instance, if point A on the part is ¼ inch thick and point B on the part is ½ inch thick, with a point between the two that is ¾ inch thick, the robot will impact the high point between the two because it is in the way of the straight-line motion. To avoid these kinds of issues, you may have to program extra points along the straight-line distance to help the robot work around changes in thickness or other obstructions such as clamps or fixturing.

The line the robot draws with linear motion is based on the tool's center point and the orientation of the tooling remains constant unless we alter the tooling angle at the destination point. Linear moves can create singularity issues.

Circular motion is the formation of arcs and full circles as described by no less than three points. For movement in an **arc** (i.e., part of a circle), you need to teach the robot at least three points: the start point, a middle point, and an end point. For a full circle, you will need at least four points, but five or more is better, including a start point, a point every quarter of the circle (i.e., 90°), and an end point that is the same or just past the start point along the circle. In some systems, you set the initial point for the circular motion in the circular command; in other systems, such as in FANUC systems, the first point of the circle is the last programmed point above the circular command (Figure 6-7). Be sure you know the specifics of the robot you are using. I recommend a few trial runs to make sure you have a good understanding.

For all languages level 3 and higher, the robot controller handles the math necessary for linear or circular motions based on the points defined. Level 1 and 2 languages require you to figure out how to make this happen, which is another reason for their

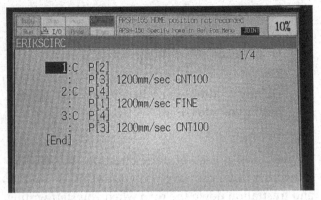

FIGURE 6-7 This example shows how a circular program is set up in a FANUC system.

rarity in industry. The higher-level languages will calculate the best arc or circle based on the points you create to complete the motion. If you teach one of the points in an odd location, your arc or circle may end up a bit distorted, so make sure you test these motions thoroughly.

Weave motion is straight-line or circular motion that moves from side to side in an angular fashion while the whole unit moves from one point to another. Think of it as a stitching-type motion across the normal line of motion. To program this kind of motion, you must set two additional points besides the normal required points for linear or circular motion. These two points determine how far to each side of the normal motion the robot weaves as well as the distance it moves forward between weaving motions. We can program any application or robot system to use weave motion, but welding applications are the most common place you will encounter this motion (Figure 6-8).

The second part of the motion command refers to position:

JOINT P[X] 100% FINE

FIGURE 6-8 This example shows what a weave weld can do. Notice how wide the weld path is and the wavy patter it creates.

A position is handled internally by the robot as a combination of numbers, one for each axis, representing location and orientation of the end-of-arm tooling. It may also include four Boolean terms representing the configuration of the arm, depending on how the manufacturer designates orientation. The internal breakdown of a position is

XYZWPR

where

- *X* is the distance in millimeters along the *x*-axis from the origin of the frame.
- *Y* is the distance in millimeters along the *y*-axis from the origin of the frame.
- *Z* is the distance in millimeters along the *z*-axis from the origin of the frame.
- *W* (yaw) is the rotation in degrees around the *x*-axis of the tool.
- *P* (pitch) is the rotation in degrees around the *y*-axis of the tool.
- *R* (roll) is the rotation in degrees around the *z*-axis of the tool.

Next to the position there is a speed designator that determines how fast the robot will move along the prescribed path. There are many different ways to express the speed of the robot:

- A percentage of the maximum speed of the robot

JOINT P[X] 100% FINE

- Millimeters per second (mm/s)
- A process-related parameter (e.g., in./s of weld)
- Something else robot, application, or manufacturer specific

Several factors determine the actual speed of the robot. For example, the programmed velocity or the setting we have in the motion instruction obviously affects the speed. In addition, an override setting may be used to limit the maximum speed of a robot, similar to the way a governor prevents an engine from going faster than a certain speed. For instance, in manual mode, most robots move at a set speed cap that is considered safe for humans to be around. Of course, any axis of the robot can move only at the maximum speed it is capable of. The main limiting factor on maximum axis speed is either the hardware of the system or certain types of moves. For example, in joint motion, all of the

axes move at the speed of the slowest axis. These factors, along with anything unique about the robot in use, determine the actual speed of robot motion.

The next part of the motion instruction is the termination type. The termination type may significantly influence robot motion path, but is not used by all manufacturers, and differs from manufacturer to manufacturer when it is used. If a termination type is specified, it will be something along the lines of the following:

- FINE, which stops robot motion at the point

JOINT P[X] 100% FINE

- CONTINUOUS, which slows but does not stop robot motion at the point

Both FANUC and ABB use fine termination to stop the robot at the end of the motion, but each uses a different termination type for continuous termination. FANUC software uses the expression CNT** for continuous termination. An ABB motion command uses Z*** for continuous termination. It is up to the operator, programmer, or process engineer to be familiar with the terminology used by the particular robot.

Fine termination stops the robot at the end of the move and adds approximately 500–750 milliseconds to the cycle time every time the robot stops. This additional time can significantly increase the overall cycle time of the program. The exact amount of time added will depend on several factors, including robot payload, the programmed velocity of the motion, the acceleration time of the servo, and the distance between the programmed positions. Because of this, most programmers use fine terminations only when necessary—that is, when the robot must hit the programmed point.

Continuous terminations are advantageous for several reasons. Since the robot does not come to a complete stop, continuous terminations are less likely to significantly increase cycle time. Moreover, the impact on cycle time is reduced. Continuous terminations also affect the trajectory that the robot will follow, with the exact trajectory determined by several factors:

- The programmed velocity of the robot
- The location of the positions
- The amount of deceleration
- Any override speed change settings

If your robot has the option of stopping or just slowing down at the points, make sure you understand all the dynamics of these two commands and use them properly. A few seconds saved on each cycle can translate into thousands of dollars saved each year!

While on the topic of motion, this is a good place to talk about kinematics, as you will likely hear this term used with robot motion at some point. **Kinematics** is a branch of mechanics that uses math to describe motion without considering either the cause of that motion or the masses involved. In physics, kinematics is typically defined mathematically by the formulas for displacement, time interval, initial velocity, final velocity, and constant acceleration. In the world of robotics, kinematics is the analytical study of motion of a robot manipulator and the basis for the complex math that goes into the movement of modern robotics with multiple axes or unique configurations, such as the delta-style robots.

In the robotics world, kinematics is further broken down into forward kinematics and inverse kinematics. Forward kinematics is based on calculating the position and orientation of the EOAT in terms of joint variables, often based in a Cartesian plane. In inverse kinematics, the position of the tooling is converted from Cartesian space to joint space. This conversion involves numerous algebraic calculations that take a fair amount of processing time, compared to the forward method. Inverse kinematics is often used with delta-style robots and hexapods that use six legs to travel due to their unique movements. The good news here is you can run and program an industrial robot without having to perform these calculations, thanks to the software doing all the math for you in the controller. If you choose to go into the field of robot development, kinematics is something you will need to learn more about—and I recommend brushing up on your algebra before you start.

FOOD FOR THOUGHT

6-1 Example of Movement Planning

This image shows a student-built work cell to simulate a manufacturing process by drilling a hole in a short section of PVC pipe. The work cell consists of the following components:

- Two conveyors, in feed and out feed
- A small six-axis FANUC robot
- Two variable-frequency drives to control the conveyor speed
- Operator panel
- Main control panel
- Drilling station with an air-driven drill
- Work piece fixture

The operator places the piece of pipe on the input conveyor on his or her left, and presses the two start buttons to begin the process. The robot waits for the pipe to reach the end of the conveyor, where a signal from an optical switch provides digital input to the robot telling it the part is at the end of the conveyor ready for pickup. The program then initiates the pick process, in which the robot moves to pick up the work piece and place it in the fixture in the center of the cell. A joint motion moves the robot to a position above the work piece; it is followed by a linear motion that moves the robot to the work piece. A 0.2-second pause allows the gripper to grasp the part. This is a fairly common procedure, as robots often move faster than the gripper can physically close.

A linear motion moves the robot to a position 300 mm above the conveyer, and then a series of joint motions take the work piece to a position just above the fixture. A linear motion places the part in the holder, and a 0.2-second delay allows the gripper time to release the part. The robot then makes a short linear move to clear the fixture, and then a joint move to the home position. The fail-safe condition is for the robot to stop or stay in position if the part is not detected. A proximity switch detects when the part has been placed in the holder and signals the controller to close the jaws of the work fixture. Once the system detects the part in the fixture, the robot returns to the home position.

When the robot hits the home position, it sends an output signal to the program. The fail-safe condition here is that the program will wait until it receives the at-home signal. Only then does the software send a series of signals to raise the drill, send the drill forward to drill the hole, return the drill back to its start position, and lower the drill to finish the process. The drill is tracked as it moves to each position by a proximity switch that detects the position of the drill. The fail-safe condition is for the program to wait until the drill has returned to the full back, full down position.

When the drill has returned to the full back, full down position, the robot moves to the work piece holder to grab the part. A signal from the controller tells the work piece holder to release the part; it is important that work piece holder does so before the robot gripper grabs the part. If the robot grabs the part and

attempts to remove it before the part is released by the holder, the system generates a collision detection alarm and faults the controller, resulting in the need for operator assistance.

The robot takes the part from the work piece holder and places it on the out-feed conveyor. There, a proximity switch stops the conveyor when the part has reached the operator so that the operator can safely remove the part from the conveyor. A light curtain at the operator's station is a backup signal to stop the conveyor if the proximity switch should fail. As an additional protection, time-outs are programmed with some of the outputs. In the event the robot does not return home or the drill does not return to the full back, full down position before the time-out, the program would initiate an error routine to prompt the operator to investigate and take corrective action.

This simple example gives you a glimpse into the kind of planning that goes into a program and the things one must consider in the process. Even though relatively little was happening in this case, there was still a need to address safety protocols, delays, route planning, and positioning considerations. Many of the tasks in industry require a multitude of functions to occur in a precise order at precise times, all while taking operator safety into consideration. In turn, new programs often require tweaking and adjustment for days, weeks, and even months in some cases to get all the bugs out.

Does the Robot Need to Avoid Any Obstacles?

Once you have decided how the robot moves between points, look for anything the robot might hit along the way (Figure 6-9). The commonly overlooked item here is the fixturing. Fixtures hold parts in place for various industrial processes by clamping or holding the part in some manner. The way we hold parts in place can become an obstacle to the robot, especially with tasks other than pick and place. On several occasions, I have seen a robot run into fixture clamps during a welding operation, but not alarm or stop. Instead, the robot shifted out of its programmed path slightly due to the flexibility of the tooling and created a weld that did nothing to bind the parts together in that spot.

You should also make sure not to leave tools and other objects in the work envelope, as they may be in the robot's path as well. Many machines that the robot works with and around have a monitor and/or operator control system that is movable. In one case in my experience, a spectacular crash occurred when the operator changed out a tool in a CNC machine,

FIGURE 6-9 Sometimes there are many potential obstacles that may become problems when programming the robot.

but forgot to push the monitor back into the safe area before restarting the robot. The robot smacked the display and shattered it, while merrily performing its program without so much as an alarm or warning. That several-hundred-dollar mistake shut the CNC machine down for two days while the company waited for a new monitor to arrive. Loose objects, such as hammers or wrenches, have the potential to become projectiles if hit by the robot, spreading the destruction to new and scary areas. Imagine being at your workstation or somewhere on the production floor when a hammer suddenly flies by your head or, even worse, hits you! A moment's inattention is all it takes to create a disaster, so make sure you always take a moment or two to verify everything is clear before you hit the go button.

What Is the Robot Doing at Each Point?

Does the robot stop at a point, or can we use continuous motion? Does it open or close a gripper at that point, and do we need to add in a delay to accommodate this action? Do we need to turn on a welder, paint sprayer, glue dispenser, or other device, and are there any special settings involved with that tooling? You have to know why you set each point to make sure the robot does what you want. Recall that cycle time—the time it takes to complete the program—is greatly impacted each time we have the robot stop at a point or add in a delay, so we try to keep those intrusions to a minimum. Adding in points that are not needed is one of the fastest ways to waste cycle time and often can be avoided by proper planning and analysis of the task performed by the program (Figure 6-10). If you just start adding points with no clear-cut goal in mind, the resulting program is likely to be slow, inefficient, and quite possibly unable to perform the desired task.

What Is the Robot Doing Between Each Pair of Points?

Often this question determines the motion type between points and other program specifics such as whether the robot is welding (or not) and how fast it can go between the points. As a rule, movement between points for staging purposes occurs at full speed, whereas movement during tasks such as welding, painting, and part pickup is slower. Staging points are positions that get the robot close to the desired point, but still a safe distance away, allowing for clearance and rapid movement. I highly recommend adding a staging point before and after any precision movements, as coming in too fast or leaving too abruptly can cause problems. Although it might seem as if these extra points will slow the program, but you can always remove staging points later if you determine they are unnecessary. I have seen more than one

Photo Courtesy of FANUC America Corp. www.fanucamerica.com

FIGURE 6-10 Here is a great example of the need to think about what the robot is doing at each point. If the programmed point for pickup is off, the metal prongs on the gripper will impact the rollers on the conveyor. If the gripper closes too early, the prongs could rip the bag open and spill out the chocolate.

crash resulting from a lack of staging points when the robot ran at full speed.

Do Any Conditions or Other Factors in the Process Need to Be Addressed?

This is the step in the planning process during which you factor in things such as sensors, waiting for machines to finish their part of the process, singularities, and anything else that might affect the program. Singularity is a condition in robotics in which there is no clear-cut way for the robot to move between two points. It results from lining up two axes, such as 4 and 5, in a straight line, with one of the axes (usually axis 5) at zero degrees, so that the robot could go two different ways to reach the programmed point. Figure 6-11 shows an example of the dreaded singularity. When this happens in a program, the robot may move erratically and faster than the designated speed. Some systems alarm out or warn the programmer when these conditions occur, but others just do the best they can—which can result in damage to the robot, the part, or both.

Luckily, we have several options to prevent or correct singularity. First, careful planning of robot motion that avoids changes in wrist orientation goes a long way toward avoiding singularity. Second,

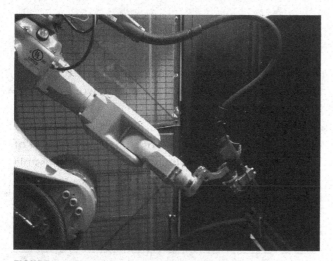

FIGURE 6-11 Because axes 4 and 5 are lined up, the robot can run into computational trouble during motions that pass through this point.

mounting the tool so that it is offset or does not come straight out of the faceplate helps to keep a bend in the arm and avoid the zero-degrees point transition. If a singularity arises within your program, moving the problem axis by approximately 10 to 15 degrees will usually correct the situation. In other words, you want to put a bend in the axis so it does not hit the zero-degrees point. If the robot absolutely must pass through the zero-degrees point, such as during a tooling angle transition, you can use joint movement to get there. Joint motion is the only type of motion that is immune to singularity and, therefore, your only option if none of the other fixes is possible. You may have to perform the move above the work area since joint motion often causes arcing or other unexpected movement.

Is the Process Logical?

This is a simple question, but it can cause major trouble if you do not take the time to ask and answer it. Remember, a program is a set of logical steps; thus, if you have an illogical plan of operation, your program is likely to fail. Make sure the robot's movements flow in a way that makes sense and are sequenced properly, rather than just however you happened to think of them first. Make sure you added in the necessary steps for any data collection or looping of sections.

This planning phase is where you make sure your program consists of a viable set of commands to complete the tasks you have in mind. For example, it makes no sense to start the movement to glue a part before you turn on the glue application process, but

it does make sense to unload the finished part from a machine before you try to load in a raw part. Take a few minutes and look for anything that is out of order or other problems in the flow of the tasks. In other words, channel your inner Vulcan, roboticist, programmer, engineer, or however you prefer to think of it to make sure the process does not have illogical steps.

If you take the time to think about and answer all of the questions posed in this section, chances are you will have a solid plan for your robot program. The first few times you go through this process it may seem odd, but as you write programs and work with robots, you will find this kind of planning becomes easier and evolves into a natural part of your programming process. You may find other questions that help you with the process, or you may find that changing the questions' order makes it easier for you. These are guidelines, not iron-clad rules, derived from my years of programming and teaching programming for robotics. Feel free to adapt them in a manner that works best for you.

SUBROUTINES

Now that you have a plan in place and have mapped out all the various tasks involved, this is a good time to talk about subroutines. **Subroutines** are a sequence of instructions grouped together to perform an action that the main program accesses for repeated use. The purpose of using subroutines is to reduce the lines of code in a program and to make it easier to write programs. Some examples of subroutines are instructions for opening or closing grippers, tooling changes, alarm response actions, and other actions that support the operation of the robot. There is not a specified limit on the lines of code in a subroutine, but often these programs are much smaller than the operating program and some may have as few as three or four lines of code. Depending on the controller, you may need to add a specific line of code to return to the main program at the end of a subroutine, although some controllers require only an end statement to exit the subroutine. No matter how it is accomplished, there must be some way for the system to return to the main program once the subroutine is complete; otherwise, the system will not function properly.

By taking a few minutes to identify repetitive actions in your program that can be turned into subroutines, you can save yourself a large amount of time writing the program. For instance, if it takes three lines of code to open the gripper and another three lines of code to close the gripper, a program that requires 10 changes in gripper state would require 30 lines of instruction to complete this action. If we turn the opening of the gripper into a subroutine and the closing of the gripper into a second subroutine, then we would need only 10 lines of instruction to complete the 10 changes in gripper state. Not only does this save a lot of time in writing the program, but it also reduces the chances of making a mistake. Once we have a subroutine working, it will continue to function properly until we change the program or something changes with regard to the basic operation of the robot.

When creating a subroutine, make sure you create it as a global function instead of a local function. **Global functions** are variables, subroutines, and other code or data that is accessible by any program you create on the robot. This definition holds for all systems that offer global data sorting and saves you from having to recreate subroutines or record key points when you create new programs. In turn, global functions translate into a huge savings in time and effort for the programmer, as they can be used with any program in the system. Some commonly used global functions in robotics are open and close subroutines for grippers, the return to home position for the robot, tool change subroutines, and other positions or actions that are useful to multiple programs.

With **local functions**, in contrast to global functions, only one program can access the data. Most of the positions you create while writing a program are local data. If you create a subroutine and make it local, then only the program in which you created it can use the subroutine.

To determine whether you are creating a local or global subroutine, you must understand how your controller sorts and stores data. You may have to create the subroutine as its own small program for global access. Some systems call global subroutines **macros**— an adaptation from the computer science world, where macro instructions invoke a macro definition to generate a sequence of instructions or other outputs. The *how* behind setting up local or global subroutines depends on the system you are working with and the way its software is set up.

Ideally, you will determine which tasks you want to turn into subroutines *before* you begin writing the program. The whole point is to save time and effort, so it makes sense to figure out this answer during the planning phase. If you forget this step or realize as you write your program that you could turn some

functions into subroutines, there is no rule that says you cannot do so mid-program. In fact, if our earlier example used the three lines of code to change the gripper state five times and a subroutine to change the state the other five times, it would work just fine. The danger is that we may have made a mistake while inputting one of the 15 lines of instruction we used for the first five changes and, of course, we lost a bit of time putting in the extra lines. The only other possible concern is usage of memory, though this is a rare problem in most modern systems.

WRITING THE PROGRAM

Now that you have settled on your game plan, mapped out the various tasks and their order, and decided what to turn into subroutines, it is time to write the program. This is *where* you have to know how your particular robot works, how the controller organizes information, and, most importantly, how to build a program for the specific robot you are using. Panasonic robots use an operating system similar to Windows, so programming on those systems is similar to creating a Word document (Figure 6-12). FANUC and ABB use their own software that has evolved with their robots, but both rely on user-friendly, level 4 languages. The Baxter robot, due to its level 5 language, has you physically move the robot to create a program. As these examples suggest, writing a robot program is model and controller specific—which is why we do not focus on any specific system for this chapter.

On a positive note, it has been my experience that once a person learns to program one type of robot, that individual tends to learn new styles of programming

FIGURE 6-12 The icon in the top-left corner that looks like a file allows you to do the same things that the file tab does in most Word documents. The icon next to it provides editing functions such as cut, paste, and copy. Several icons are related to the robot only, such as the robot icon and the icon that looks like a circle with two arrows.

quickly. The process of writing a program is fundamentally the same from robot to robot; it is the specifics or syntax that changes from system to system. If you learn a level 3 language first, it is often easier to learn how to program a level 4 or 5 system. Going from a higher-level style of programming to a lower level may prove a bit more difficult, but having programming experience of any kind under your belt will be a definite plus.

Start with your global subroutines if your system allows for that programming approach. Since we will use these subroutines as part of the main program, it is easier to have them in place before we write the program. Often times this step has already been done and the subroutines provided as part of the initial package of the robot, especially in the case of tooling subroutines. In most cases, you will simply need to check the labeling of the subroutine in your system and then write that down to help with the programming process. If this is the initial setup of the robot, you are using new tooling, or you have a new function in mind, you will have to create these subroutines. When creating subroutines, make sure to test them out before you just assume they are good to go. (We will talk about the testing process shortly.)

Now it is time to create the program. The specifics of this effort depend on the system, but the common steps at this point are naming the program, determining the program type, designating the frame, picking the proper tooling, and designating the offset tables to reference. Again, not all systems will have these specific options, and some systems will have different options added to the mix as well, but you should be able to find the basics in the programming manual or learn them via specialized training.

When you name the program, make sure you pick a name that fits within the length constraints of the system and does not duplicate a name that has already been used. Using a too-long name may make your program unrecognizable to the controller or alarm out the system. Using an existing name is likely to erase the previous program—and you may or may not get a warning before that happens.

Frames

A frame provides both the robot and the programmer with a reference point in establishing positional data. Choosing the frame is often one of the next steps completed after naming the program. A **frame** is a set of

three planes at right angles to each other, and where the three planes intersect is the origin of the frame.

Modern industrial robots come standard with a **World frame**—that is, a Cartesian system based on a point in the work envelope where the robot base attaches. Within this frame of movement, the robot moves in straight-line motions we can predict with the right-hand rule, allowing us to anticipate robot movements during programed or manual operation. The **right-hand rule** is a way to determine robot motion while using a Cartesian-based frame (e.g., the World frame), with the thumb, forefinger, and middle finger (Figure 6-13). Orient yourself in the same direction that the robot is facing and, using your right hand, stick your thumb up, point your forefinger straight ahead, and point your middle finger toward the left side of your body at a ninety-degree angle to the forefinger. Your thumb is pointing in the positive direction for the z-axis, your forefinger is pointing in the positive direction for the x-axis, and your middle finger is pointing in the positive direction for the y-axis. Bear in mind that this rule works with the World frame and that you must orient yourself in the same direction as the robot or looking from the robot toward the work envelope, not facing the robot from the other side of the work envelope. If you try to use the right-hand rule while facing the robot, you will actually be pointing in the negative x and negative y directions, which could cause a crash if you think of these as positive directions.

While we are on the topic of frames, this is a good time to talk about the *differences* between the Joint, World, and User frames or coordinate systems.

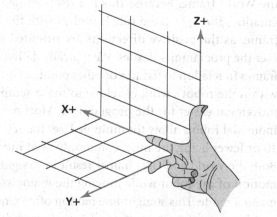

FIGURE 6-13 Your index finger should point in the same direction that the robot faces, which is the same direction as the main part of the work envelope.

The terms frames and **coordinate systems** are interchangeable when used in reference to robot movement, so do not be alarmed if you see frames in one place and coordinate system in another throughout this book or in the reference material for a robot. The World frame, as noted earlier, is a Cartesian-based system that uses the origin or zero point for the robot as the zero reference of the frame (Figure 6-14). The zero point for industrial robots is usually a point in the middle of the base that we bolt to the floor or other surface.

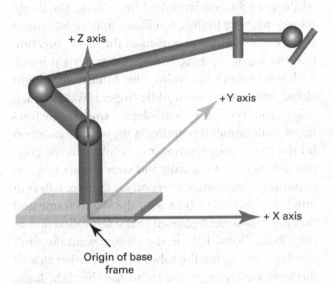

FIGURE 6-14 The vectors of travel based on the zero point of the robot.

The **User frame** is a specially defined Cartesian-based system in which the user defines the zero point and determines the positive directions of the axes, including the option of straight-line motions moving at an angle instead of along the flat lines of the World frame. Because this is a user-defined orientation, the right-hand rule is useless with the User frame, as the positive directions are oriented however the programmer desires. We typically define User frames in relation to fixtures or other points of interest within the robot's work envelope to make setup and movement easier for the programmer. Most modern industrial robots allow for multiple User frames, with 10 or fewer being the most commonly used number. Both the World and User frames result in straight-line motion of the robot while moving the major axes in manual mode. This straight-line motion often requires multiple axes of the robot to work in unison under the direction of the robot controller.

One type of user-defined frame is the **Tool frame**, which establishes the location and orientation of the

end-of-arm tooling (EOAT) for the robot software and the direction of motion for the x-, y-, and z-axes; it also affects payload and inertia calculations. A critical element in setting a Tool frame is the location of the **tool center point (TCP)**. The TCP is the reference point where the robot is moving through space. When you set a point, you are basically storing information that records the location of the TCP. Most robots use the center of the faceplate at the end of the wrist as a default TCP (Figure 6-15). When we set the TCP, it is normally expressed as an x,y,z distance and w,p,r angle from the (0,0,0) point in the center of the faceplate. For a material-handling robot, the TCP is usually set at some point in the center of the gripper's travel path (Figure 6-16). In an arc-welding process, the TCP is usually established at a distance of ½ to ¾ inch from the end of the torch, as this distance allows enough wire sticking out to prevent the torch from welding itself to the parts being joined. With a spot-welding gun, the location of the TCP depends on the type of gun used. If both tips of the gun move, the TCP is established where the two tips join. If the gun has one moving tip and one fixed tip, the TCP is established at the location of the fixed tip.

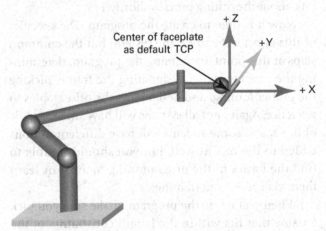

FIGURE 6-15 The default TCP for most industrial robots.

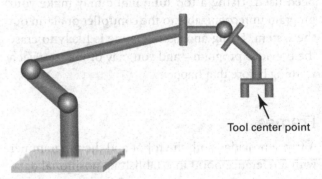

FIGURE 6-16 With clamping-type tooling, the TCP is often adjusted to the middle of the part of the tooling that closes.

Unlike the World and User frames, the **Joint frame** moves only one axis at a time while in manual mode, with the positive and negative directions determined by the setup of each axis's zero point (Figure 6-17). This is the perfect mode to test each axis of the robot independently and, in the case of problems, determine which specific axis is making noise or moving oddly. The downside of using the Joint frame is the fact that the only linear movements of the robot involve those axes that are linear in nature. Since most robots primarily have rotary axes, the robot will move in a circular fashion during manual motion. On top of the nonlinear motion, the multiple-axes movements that the controller performs for the other frames become your responsibility in the Joint frame. The result is a near inability to perform straight-line manual movements in the Joint frame and the necessity of switching among multiple axes to get the robot to a desired point in the work envelope.

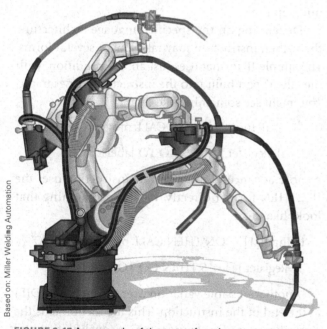

FIGURE 6-17 An example of the type of motion you can expect while moving in the Joint frame mode.

Based on: Miller Welding Automation

Picking the right reference frame and tooling for the program is crucial for operational success. In many cases, the default frame is the World frame, which uses the robot base as the origin point. If you change the reference frame after creating the program and saving points, there is a high probability that your taught points will change—which could be a recipe for disaster. In systems that use only one type of tooling, you just need to verify the tooling is present in the program

parameters; this is often a default setting. When working with a system that has multiple tooling, make sure to select the proper tooling or tooling group as well. Changing the selected tooling after you have created a problem usually does not cause problems with positioning, providing you taught the points with the proper tooling in place or the TCP is set up correctly for the tooling used.

Tech Note

We use tooling groups when a robot has more than one set of tool position data. This is how the robot keeps track of where the positional point or tool center point is for each unique tooling. You can set a tooling group as a default on most systems, ensuring new programs use this data automatically.

Final Touches on Program Creation

Once you have set the main parameters and double-checked any other variables in the program creation process, such as offset tables, you are ready to finalize the creation of the program. Finalization saves the parameters set during program creation and allows you to start teaching the specified points and set the logic of the program. Before you finalize the creation of the program, it is a good idea to add in descriptions, if possible. Descriptions change nothing about how the program operates, but are a great way to log the frame and tool group used to set the points of the program. I have my students log with the User frame and the Tool frame (whichever they are using) when they set the points for a program, as these points are required to run the program again down the road. My students and I have discovered the hard way that a program for a FANUC robot will not run if you have a User frame or Tool frame selected for the system that is different than the one used to create the program.

Once you have finalized the program, usually by scrolling to the bottom of the creation options and clicking Enter or similar, you are ready to set points and create the logic of the program. This stage is where you take all the information you gathered and outlined during the planning process and create your program. When you set the points of your program, make sure you select the proper motion type to reach that point. Pay attention to robot speeds and tooling conditions as well. Make sure you add in any subroutines or other logic functions at the appropriate points.

In short, finalization of the program is the main part of the programming task. In standard line-based programming, each line will have a unique number that dictates the flow of the program and allows you to reference specific parts of the program with various logic commands.

Program Instructions

A program consists of two kinds of instructions: motion control instructions and program control instructions. As described earlier in this chapter, the motion control instructions are developed as part of the motion planning process and are incorporated into the program to establish the path or trajectory that the robot will follow. Program control instructions allow the robot to communicate with the work cell, monitor the production process, and control program execution.

Several different types of program control instructions may be used that perform specific functions in the programming process. We can combine these commands in nearly limitless ways to create a program that far surpasses a simple "move here and do that" kind of logic sorting. The specific options available for program control vary from manufacturer to manufacturer and from process to process. Nevertheless, some instructions are used by most robots. The various instructions available usually fall into the following categories:

- Branching
- Looping
- Register
- Input/output
- Arithmetic
- Call
- Data structures

Branching

Branching instructions cause the program to jump to another point in the program and are classified into two broad types: conditional branching and unconditional branching. Unconditional branching instructions simply cause the program to go to another instruction to continue execution. The most common types of unconditional branch instructions are the jump (JUMP) and go to (GOTO) instructions. These are usually two-part instructions that send the program to a label or marker that the instruction has to find to complete the instruction. The software must have both parts of the instruction to properly execute it, and the program interpreter or compiler will check for both parts. You will usually get a warning message or error if one of the parts is missing.

Conditional branching instructions test for specific program conditions before executing a branch. The two most common types of conditional branching instructions are if/then (IF...THEN) and select (SELECT). The if/then instruction is basically a true-or-false question in which a specific program condition is examined, and the branch is executed based on the condition. Typical program conditions that could be evaluated include program inputs, register values, system variables, and inputs from external sources. These instructions may be combined with other instructions to further expand the range of programming options.

Depending on the specific language architecture, the if/then instruction may take any of several forms. The simple IF format uses just an "if" condition, with the "then" part built into the instruction. For example, you might see something like this:

IF Input[1] = ON CALL program XXXXXX

IF Register [1] = 5 GO TO label1

A more advanced format of this instruction uses the IF...THEN format directly, creating something that looks like this:

IF Input[1] = ON THEN CALL program XXXXXX

IF Register [1] = 5 THEN GO TO label1

Another possible variation uses the endif (ENDIF) at the end of the instruction. This instruction tells the controller to stop looking for the "if" qualification and move on to the next line. If this format is used, the endif must be included as part of the instruction and may look something like this:

IF Input[1] = ON THEN CALL program XXXXXX ENDIF

IF Register [1] = 5 THEN GO TO label1 ENDIF

Yet another variation of the if/then instruction has an optional else (ELSE) clause. In this case, if the variable examined for IF is true, the first event happens; if the IF variable is false, the second instruction is executed; and if neither of those events can happen for some reason, the endif tells the controller to move on. Possible variations of this instruction may take the following forms:

IF Input[1] = ON THEN CALL program XXXXXX
    ELSE CALL Program YYYYY ENDIF

IF Register [1] = 5 THEN GO TO label1 ELSE GO TO label2 ENDIF

In many software applications, it is possible to logically connect IF instructions using either an AND or an OR function. When using the logical AND function, all IF conditions must be satisfied for the condition string to be evaluated as true. When using a logical OR function, if any one of the IF conditions is true, then the entire condition string is true. This may look something like this:

IF Input[1] = ON AND Input[2] = ON CALL program XXXXXXX

IF Input[1] = ON OR Input[2] = ON CALL program XXXXXXX

In the case of the first instruction, both Input[1] and Input[2] must be ON for the CALL instruction to be executed. (We will look at the CALL instruction soon.) In the case of the second instruction, either input being ON will cause the CALL instruction to be executed.

The select (SELECT) instruction is a multiple-choice question that may appear as a TEST, SELECT CASE, or TEST CASE instruction. Instead of checking for a single condition, this instruction checks for multiple conditions with an optional ELSE (none of the above) clause. If any of the conditions in the select clauses is satisfied, the program will execute the appropriate set of instructions. The optional ELSE clause is a set of instructions that the program executes if none of the select conditions is satisfied. A typical select instruction may look like this:

SELECT REG[1] = 1 CALL PROGRAM AAAAA

    REG[1] = 2 CALL PROGRAM BBBBB

    REG[1] = 3 CALL PROGRAM CCCCC

    ELSE CALL PROGRAM HOME

Looping

Looping instructions cause the program to repeat a series of instructions either for a specified number of times or until a specific program condition is met. There are three common types of looping instructions:

for/do (FOR...DO), repeat (REPEAT), and while (WHILE). The for/do loop causes a program to repeat a series of instructions for a specific number of cycles or times. The number of iterations is determined by a program function, and the loop can either increment (count up) or decrement (count down), depending on the program function. The software should automatically increment or decrement the count after each iteration, but if for some reason it does not you may need to add in a math command to help things along. The for/do may look something like this:

FOR A = 1 TO 5 DO

    MOVE TO P1

    MOVE TO P2

    MOVE TO P3

    MOVE TO P4

    ENDFOR

The repeat instruction is a do-then-test or bottom-testing loop. It performs a series of instructions, and then tests a program condition to determine whether it should repeat the instructions. If the condition is satisfied, it will repeat the sequence, and then test the condition again. As long as the condition is satisfied, it will repeat the instruction set. Once the condition is false, it exits the loop and moves to the next line in the program.

The while instruction is a test-then-do or top-testing loop. It tests a program condition to determine whether it should execute a series of instructions, and executes the instructions if the condition is satisfied. After execution, it tests the condition and repeats the series of instructions as long as the condition is satisfied. If the condition is false or not met, the program will jump to the line below the loop and continue on.

Register

Register instructions allow the programmer to use arithmetic registers to store and manipulate data. They may be used as part of other instructions, such as conditional branching instructions or looping instructions, and are sometimes included within arithmetic instructions. Basically, registers are a place to store numbers that the program can manipulate, usually for logical sorting or data reporting reasons. Some common applications include part counts, loop counts, numeric input from other sources, and calculation results.

Input/Output

Input/output instructions are used to monitor conditions and cause actions in the robotic system as well as whatever else it is connected to. **Input** instructions check the value of program input data and are often tied to sensors, registers, and other places where data is stored in the controller based on either real-world situations or programmed outcomes. Inputs can be either **digital**—which means they have only two states: 1 or 0, one or off, yes or no—or **analog**—which is a range of voltage or current values scaled to mean something in the system. Inputs can come from proximity or limit switches, process sensors, signals from a programmable logic controller (PLC), feedback data from process-related equipment, operator inputs, or any other recordable source. Input instructions can be combined with other instructions to create loops or monitor process conditions.

Output instructions are used to cause an action or write data somewhere. Whereas inputs look for something, outputs do something. We use output instructions to initiate robot motion; start, stop, and control production processes; write data in registers; send information to other systems or machines; illuminate lights; start or stop motors; or do anything else in the realm of taking action. As with inputs, outputs can be digital or analog. They are used in many of the advanced programming commands.

Analog input/output signals use either a voltage range (often 0 to 10 volts DC or -10 to $+10$ volts DC) or a current range (usually 0 to 20 milliamps or 4 to 20 milliamps) to represent a process parameter, such as air pressure, fluid flow rate, fluid volume, or any other condition that is not measurable by a signal being either ON or OFF. When analog signals are used, the controller input/output section converts the analog signal to a digital signal with a specified number of data bits and a sign ($+$ or $-$) bit. The controller software processes the information digitally, and then converts it back to an analog signal before sending it as an output to the work cell or robot. Some software applications allow analog inputs to be read from or loaded into arithmetic or data registers, or other system parameters.

Arithmetic

Arithmetic instructions are used to perform arithmetic operations. Although the set of arithmetic instructions depends on the type of software, all basic functions are available in most robots—addition, subtraction, multiplication, and division. Some software applications may include the additional function of division with whole number and division with remainder. Some systems may offer trigonometric functions, roots, and powers as either an option when programming with the teach pendant and/or when using offline software.

Call

A **call (CALL) instruction** is used to call other programs from a main program and can be considered a type of unconditional branching instruction. Such an instruction calls another program and turns control of the robot over to that program, which then executes its instructions. When the called program ends, the controller returns to the original program and continues with line of instruction under the call command. This is how we utilize the subroutines created previously as well as programs that are only used on special occasions such as tool changes or welding gun tip cleaning.

Data Structures

A **data structure** is a collection of related items used to control a process, often arranged in a table format. The individual programmer may or may not have to create the structure or enter the data into the structure to create a process. An example of a data structure is a weld schedule for an arc-welding robot, which may contain the following information:

- Weld voltage
- Weld current
- Wire feed speed
- Weld travel speed

In this case, the data structure contains several data items that would be set up using analog output information or arithmetic register data. Placing all of this information in a single data structure simplifies programming, as the programmer calls the weld schedule as part of the program rather than having to enter each item individually.

When we use structures to control robot motion and/or other process parameters, it is easy to make adjustments to the system. If we had to go into each program and alter each instance of these variables every time we wanted to change something, we would lose a large amount of production time on top of likely missing instances of the variables in a few spots, with all the havoc that could cause. While it may take some time to change structures or create new structures as the demands of the task change, it is time well spent in the long run. I also recommend creating a new structure when most, if not all, of the variables need changed—you never know when you might want to use the old structure again.

You may encounter other logical filters within the robotic systems you program, but at least now you have an understanding of some commonly used commands. Often you can consult help files on the teach pendant or reference manuals to help you figure out the specifics of anything not covered in this section. A quick Google search may also yield a wealth of information about programming instructions. As your programming skills improve and you begin to understand the robot and the task better, try adding in some higher-level commands where they would benefit the process the most. The best way to learn how a command works is to use it in a program and then verify it performed as you expected.

Saving Your Program

Another crucial part of the program creation process is saving your data. I always tell me students, "Save early, save often, save your data!" One of the most frustrating events when programming is spending a large chunk of time writing a program, only to lose all of your work due to momentary power loss, a missed action on your part, or a glitch in the system. I cannot count the number of times my students have spent the entire lab time working on a program, only to lose their data at the end of class due to a slight oversight. In particular, shutting down the robot before saving the program is a great way to lose your data and bring a tear of frustration to your eye. Until the program is saved into the memory of the teach pendant or controller, what you have is a theoretical program. Sure, it has code, logic, and positions, and you may be able to step through it for verification. (which is a process we will cover soon). Until you actually save it in memory, however, it is vulnerable to human and system error.

For long programs, I recommend saving the program every 10 minutes or so when you are creating the program. At the very least, you need to save the program once you have finished entering it and before you begin testing it. Many of the modern industrial systems save the program automatically before you enter auto mode, but you should not get lazy and rely on this method for saving all your programs.

The specifics of creating a program, saving points, setting the motion types, creating the logic filters, calling subroutines, and all the other bits and pieces that make up a program depend on the system you are working with (which is why this chapter does not focus on a specific programming style). Most likely you will have the chance to do some programming in your class, and your instructor will provide the specifics needed to write a program for the robot(s) used in the classroom. If this is the case, learn everything that you can during this activity: This knowledge will greatly help you out in the field when you encounter new systems and new programming styles.

TESTING AND VERIFYING

Once you complete your program, there is still one crucial step before you place the system in auto and give it a go: You must test your program. Even programmers with years of experience need to verify the operation of their program before declaring a program ready to go or viable. Helmuth von Moltke, a German military strategist, said it best: "No battle plan survives contact with the enemy" (Levy 2010). While programming is not war, it can certainly seem that way when things go wrong and the boss is breathing down your neck for results. Moreover, there is a good chance you will need to make changes to your original plan once you start writing the program or when you test it out for the first time; this is a natural part of programming process. Do not hang on to your planned program so tightly that you cause yourself grief during the testing process, as there is a high probability you will need to make some modifications to tweak the program and get the desired results.

The system you are working with will determine which testing options you have at your disposal, but most industrial systems offer a manual, step-by-step testing method as well as a continuous testing mode (Figure 6-18). For most systems, manual testing requires you to hold the dead man's switch on the teach pendant as well as press and possibly hold some of the keys on the pad. If you release the dead man's switch or the key you have to hold down, the system will stop immediately, giving you the control you need to keep the robot from crashing into something or damaging parts. In addition, the system typically runs at about half of the normal speed during manual testing. Robots can move faster than human scan react, so this slower movement speed allows the operator to

Image courtesy of Miller Welding Automation

FIGURE 6-18 Welding systems often give you the option of welding in manual mode during the testing process. If you are working with a system that performs an operation during testing, remember to use all the proper safety gear.

keep up with everything that is going on and gives the operator the chance to respond before disaster strikes or get out of the way if the operator is in the danger zone (Figure 6-19).

Image courtesy of ABB Inc.

FIGURE 6-19 Sometimes you may be close to the action during program verification. This is why the robot moves more slowly when in manual mode and why you must always take care when proving out your programs.

In **step mode**, the robot advances through the program one line at a time and requires you to press a button on the teach pendant keypad before it reads the next line of the program and responds. Most systems will let you step forward and backward in the program so you can bounce between two lines of code if you

need to watch the motion over again. Of course, just because the robot is executing only one line of code at a time, that does not mean it will not travel a large distance. If you set two points 10 feet apart, the robot will travel that 10 feet as you execute the line of code that triggers the movement to point B. You can stop the motion by releasing the dead man's switch or, in the case of systems where you have to hold down a button on the key pad, by releasing the button. When testing, it is my recommendation to start with step mode as this mode allows you to determine exactly what the program does line by line and makes it easier to correct the proper portion of the program as needed.

Continuous mode works as the name implies. Once you start the program in manual continuous mode, it continues to run until you stop the program by releasing a button or dead man's switch, the program reaches its end, or something causes the system to alarm out. Many industrial systems let you perform this method of testing backward as well, so make sure you know which way the program will flow. I recommend running the program in continuous mode once you have verified and are happy with the operation of the program in step mode. This will give you a good idea of how the program looks as a whole and lets you see the big picture.

Some motions in your program cannot happen in the space allotted if the robot moves slowly. If you run into this situation, most systems will give you a warning alarm. Once you accept or do whatever is required to authorize the movement, the next movement the robot makes will be at or near full speed. This is often the case when you make a radical change in the tooling orientation between points or other moves where the axes involved have to turn a great distance (Figure 6-20). If you just select yes and do not keep in mind what is about to happen, you could get a nasty surprise—especially if you are close to the robot when this move happens. Whenever you get an alarm for the robot, take the time to read it and determine what the alarm is telling you *before* you go full tilt ahead. I have had students crash the robot, damage parts, have near misses, and mess up their programs by getting in a hurry and ignoring what the alarm said. This is a common error that many experienced programmers make: They get used to seeing the warning messages and fall into the habit of selecting yes until the system runs again.

The step and continuous manual modes are also a great way to advance through an existing program for rework and post-crash operation. When these modes are used this way, you should take special care to ensure

Image Courtesy of Miller Welding Automation

Image Courtesy of Miller Welding Automation

FIGURE 6-20 In these two pictures, you can see the types of movements that often require a rapid move in manual mode. Switching the orientation of the tooling in such a radical manner over a small distance usually requires a rapid movement with a break in the welding, which can be detrimental to weld quality.

the tooling is functioning as needed to finish the parts. For instance, welding guns have to strike an arc before they begin the weld, and you may have to do something extra to make this happen mid-weld. In another case, you may have to open grippers before stepping through to the part pickup portion of a program. The key point is to make sure you understand what is supposed to happen when using the manual modes for crash recovery or manual operation of the system.

Once you have tested the program in both step and continuous modes and are happy with the robot's operation, make sure to save the program once more. This ensures the program in memory is the one you want, rather than the pretesting program.

Sometimes testing reveals that the program is so laden with problems that the programmer is better off starting over rather than trying to fix the existing program. This may sound crazy, but sometimes it really is easier to start over instead of devoting time to fixing a hot mess. Some of the warning signs to look for are

multiple alarm conditions, problems that get worse instead of better after several attempts to correct them, code that causes the program to stop for unknown reasons, and a program that has been corrected so many times that you no longer understand the logical flow. Some programmers also find it easier to create a completely new program than to make changes to an existing program written by someone else. Just as entering into battle may require generals to scrub their strategy and start over, so testing may send the programmer back to the planning stage to create a new program.

The whole point of testing your program is to get the bugs out of the system before you put the program in automatic mode and let it run. The more effort you put into this step, the greater the chances of successful robot operation in automatic mode. Rushing through testing may cost you a lot more time in repairing damage than you would have lost by doing it right, so keep that in mind. When it comes to programming, quality is much better than quantity!

NORMAL OPERATION

Once you have completed writing and testing the robot program, the next step is to press the green start button and let the robot get to work. Even though you are now ready to begin normal operation, this does not mean you can just walk away with no concerns. As a safety feature, most systems run at a slower speed in manual mode than the robot is capable of running in automatic mode. Unfortunately, this feature may mask certain problems that become evident only when the robot is running at full speed. On top of this, the changes that naturally occur in the world around the robot as well as changes in the robot's internal systems can cause problems with normal operation. It is the job of programmers and operators to monitor the robot's behavior, make the necessary adjustments as needed, and, in some cases, prevent disaster before it happens.

When running a new or edited program for the first time in automatic mode, make sure you watch the robot go through the program a few times before you consider it ready to run with limited supervision. Many programmers have skipped this step, only to discover problems with the program after the robot has ruined parts or damaged equipment. By verifying that the program is working correctly during a couple of cycles, you can avoid the expense and frustration of down time as well as the likelihood the boss will ask you some uncomfortable questions: "Didn't you test this thing?" or "What do you mean it crashed?" You should also keep your hand near the E-stop or some other stop button during that first run in automatic mode, in case something does go wrong. You cannot always stop the robot in time, given that robots move very fast and sometimes leave too little time for mere mortals to react, but this precaution will at least give you a fighting chance. After you have verified the operation by watching the robot run through the program a few times or cycles, then you can truly declare it ready to run (Figure 6-21).

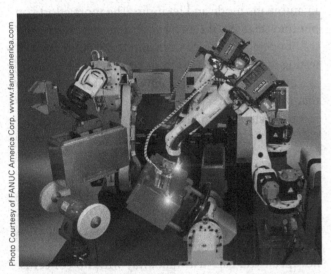

Photo Courtesy of FANUC America Corp. www.fanucamerica.com

FIGURE 6-21 A properly designed and tested program allows multiple robot systems, like the setup pictured here, to work together in fluid elegance.

Once you are running a verified program, that does not mean the robot no longer needs human help. Variances in parts can cause problems with pickup and placement. As tooling and fixtures wear, the positioning can change, parts can move, and the quality of work may decrease. As the robot wears out due to normal use, the positioning of the various axes may change or slop in positioning may develop. These and many other changes may make enough of a difference in the working conditions of the robot that it needs human intervention, and it is common in industrial applications for operators or programmers to tweak robotic programs from time to time to deal with these changes. Often this procedure takes just a few minutes and can be as simple as adjusting one or two points in the program and saving the changes. In these cases, there is no need to reinvent the wheel and write a new program. In addition, the testing is typically minimal, such as stepping through a few lines of code to verify that the new positions work.

Whether the robot has an operator watching it or a complex sensor array that looks for problems, some system for monitoring the robot is needed. Remember, the robot can do only what we program it to do. It has no intuition. It has no feelings. It has no problem-solving skills other than those we give it through programming. As we figure out new and inventive ways for robots to solve tough problems, it may become easier for systems to respond to unexpected circumstances, but these responses are still program driven. Eventually, the robot needs a human to solve the tough problems, find better ways to perform tasks, and make the determinations that are simply beyond the scope of its programming. This is part of the partnership between people and robots, where each party has its own strengths and weaknesses. We capitalize on the strengths of both parties to compensate for the weaknesses of each and to perform tasks that would be difficult, if not impossible, for either party to do alone.

FILE MAINTENANCE

Robot controllers are specialized computer systems and, as with computers, you need to perform data or file maintenance on them from time to time. In many ways, this work is similar to managing the various documents you might have in your word processor program or the music you might have on your phone. You want to save the data you like, make sure you can find and use it when you need it, and delete the data you no longer want. Let us take a closer look at what each of these functions entails in the world of the robot.

Earlier I mentioned my mantra for saving: "Save early, save often, save your data." This goes beyond just saving the program into the teach pendant or controller of the robot. Most robot controllers do not write programs to a storage device unless specifically instructed to do so—and that means once the power is gone, so are all your programs! While most controllers have battery backup, batteries last only so long and people are notorious for forgetting to change them when prompted or per the manufacturer's recommendations. If you do not have your programs written to some form of backup such as a flash drive, SD card, hard drive, or other storage medium, all of your programs are in peril of deletion! To help you avoid this unhappy outcome, most manufacturers provide some way to backup data and programs from the robot, and often you can set a timed interval for when the backups should occur. For instance, Panasonic welding robots use SD cards as their backup media; these cards plug into a slot inside the teach pendant for easy access. A setting in the controller allows you to select how often the system backs up all the programs as well as how many backups are put on the SD card before overwriting the oldest backup. This backup scheme allows you to return to previously saved programs in case you edit and save a program that is not to your liking. The backup process varies from robot to robot, with some giving automatic options and others requiring manual saves. Make sure you understand how your robot works and regularly save all your programs and data.

Another data trick you might have available for use is sorting. Imagine for a moment that all your documents or music thrown into one folder, with only the title of the document or song being used to sort things. This would make it difficult to find specific items and places a large amount of data in one spot. To get around this sort of chaos in the computer world, we create folders within folders, which we then use to sort data into groups that make sense to us. Depending on the system, you may have this option with your robot as well. It is a very handy feature, especially if you are looking for something specific in a backup file. If you have this option with your robot, make sure you create folders that are meaningful and sort the data appropriately. Failure to do so could lead to problems when both you and the robot need to find certain data. Many industrial systems create specific folders automatically during the backup process, in which case you simply need to learn how the robot organizes the data for efficient navigation.

Do you have everything you have ever written saved on the hard drive of your computer? The likely answer is no. Just as we discard documents that are no longer useful or relevant, so too must we delete programs that are no longer beneficial. With modern hard drives, computers have massive amounts of storage and deleting a word processor document has very little impact on the storage capacity of the system. Unfortunately, the world of robotics tends to be different. Many robots, especially older ones, have only a small amount of memory. Thus, if several complex programs are stored in the robot, you may not have enough storage space for another large program. To get around

these kinds of limits, we save programs to external storage devices, as mentioned earlier, or we delete programs that are no longer of use.

Before you delete a program, I would offer one caution: Make sure to save an external copy, if possible. Obviously, if you are deleting a program to make space on the robot's memory, you cannot save the program on the robot. If you can save the program externally, make sure you label the storage device properly or in some way document where to find the program so you can retrieve it if needed. If saving the program externally is not an option, you will have to live with the fact that deleting a program now may require you to rewrite it later.

Make sure you keep track of what the automated backup systems are doing. You may have to manually go in from time to time and delete old backups so there will be room for new/current backups. In the fast-paced modern world, it is easy to ignore the automatic backups, as they are "out of sight and out of mind." You may need to create a checklist or some form of automatic reminder to help you monitor the storage space for backups if the robot does not generate an alarm message or other warning that backup system has used all the available space.

You may run into tasks specific to your robot when it comes to file maintenance, so make sure you understand how your robot deals with data and your responsibilities in this respect. If you hit a data snag, take the time to come up with a corrective action that accounts for all the variables and needs. Experience is a great teacher, and there is no greater teacher than hours of tedious work that you could have avoided with a few minutes of proper file management. When dealing with files and data, do not forget about issues such as home positions, user frames, offsets, data structures, and other information that affects the operation of the robot. Just because they are not a program, it does not mean they are unimportant. In fact, loss of the zero point for one or more axes can cause you a great amount of downtime and frustration as you correct the situation, whereas saving this data beforehand might make correcting the problem as simple as uploading a file. File management is another place where the *how* and *what* varies from robot to robot and sometimes model to model, so make sure you take the time to learn the details for your robot.

REVIEW

While we did not cover the specifics of programming any one system, the information from this chapter will help you in programming the robots encountered during your exploration of the field. Once you learn the basics of programming for one system, it becomes much easier to learn how to program different robots. Whether you are working with a FANUC, Panasonic, ABB, Yaskawa Motoman, or other robotic system, the process is the same as that described in this chapter, with only the syntax or how you create the program changing. The language level does have an impact on how you write the program: The lower the level, the more work required on your part. During our exploration of programming and file management, we covered the following topics:

- **Programming language evolution.** We looked at the five levels of programming languages and considered how each of them works. We also explored the evolution of digital programming.

- **Planning.** Planning is intended to make sure we know what we are doing before we write the program.

- **Subroutines.** Subroutines are small programs and sections of code that we use repeatedly during the course of operation.

- **Writing the program.** We looked at the process of writing a program, including the types of motions you can expect to use and some of the common logic instructions.

- **Testing and verifying.** It is essential to double-check your program for proper operation during and after the process of writing the program, but before running the system in automatic mode.

- **Normal operation.** This section covered normal operation, including the human responsibilities during this phase.

- **File maintenance.** Just as we manage the data on our computers, so too must we manage the data and files in the robot.

KEY TERMS

Analog	Digital	Looping	Step mode
Arcs	Fixtures	Macros	Subroutines
Arithmetic instructions	Frame	Output	Tool center point (TCP)
Branching	Global function	Programming	Tool frame
Call instruction	Input	languages	User frame
Circular motion	Joint frame	Register	Weave motion
Continuous mode	Joint motion	Right-hand rule	World frame
Coordinate system	Kinematics	Robot program	
Cycle time	Linear motion	Singularity	
Data structure	Local function	Staging points	

REVIEW QUESTIONS

1. What are the five levels of robot programming?

2. Describe programming a robot with a level 3 language.

3. How do you write a program in a level 4 robotic language, and what are some of the benefits?

4. What is the difference between a level 4 and a level 5 programming system?

5. What are the seven questions you should answer during the task mapping phase of planning?

6. What are the four main motion types, and how does the robot move during each?

7. Which types of motion can joint movement generate between points? What is a potential danger from this motion?

8. How do you create an arcs and circles using circular motion commands?

9. What is the difference between a fine termination of a motion command and a continuous termination of a motion command?

10. What is robotic kinematics?

11. What are staging points, and how are they typically used?

12. What is singularity, and how can we avoid this condition?

13. What is the purpose behind using subroutines? Give some examples.

14. What is the difference between a global function and a local function?

15. What is a frame, and where is the origin point for the World frame of a robot?

16. How does the right-hand rule work?

17. What is the TCP? What is the default position for the TCP? Which of the frames use the TCP as the origin point?

18. Which problems may potentially arise when you change the frame of a program after saving points?

19. What are the seven common categories of program control instructions, and what is the basic function of each?

20. What is the difference between step mode testing and continuous mode testing?

21. What typically happens when you are testing in manual mode and the robot has to make a large axial movement? What is the danger associated with this?

22. What should you do when running a program in automatic mode for the first time?

23. What are some common tasks that fall under file maintenance?

Reference

1. Levy, Steven B. *Lexician*. November 1, 2010. http://lexician.com/lexblog/2010/11/no-battle -plan-survives-contact-with-the-enemy/. Accessed August 11, 2013.

CHAPTER 7

Automation Sensors

WHAT YOU WILL LEARN

- How limit switches work and how to select the right one
- How inductive proximity switches work
- How capacitive proximity switches work
- How photoelectric proximity switches work
- How tactile sensors work and the information they give robotic systems
- Different ways to detect unexpected impacts of robotic systems
- Some of the different ways we monitor temperature

- How to monitor fluid flow
- The basics of fluid level monitoring
- How a pressure gauge works
- Which types of encoders are available and how robots use them for positioning
- How robots use microphones to track sound
- How robots use ultrasonic sensors to see
- Things to keep in mind when connecting sensors to the robot
- How to pick the right sensor for the job

OVERVIEW

Without sensors and the information they provide, modern manufacturing equipment would have zero information about the world around it and what is occurring. Safety protocols, production sequences, operational responses, coordinated actions between equipment, and many other actions performed by industrial equipment would be extremely difficult, if not impossible, to carry out without the information provided by various sensors. In this chapter, we drill down into the field of sensors, looking at the types of sensors used by robots in industry and some of the basics of their operation. As you read through this chapter, keep in mind that sensors are used by all types of machines and processes in industry—not just robots. Of course, given the diversity of this field, we cannot cover all the different types of sensors utilized in industry. Instead, we will limit our attention to the following issues:

- Limit switches
- Proximity switches
- Tactile and impact sensors
- Temperature sensors
- Fluid sensors
- Position sensors
- Sound sensors
- Connection to the robot
- Sensor selection criteria

LIMIT SWITCHES

A **limit switch** is a straightforward basic device that has been helping robots gather information about the world around them for many years. We use limit switches to make sure safety doors are closed, warn the operator that the robot is moving too far in a particular direction, and even track part presence, to name but a few applications. Not all robotic systems use limit switches, especially internally, but they are very useful for providing information about the state or position of objects in the production environment.

Limit switches are activated by contact with an object, and change the state of their contacts when the object exerts a certain amount of force. These devices consist of a body, which houses and protects the electrical contacts, and some type of actuator, which moves with physical contact. Actuators can use pressure or rotational motion to change the contacts of the switch.

In many of the larger units, the actuators are attached as a separate unit on the limit switch body, known as a head unit, allowing for options in actuator orientation (Figure 7-1). When changing out the head unit on a limit switch, make sure the new switch unit matches the limit switch in question as well as the task. Another great feature of this type of limit switch is the ability to rotate the head unit of the switch by removing the appropriate screws and then turning the actuator as need, thereby adapting the switch to the various situations in the field without needing a separate switch to handle each mounting situation. The range of motions and force needed to activate the switch can range from slight movements with 0.2 newton of force to large movements requiring around 20 newtons of force. (Newtons are a common measurement of force and the metric equivalent of the English pounds.)

FIGURE 7-1 A Cutler-Hammer limit switch with the arm mounted for contact on the right side of the unit.

When selecting a limit switch, you will need to know how much amperage the contacts must handle, how fast the switch needs to respond, and how much force is generated by the contacting object. Standard limit switches and their smaller cousins, micro limit switches, can usually handle up to 10 or 20 A of current, while the smallest members of the family, subminiature micro switches, are good for 1 to 7 A on average. Each switch takes a certain amount of time to make (activate) or break (deactivate), so make sure the switch can respond fast enough for the conditions you will use it in. For instance, if the switch is intended to help the robot keep track of parts on a conveyor line, make sure there is enough space and time between parts for the limit switch to reset. Some limit switches require a fair amount of force to activate, making them a poor choice for lightweight parts. If you are using a micro limit switch to prevent the robot from going past a certain position and causing damage, the best option is likely to be a small-movement, low-force, limit switch that can react quickly and shut down the robot before any mechanical damage occurs.

When used properly, limit switches are a valuable information-gathering tool for robots and other applications. In early robotics, they were one of the primary sensing systems, and they remain a staple part of the modern robot's array of sensors. Today we use limit switches to prevent the robot from traveling too far, monitor the doors on safety cages, confirm raw parts are ready for loading, verify that the machine the robot works with is in position, and answer other yes/no-type questions. While there may not be as many limit switches today as there once was in industry, they are still a vital part of many guarding and sensing systems that work either directly or indirectly with the robot (Figure 7-2).

PROXIMITY SWITCHES

Proximity switches are solid-state devices that use light, magnetic fields, or electrostatic fields to detect various items without the need for physical contact (Figure 7-3). **Solid-state devices** are units made up of a solid piece of material that manipulates the flow of electrons. Because they do not have any moving parts, solid-state devices are capable of performing millions of operation without wearing out. The downside is that most solid-state devices never fully stop the flow of electrons, although often only a very small amount leaks through. They are also vulnerable to electromagnetic pulses (EMP) and other

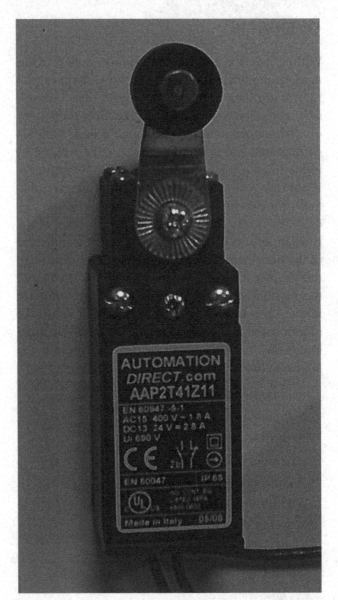

FIGURE 7-2 An Automation Direct limit switch with a much smaller arm and a wheel on the arm to help reduce the connection impact. This type of switch would work well with a door or other movable guarding.

strong magnetic fields. Because the proximity switch (also called simply a "prox") does not touch the object it is sensing and has no moving parts, these switches can work for years with minimal preventive maintenance.

Proximity switches send low-amperage signals to controllers and often have only one or two signal outputs, but the outputs come in a variety of voltages. A common range for DC prox switches is 10–30 V; a common range for AC prox switches is 20–264 V. Some prox switches can work on either AC or DC voltage, in which case the common range is 24–240 V. The main reasons that prox switches fail

Courtesy of Balluff, Inc.

FIGURE 7-3 Some truly small prox switches, which would be perfect for robotic systems.

are either physical damage to the switch or incoming voltage that is greater than the prox's voltage rating. During my time in industry, I have seen prox switches that were damaged to the point you could see the internal components, yet still worked to some extent—proof of the tenacity of this type of sensor. See Food for Thought 7-1 for a story about one tough Balluff prox.

An **inductive proximity switch** uses an oscillating magnetic field to detect ferrous metal items. An oscillating field goes on and off. In this case, it creates a wave that looks just like the sine wave of AC voltage. Unlike AC power, inductive prox switches create only the positive portion of the waveform. They have thousands of cycles per second, whereas the AC power found in the United States has only 60 Hz. The oscillator—the part of the prox that creates this rapid positive sine wave pulse—connects to a coil of copper in the end of the prox that generates a magnetic field extending from the end of the prox switch (Figure 7-4). This magnetic field interacts with ferrous metals (i.e., those that are magnetic), which is how the prox switch changes states. Prox switches with a low-strength magnetic field can sense items at a maximum distance of 0.06 inch (1.5 mm), whereas the stronger inductive prox switches can sense metal items up to 1.6 inches (40 mm) away.

When a metal part enters the magnetic field of the inductive prox (Figure 7-5), it generates eddy currents in the part, which takes energy away from the magnetic field created by the oscillator in the prox. **Eddy currents** are flows of electrons created by a magnetic field moving across a ferrous metal item; they in turn generate their own magnetic field. The movement of the magnetic field in this case arises primarily because the sine wave of the oscillator creates a magnetic field that builds to a maximum and then collapses back to zero thousands of times per second. The eddy currents and their induced magnetic fields draw enough power from the oscillator that eventually it can no longer create an oscillating field. A separate unit in the prox known as the

FOOD FOR THOUGHT

7-1 Balluff Prox Toughness

We often hear companies' claims about how great their products are or how well they hold up under adverse conditions, but many of us wonder just how much of those claims is hype. While I cannot vouch for every proximity switch you might encounter, I can relate my own Balluff prox story and testify about the conditions in which I have seen proximity switches survive.

The Balluff prox in question sensed the motion of sanding tape in a polishing machine that polished the part in two directions. Its purpose was to ensure that the proper amount of sandpaper advanced with each direction change. The sandpaper was on rolls, which were threaded through some rollers, including a plastic roller that had a piece of metal embedded inside. The prox was set up to monitor this plastic roller and register when the metal piece rotated by, thereby indicating to the system that the sandpaper had advanced. If this prox did not change states, the system would fault out and turn on an error/alarm light. To reset the alarm, the operator had to press a reset button and then start the system again.

On the day in question, the operator was having trouble with the system faulting out intermittently. As improperly routed/installed tape rolls were a common

(continued)

problem with this operation, I generally started there when troubleshooting the system. This time I found that not only had the operator routed the tape incorrectly but also the roll was placed in such a way that the abrasive side of the sanding tape was running across the Balluff prox that was used to ensure tape advancement. The entire front of the prox had been sanded down to a 30° to 45° angle, with very little of the flat tip remaining. At best, there was perhaps 1/16 inch of the flat-end surface left. Now here is the amazing part: The prox was still functional! The operator was having only an intermittent problem with the machine, which means the prox was still functioning most of the time. In fact, I verified this functionality by advancing some of the tape by hand and watching the indicator light change state on the prox. Of course, I recognized that this was the problem with the system given the damage to the prox, so I began to remove the prox from the system.

This is where I got my second surprise of the day. I had changed many of these prox switches in the past, and knew that it was common practice to do so under powered conditions: Most of the prox switches we used had a handy screw on the connectors that

made replacement a breeze. The proper sequence was to unscrew the connector, thereby removing power, and then remove the prox from the bracketing. On this day, I was not thinking and started by removing the prox from the bracket first. At this point, I discovered that not only was a large chunk of the tip of the prox gone but also the live wires inside were exposed. Yes, I discovered this in the way no maintenance person should—by being shocked. Luckily, it was a low-voltage prox and the shock was minimal. While this is the entertaining portion of my story, it demonstrates how much damage a Balluff prox can take and still function at least intermittently. Once I realized the error of my ways, I removed the connector, and the power from the prox, before finishing the removal of the prox from the bracket. The rest of the story is much more boring and standard as far as replacing a prox goes.

Safety disclaimers: Balluff prox switches may still function with exposed internal electrical components, so always kill or disconnect the power before attempting their removal. Also, remember a current of only 0.05 A can be fatal, so always, always, be careful with electricity!

Courtesy of Balluff, Inc.

FIGURE 7-4 A cutaway of a prox switch, showing both the internal solid-state components (surrounded by the red resin) and the coil in the tip (surrounded by the black resin). The coil is the copper color in the black resin.

trigger unit looks for the oscillation to die down to a certain level and then activates the output(s) of the prox accordingly.

Capacitive proximity switches generate an electrostatic field and work on the same principle as regular

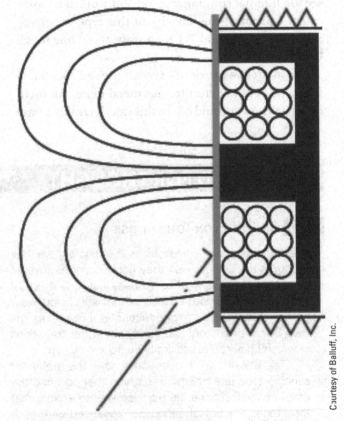

Courtesy of Balluff, Inc.

FIGURE 7-5 This drawing represents the magnetic field generated by an inductive prox that interacts with magnetic metals.

Tech Note

The oscillation of magnetic fields is a key concept and the heart of many electrical components. For example, transformers use this principle to step up, step down, or clean up and isolate electricity in systems. In addition, the induction motor—one of the main motors found in industry—uses induced magnetic fields to cause the rotor to spin. As you learn more about electricity and motors, you will delve deeper into how induced magnetic fields are crucial to the modern electrical world.

capacitors. The switch provides the electrostatic field, air is the insulating material or the dielectric, and the object being sensed provides the conductor that finishes the capacitive circuit. Unlike an inductive prox, a capacitive prox will interact with materials that are both magnetic and nonmagnetic (Figure 7-6). The capacitive prox works in the opposite manner as the inductive prox, in that the oscillation of the circuit *begins* when the item sensed enters the electrostatic field and *ends* when no object is present. Another important difference is that the field on the inductive prox does not change, due to the construction of the switch. In contrast, the capacitive prox allows adjustment to the switch, such that the user can vary the electrostatic field and change the distance it interacts with objects. As a consequence, it is possible to set the switch to detect items inside cartons, boxes, or other packaging materials.

FIGURE 7-6 A capacitive prox switch. Notice how the body is plastic, rather than metal. It is common to find inductive prox switches with metal bodies and capacitive prox switches with plastic bodies, but do not assume that all metal prox switches are inductive or that all plastic prox switches are capacitive.

Capacitive sensors generate small radio-frequency energy fields, typically in the 100 kHz to 1 MHz range—a point to keep in mind, as this field might interfere with

some equipment. In terms of targets, while the inductive prox needs a magnetic metal, the capacitive prox can sense liquids, wood, plastic, ceramics, certain glass materials, and metals. The preferred target is a flat, smooth object that has the capability of interacting with the electrostatic field, with common sensing ranges for the capacitive prox being 0.1 inch (3 mm) to 2.4 inch (60 mm). This kind of prox is often used to sense the material within a package while ignoring the outer container.

Photoelectric proximity sensors or switches detect levels of light, sense objects via reflected or blocked beams, and detect colors via an emitter that sends out a specific wavelength of light and a receiver that looks for this wavelength to return after it has interacted with the surrounding environment. Several configurations and operational modes for this type of sensor are used. The most commonly encountered components are the **transmitter**, which sends out a specific type of light; the **receiver**, which receives the light and looks for specific light information; and the internal process/controls, which takes the incoming signal and either turns it into information to send on or changes the state of contact(s) within the sensor. Often these sensors use infrared light-emitting diode (LED) light that is outside the visible spectrum, but they can also work with laser light, fiber-optic light, or visible light depending on the application.

The typical sequence for a photoelectric switch is as follows:

1. An oscillator module turns the transmitter, usually an infrared LED, on and off at high speeds.

2. The light either travels in a straight line to the receiver or is reflected by some surface.

3. The receiver, when the correct light strikes it, transmits signals to the internal components of the switch.

4. If a threshold is met, the switch changes the state of contacts or passes along information to another process. This threshold value is adjustable in many sensors, and this adjustment is crucial for small, fast-moving parts.

5. If the wrong frequency or type of light enters the receiver or if not enough of the transmitted light makes it to the receiver, the switch does not change conditions or passes on only partial information.

Through-beam photoelectric sensors separate the transmitter and the receiver into different units, placed opposite each other. When this configuration is used, the system is often set up to change state when the signal is broken, indicating that something is blocking the light. With the use of a laser beam, this system can cover distances of approximately 300 feet (92 m), and heavy dust or other airborne particles are less likely to interfere with the sensor. The light curtain (mentioned in Chapter 1 in the discussion of safety) utilizes this setup, with infrared LEDs in a line configuration covering a broad area where people might get close to a dangerous part of the machine (Figure 7-7). While this system is fairly forgiving of the environment in which it works, the transmitter and the receiver must be in a straight line with each other. Also, if the environment includes a lot of airborne dust or other contaminants, you will need to clean the covers on both sides from time to time.

FIGURE 7-7 One side of a light curtain set. Often light curtains are used to monitor the area where parts are loaded and unloaded to protect operators from dangerous motion as the finished parts cycle out and raw parts cycle in.

Retroreflective photoelectric sensors place the transmitter and the receiver in one assembly, and often use something like a bicycle reflector to return the emitted light to the receiver. Similar to the light curtain, this switch often changes state when the beam does not make it back to the receiver, but the range is reduced in this application to a maximum of approximately 165 feet (50 m). Shiny parts may return enough light in some applications to keep the sensor from changing state; in such a case, we need to either use a different sensing system or apply a polarizing filter over the sensor. The polarizing filter projects the transmitted light in one plane only, so it requires a corner-cube type of reflector to return the light to the receiver—something the shiny part cannot accomplish.

Diffuse photoelectric sensors also house both transmitter and receiver in one unit, but in this case only a small amount of returned light is needed to activate the sensor. Instead of a reflector, the part is what returns the light. Thus, this system tends to change state when the part is present and enough light reaches the sensor. We can further dial in this property and use a diffuse photoelectric sensor to detect a certain feature or marking on the parts via a specific amount of reflected light. Obviously, this plan works only if the feature we are looking for differs from the rest of the part enough to reflect a unique amount of light back to the sensor. Because such a sensor works with just a partial reflection of the emitted light, the maximum range is approximately 40 inches (1 m).

With the **light level detection** photoelectric sensor, we can tune out or ignore the background, differentiate between parts on a background, and even detect transparent objects. When using a simple photo eye, the beam is either returned or not. The sensor, however, analyzes the light coming back to the receiver and determines whether it has been affected or altered in any way. A light level detection prox could differentiate between a rough, raw part and a smooth, finished part as it passes by, based on the differences in the light returned to the switch. We can use this type of prox to make sure labels are in place, monitor empty spots on the conveyor, or detect any other condition where there is a clear and consistent difference in the light returned to the prox between the two states monitored. Proper calibration is a must with a light level detection sensor, however, and dusty conditions or changes in ambient light may be a problem.

Color detection photoelectric sensors analyze the light coming into the receiver to determine the color or a color difference in parts. Such a sensor can find the edge between parts of different colors, thereby providing important positioning information to robotic systems. Roboticists have used color detection sensors to tell the difference between blue and red balls, follow colored lines, and perform other such tasks involving color recognition.

Color detection photoelectric switches do have some drawbacks that you must take into consideration. Changes in the ambient light will change the light returning to the switch and the reading, so the sensors require recalibration. For example, this is why a robot might be able to tell the difference between red and blue balls in the lab, but fail to work at the science fair in the gym. Also, these sensors have an emitter that sends out a specific wavelength of light. Depending on the light emitted, it may be impossible for the receiver to detect some colors or the difference between certain colors. Clearly, then, there is not a "one size fits all" color detection prox.

Hall-effect sensors respond to magnetic fields and can produce either a digital on/off-type signal or a varying analog signal, depending on the sensor and application. When used in a digital application, such sensors are "off" when there is no magnetic field and "on" when a magnetic field of sufficient strength passes over the sensor. With the analog variety, the output increases with the strength of the magnetic field interacting with the sensor. The Hall effect was named after Edwin Herbert Hall. In 1879, he figured out that when current flows through a conductor in a magnetic field, the magnetic field tends to push the negatively charged atoms to one side of the conductor and the positively charged atoms to the opposite side, when the magnetic field is perpendicular (at 90°) to the conductor. Today we use a semiconductor to allow a voltage to pass when a magnetic field lines up with the sensor, and we apply this principle to detect position, monitor rotation, measure current in conductors, and gather useful information about systems. The magnetic field can come from either a magnet added to the system or via magnetism created/induced by current flow.

As you get into the field and work with various proximity detection systems, you will have the chance to learn the ins and outs of the specific sensors in your application. When replacing these sensors, take special care to replace the old device with the same type of sensor, based on factors including how much input and output voltage there is, how it responds to a change in state, and which type of contacts it has. Many prox switches look identical, but operate in a completely different fashion, which could lead to false signals and incorrect operation at best, or damage to the system at worst.

TACTILE AND IMPACT SENSORS

Another sensing field for robotics is **tactile** sensing, or the ability to sense pressure, and **impact** sensing, or sensing when the robot contacts an object in the intended movement path. Both of these reactionary types of sensing are based on force, but the applications and techniques for each differ significantly. Tactile sensing seeks to determine how much force is being applied, what the shape of the part is, how it is gripped, and whether the part is hot or cold—basically the same kinds of sensing tasks that the skin performs for the human body. Impact sensing is concerned with detection of collisions, determining whether forward movement is impeded or stopped, and, perhaps most importantly, shutting down or modifying the motion of the system to prevent damage to both whatever is hit and the robot. Our exploration here will cover only a few of the many ways possible to perform tactile and impact sensors; even as you read this text, new sensors and new methods are undoubtedly under development.

Tactile Sensors

Tactile sensors, as just noted, focus on how much force the system applies to an object and which area that force covers (Figure 7-8). These systems range from simple buttons that become active when depressed and disengage when released, to complex arrays of sensors that closely mimic the function of human skin in terms of the amount of information they provide. For the simple "Is it gripped?" system, all that is required is a switch that will fit in the desired space, usually the gripper; that is capable of activation by making contact with the desired object; and that is strong enough to handle the forces involved. A subminiature micro limit switch or push button is often the switch of choice for this type of operation. If you have experience with robots, you might ask, "What about sensors that detect whether a gripper is open or closed—would they be considered tactile sensors?" The answer is no. Tactile means "of or connected to the sense of touch," so a tactile sensor must "touch" the part it is sensing. Otherwise, it is a different type of sensor application, even if we use the same type of sensing device in both applications.

Source: NASA

Source: NASA

FIGURE 7-8 These two pictures demonstrate the principle behind tactile or force sensing. The astronaut shaking the Robonaut's hand would likely not have enjoyed the experience if the Robonaut had used the same amount of force as it did lifting the 20-lb dumbbell.

Complex tactile sensors use **sensing arrays**—that is, organized groups of sensing elements—to gather information about contact with objects. The construction of the robot and tooling, along with the application, determines how much area is available for installing the array. Each element of an array is an individual sensor that provides information about how it is interacting with the part. The smaller the sensing element, the more elements we can place together in an array. Just as more pixels in a camera improves the image quality, more elements in an array increases the amount of information provided to the robot.

Simple tactile sensing arrays consist of elements that have digital output (e.g., on or off, 1 or 0). With this type of sensor, the system knows if it is gripping the part and how many elements are involved with that grip, which is enough to find edges and determine whether the part is moving in the gripper during travel by monitoring for changes in element state. The downside to this type of element is that it does not provide any feedback on the amount of force generated, the surface features of whatever is gripped, temperature, or anything besides the simple yes/no output of the digital element.

High-end, complex tactile sensors can provide information about the shape of the object, the amount or pressure applied, temperature, finish, and other analog-type sensing data that is difficult, if not impossible, to convey with digital signals. This added data allows the robot to handle delicate items such as a single egg one minute and heavy items such as a case of eggs the next minute. Complex tactile sensors have a variety of options for sensing changes in force, such as changes in voltage, resistance, capacitance, and magnetic flux. Regardless of which signal generation method is used, they all share one similarity: A change in the shape of the element causes a change in the signal generated. The strain gauge element changes resistance as the wire inside is stretched or compressed. The capacitive element experiences a change in the charge it stores depending on the distance between two electrodes. Piezoelectric elements create electricity with applied force: The greater the force, the more electricity produced, though it is in the millivolt range. With each of the force sensors, the greater the pressure, the greater the deviation from the base signal or the signal generated by the element under noncontact conditions.

Often the manufacturer of the tactile sensor will provide data on the change in force necessary to change the output ratio, or some other means for making use of the data from the array. This is crucial information when you are installing and using third-party sensors. Systems that include complex tactile systems as standard components should require little more than calibration from the user. **Calibration** is a process that ensures a precision system performs properly and provides for any adjustments needed. In this case, the user would place a verified accurate pressure gauge in the tactile sensor area of the robot and then apply force. The user would compare the reading from the pressure gauge and the robot to determine the amount of difference between the two. If the difference is great enough or out of tolerance, the user would offset the robot system to correct for the error measured. You should calibrate the system during robot installation and verification, after changing sensors, after any damage to the system, and periodically as recommended by the manufacturer.

Impact Sensors

Impact detection plays a crucial role in minimizing or preventing damage to the robot as well as to people and equipment. Initially, impact detection dealt with keeping the robot and other equipment safe, as people were protected by fences, light curtains, pressure mats, and other devices designed to ensure they stayed clear of the work envelope during operation. In fact, it was not until 2012 that this technology advanced enough for the robot to work without the isolation cage and beside its human counterparts in industry. Sensing an impact with something rigid—for example, the machine into which a robot is feeding parts—is easier than detecting an impact with something lightweight and movable—for example, a person.

Today, we typically either monitor the amperage the motor uses or insert sensing devices designed to detect impacts. The most commonly used sensing devices for impact are strain gauges at crucial points, such as the joints of the robot, that detect sudden changes in the forces of motion so the robot can respond accordingly. Bosch went a different route, however: The company coated its APAS assistant with a leather-looking tactile skin that detects impact. At the writing of this book, Bosch was the only company that had gone this route in the collaborative robot family, but if the idea proves to work better and more cheaply in the long run, other companies may follow in its footsteps.

When the robot uses the amperage draw of the motors to monitor for impact, it creates a very accurate system with great sensitivity, but at the price of greater complexity—and greater complexity means more things that can go wrong. Some high-level math is involved (which we will not detail), which takes into account the programmed path, speed of travel, payload, and type of motors performing the work to determine the amperage draw of the motor(s). The result of this calculation is compared to the actual current draw of the motor(s). If the specified plus/minus error limit is exceeded, the robot responds as programmed.

Problems that can occur with this method include system shutdown due to excessive payload, moves that result in high-velocity turns/stops, and noise in the system. (We will discuss noise shortly.) Since we are monitoring the torque output of the motor via the amount of current used, there is no distinction between high current draw due to a heavy part or velocity multiplication and high current draw caused by an actual impact. This is why some robots alarm out and stop, indicating an impact, when a visual inspection shows no objects in the robot's path. Anything that causes the motor to use excessive amounts of current—such as a too-heavy load, bad bearings, friction, or caked dirt in the joints, among other things—can cause an impact alarm.

Noise in this context refers to the mathematical difficulties that arise when trying to properly calculate the torque required at the start of motor motion for robot movement. Motors require more amperage to start moving than to maintain motion, so we have to figure that surge into our calculations. The calculations also need to factor in how we will move the robot. For instance, the motor will need more amperage to raise the arm than to lower it, as gravity is a factor. In addition, we cannot forget to figure in how fast we want the robot to move, along with what happens to the motor as the robot accelerates and decelerates. Collectively, these factors create some mathematical gray areas where we have to estimate what will happen, and we call these estimates noise. If we are off in our estimates and set the alarm parameters tightly, the system may alarm out unnecessarily. The biggest danger with false alarms, whether they are due to noise or mechanical issues, is that someone will just change the alarm parameters by increasing the plus/minus window instead of fixing the problem or doing the math needed. This could lead to impacts that could damage man and machine alike without the system alarming out!

No matter how we detect the impact, the robot must do something to prevent or limit the damage caused. One tactic is to E-stop the robot and lock it in place. With this approach, the robot stops all motion as quickly as possible and then freezes in place to let the operator decide what to do next. The problem with this tactic is that the robot continues moving for a short period of time after the detected impact, before the full stop occurs. Granted, this is usually less than a second, but a fraction of a second can seem like an eternity when a robot arm is crushing you.

A secondary concern is the strain placed on the internal systems of the robot when this tactic is used. As an analogy, think about driving down the road and suddenly throwing your vehicle into park. While this action may not kill your vehicle, it surely did not do it any favors.

To avoid these problems, many robots now disengage the motor from the joints until the system can come to a full stop. To continue with the car example, instead of going from drive to park, now you put your car in neutral, take your foot off the gas, and apply the brakes to stop the car. This method reduces the force the robot has to stop, the energy of the impact, and the stress on the robot's internal systems. Such a response requires an advanced force reduction system coupled with specialized programming and sensors, but it has brought the robot out of the cage and enabled it to work alongside human workers.

You can expect tactile and impact sensor systems to continue to advance and evolve with the robot, especially in the collaborative robotics field, as these systems allow the robot to respond to a greater range of events encountered while in use. The current advancements in tactile sensing have given the robot a broad range of data to work with, creating exciting new options and tasks that were previously outside of the robot's scope of operation. We often use robots to perform difficult and dangerous tasks in which data about what is happening to the robot, including impact sensor data, could be crucial to protecting the system. As this technology continues to improve and mature, you can expect to find these sensors as a part of the newer robotic systems in applications where they make good financial sense.

TEMPERATURE SENSORS

Another type of sensor you may encounter in industry is the temperature sensor. There are a multitude of industrial processes in which a few degrees means the difference between a quality part and a smoothly flowing production process, and scrap parts and a complete disaster that could take hours to fix.

When selecting a temperature sensor, we have to look at the temperature range of the process, the

measurement procedure, and the precision requirements for the reading. Every sensor has a maximum temperature it can handle, and exceeding this limit is a good way to destroy the sensor. Sometimes we need to submerge the sensor in a fluid or production material; in these cases, the sensor needs to be watertight. If the sensor's smallest reading is to the single degree, then we can keep the system only within a degree or two of our set point. In many cases this is sufficient, but for higher precision the sensor needs the ability to register in the partial-degree range. In this section, we look at some of the common sensors used to measure temperature and the specifics of each.

Thermocouples are temperature sensors that work off the principle of a small DC millivoltage generated when the junction of two dissimilar metals is heated. These sensors are workhorses in industry, but a few points must be kept in mind to make sure they work correctly. The voltage generated by the sensors, and in turn the signal emitted, is a function of the difference between the hot junction—where the two metals are joined, and which is exposed to the temperature to be measured—and the cold junction—where the two loose ends of the sensor connect, and which sends the signal (Figure 7-9). For an accurate reading, the cold junction needs to remain at a constant temperature so we can calibrate the system correctly. If the cold junction's temperature changes, the difference between the two junctions changes and the output signal will also change. It does us no good to have a highly accurate thermocouple if the cold junction's temperature fluctuates greatly, as this will cause our reading to be off every time this temperature changes.

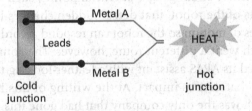

FIGURE 7-9 The basic structure of a thermocouple, including the two junctions.

To protect the wires of the thermocouple, they are placed inside a metallic tube. The walls of the tube are referred to as the sheath, and the end is called the tip of the probe. The sheath protects the wires from damage due to impact and allows for easy handling. The tip comes in three different configurations. With a grounded tip, the junction of the thermocouple wires is joined to the tip, ensuring good transference of temperature. With an ungrounded tip, the hot junction of the thermocouple is isolated from the tip and set slightly back from the end. The air space acts as an insulator and allows the thermocouple to work with higher temperatures. Because the air space is static, the output of the sensors is still predictable in nature. The third type of tip does nothing to protect the hot junction, as it is completely exposed. Extra care in placement and handling of this type of probe is necessary because any damage to the hot joint could change the output of the sensor.

We can use several different metals in the construction of thermocouples. As a result, there are different classes of these sensors, which are identified as follows:

- E: 95–900°C (200–1650°F) and −200°C to 0°C (−328°F to 32°F)
- J: 95–760°C (200–1400°F)
- K: 95–1260°C (200–2300°F) and −200°C to 0°C (−328°F to 32°F)
- N: 95–1260°C (200–2300°F)
- R: 870–1450°C (1600–2640°F)
- S: 980–1450°C (1800–2640°F)
- T: 0–350°C (32–660°F) and −200°C to 0°C (−328°F to 32°F)

The American Society for Testing and Materials has created calibration charts for these thermocouples. Thus, for each letter designation thermocouple, a specific calibration chart is available that tells you x voltage equals x temperature. This makes it easy to select and use thermocouples for various process applications. Thermocouples can achieve resolutions of 0.1°C and accuracies of ±0.30°C.

Resistance temperature detectors (RTDs) are based on the principle of a linear increase in resistance that occurs when a metal is exposed to heat. By supplying a steady voltage to this device, we can use the decrease in current flow through the sensor to determine to which temperature it is being subjected. Platinum works the best for this application, but we can also use nickel, copper, or a nickel/iron mix for the wire in this sensor. A grid of this wire is placed on a ceramic support element so there is a larger area subjected to the temperature to be measured; this element is often trimmed with a laser to create a specific resistance, thereby calibrating the device (Figure 7-10). Like the thermocouple, the RTD is encapsulated in a metal tube (a sheath); unlike

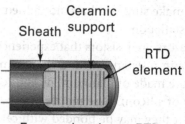

Encapsulated type RTD wound with platinum wire

FIGURE 7-10 The basic design of an RTD probe. Instead of a ceramic-based grid, some devices include a spiral of metal in the end of the probe.

the thermocouple, however, the RTD does not have varying tip types. Depending on the metal used, the following ranges of temperature measurement are possible with RTDs:

- Platinum: −260°C to +650°C (−436°F to +1202°F)
- Nickel: −100°C to +300°C (−148°F to +572°F)
- Copper: −75°C to +150°C (−103°F to +302°F)
- Nickel/iron: 0°C to 200°C (+32°F to +392°F)

In terms of accuracy, these sensors are organized into four classes:

- Class AA: ±0.10°C + 0.0017t
- Class A: ±0.15 + 0.002t
- Class B: ±0.30 + 0.005t
- Class C: ±0.60 + 0.01t

where t is temperature measured in degrees Celsius without regard to sign. For example, if we had a reading of 100°C with a class A probe, we would have a variance (inaccuracy) of ±0.15 + (0.002 × 100) = ±0.35°C.

Because of their linear nature and the fact that the math is predictable, equations are often used to scale the output of a temperature sensor. Charts are available to facilitate this kind of scaling, and it is a safe bet that the RTD vendor can also help with any scaling issues you may experience with their sensor.

One last concern with RTDs is the need to ensure the wires coming from the sensor have adequate insulation to account for the temperature in which the sensor will work as well as the distance from sensor location to signal processing. Because this sensor is resistance based, excessive heat on the wires or running them too far can increase the overall resistance of the system and skew the results. If your sensor seems off, make sure you have not added resistance with the installation.

Thermistors are resistors that experience a consistent change in resistance with a change in temperature. They are made of platinum, nickel, cobalt, iron, and oxides of silicon; these metals may be used in a pure state or they may be bonded with ceramics and polymers, so in that respect thermistors are similar to RTDs. Unlike RTDs, however, these sensors can have either a **negative temperature coefficient (NTC)**, in which case an increase in temperature causes a decrease in resistance, or a **positive temperature coefficient (PTC)**, in which case an increase in temperature causes an increase in resistance. Thermistors as a group cover a temperature range of $-100°C$ to $+300°C$ ($-148°F$ to $+572°F$), but no single thermistor covers this entire range. Their accuracy is between $0.1°C$ and $1.5°C$, making them fairly accurate sensors in their temperature range. The NTC type is more widely used, but you should not assume all the thermistors you encounter will be NTCs. As with the RTD, you must keep wiring insulation and length in mind when using a thermistor, as changes in resistance will skew the sensor's output.

When trying to choose between these three types of sensors—thermocouples, RTDs, and thermistors—there are a few things to keep in mind. Thermocouples cover the largest range of temperatures, they have good accuracy and sensitivity, their linear output is more consistent than the output of thermistors, and they tend to cost the least of the three options. RTDs cover a moderate range of temperatures (better than the range covered by thermistors), are more sensitive than thermocouples, and tend to have the best accuracy and linear output characteristics of the three sensor types, but often cost more than thermocouples and thermistors. Thermistors have the most restricted temperature range, have good accuracy and linear characteristics, are the most sensitive of the three types of sensors, and cost less than RTDs. When you are selecting a temperature sensor, I recommend starting with the temperature range of the application and then working through the process from there, looking at accuracy and cost, in that order.

Semiconductor temperature sensors, sometimes known as integrated circuit (IC) temperature sensors, use semiconductor diodes that rely on temperature-sensitive voltage rather than current characteristics to determine temperature. The base-emitter voltages of the transistors are directly proportional to the absolute temperature when we use two identical transistors and feed them a constant ratio of current at the collector. In short, by using two identical solid-state transistors, we can use the voltage output to measure temperature. The range for these sensors is typically $-55°C$ to $+150°C$ ($-67°F$ to $302°F$), although some units may work a bit outside this range; the accuracy is $\pm0.2°C$ to $0.3°C$. A semiconductor temperature sensor can send out an analog signal to measure temperature, or a digital (on/off) signal can be set to trigger at a specific temperature.

Infrared radiation (IR) temperature sensors detect the heat of an object by measuring the infrared radiation created, which is a long wavelength of light below the human scale of vision. These are noncontact sensors, so we can put a little distance between the sensor and the item measured. Unfortunately, the farther away the sensor is from whatever it is measuring, the more error the result is likely to include. The distance to spot ratio tells you how many feet away from the object the sensor must be to measure a 1-foot radius circle; the best sensors have a 300:1 ratio. IR temperature sensors can measure temperatures in the range of $-60°C$ to $1500°C$ ($-76°F$ to $2732°F$) and at best have an accuracy of $\pm1°C$, with $\pm2\%$ being the more common accuracy. They do not work well at the extremes of temperature, however, so as you approach the minimum/maximum points for the sensor, you may notice more error in your readings. The typical application for an IR temperature sensor is a handheld device with a laser that illuminates what you are pointing it at, but we could also mount this system on a robot and let it do the work for us.

There are other ways to measure temperature, but this section has described the most common approaches utilized in industry. As you encounter other methods of monitoring temperature, I encourage you to learn more about these systems. It is always better for learning when you can read about a concept and then go directly into the field and work with the sensor or device just studied.

FLUID SENSORS

By definition, a **fluid** is a substance in which the particles can move past one another, conform to the shape of their container, and continually deform under an applied shear stress. Typically, when we are working with a fluid it is either a liquid or a gas, but certain fine-grained solids may move and act like a fluid, thus requiring fluid sensors to monitor them. When working with fluids, it is common to want to verify or quantify flow, measure level, or determine system pressure. In this section, we look briefly at some of the sensing devices used to gather this information.

Flow is a term used to describe the motion of a fluid, and it can be measured with some type of sensor. When measuring flow, we have to find a way to convert the kinetic motion of a fluid into a measurable electric signal. For simple applications, the output is usually a digital signal to confirm that a fluid is indeed flowing. For complex applications, in which we need to know some additional information about the flowing fluid, the sensors generate an analog signal. Like any other analog signal, it needs to be scaled so that the voltage produced by the sensor has meaning inside the controller.

Turbine flowmeters use a propeller-or turbine-type assembly placed inside the fluid to generate a signal via a magnetic pickup sensor positioned nearby, but outside of the fluid. As the blades turn in the fluid (in the same manner a windmill or pinwheel turns in the wind), magnets positioned at the tip of the blades pass near the sensor, sending a signal pulse. The greater the fluid flow, the faster the turbine turns and the faster the pulses are generated. We can use these pulses to confirm flow or use some math to determine the flow rate of the fluid by plugging the pulse rate into a formula that takes into account the force needed to turn the turbine and the area of the space in which it is working.

Target flowmeters place a disk or similar shape in the fluid and use the force of the fluid flow to deflect the target and generate a signal. A spring sets the resistance tension, and a Hall-effect sensor is usually responsible for emitting the signal. Depending on the setup of the system, it can have a trigger point adjustable via spring tension or an analog-type output. While this system usually requires plumbing the sensor into the system, target flowmeters are known to hold up well in dirty and corrosive fluids where other sensors may fail. In some cases, the sensor may come with a built-in display, allowing those nearby to see what is happening in the system without the need to run the signal through a controller.

Sail flowmeters work on the same basic principle, but instead of a round disk for the fluid to strike, a triangular or rectangular flag is dragged along by the fluid flow. These sensors provide the same data as target flowmeters. Sail flowmeters are commonly used with gas-based fluids, whereas the target versions are good choices for liquids.

Magnetic flowmeters, also known as induction flowmeters, measure the change in voltage induced in the passing fluid. Obviously, this requires the fluid to conduct electricity, so such a system does not work with all fluids. The sensor creates a magnetic field that the fluid must pass through, which induces a voltage that is proportional to the average flow velocity. This induced voltage is detected by sensing electrodes, which send an analog signal back to the main portion of the sensor for use in calculations that determine the flow rate. Magnet flowmeter sensors should not be used around metal piping and other nonfluid items that might pick up a charge from the magnetic field and cause false readings. In case of nonconductive materials, we could add iron or aluminum powder to allow the material to interact with the sensor, but we must take two considerations into account: (1) We must make sure that the powder will not hurt the fluid or process and (2) we need to recalibrate the sensor, given that it is now interacting with only a portion of the fluid passing by. My recommendation is to utilize a different sensor in cases where the fluid is nonconductive or poorly conductive.

Level sensing is all about determining how much fluid is present in a container, usually with the goal of preventing overfilling or other less than desirable conditions. Because of the nature of level sensing and the way gases interact with their environment (i.e., gases continue to expand until they fill the space), level sensors are used almost exclusively with liquids. Some noncontact methods of measuring fluid level do exist, such as ultrasonic methods (discussed later in the chapter), but we will focus here on the two common physical contact methods.

Float sensors use something the liquid can lift to determine level. A great example of how this sensor works can be found in a household toilet. When you flush, a flapper valve opens, dumping all the water from the storage tank into the system. At this point, some type of bulb or floating cylinder drops to a lower level and water starts to flow into the storage tank. As the tankfills, the float device rises with the water and eventually closes the valve that lets the water in. We use the same basic principle in industrial applications, albeit with a few twists to the process. Sometimes the float has a set of contacts in it and a conductive medium, such as liquid mercury, that makes the contacts when the switch changes from the vertical to horizontal position. With this type of float sensor, wires leading to and from the contacts need protection from the liquid as they will contact it; this protection usually takes the form of a thick rubber or plastic hose material. Another method is to place metal or magnetic material inside the float and position something like a prox switch or Hall-effect sensor nearby to interact with the float when it reaches the desired point. While this method is versatile, it may not be well suited to applications involving hot and/or harsh liquids.

Capacitive level sensors work on the same basic principle as the capacitive prox switch, with one exception: Instead of the fluid acting as the second electrode, the liquid is the dielectric (Figure 7-11). Some sensors utilize two probes or plates in the fluid, whereas other may have three or more, providing the ability to create a high- and low-level type of system. With two plates, a change in the fluid creates a change in the dielectric property of the sensor and, in turn, a change in the signal generated, leading to an analog output. Of course, we can set a threshold level just as with a capacitive prox switch and produce a digital output instead.

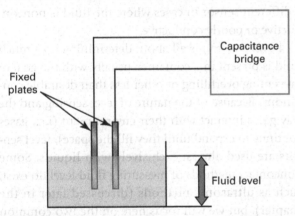

FIGURE 7-11 The basic operation of a two-probe capacitive level sensor.

When three or more probes are used in capacitive level sensors, one probe is longer than the rest, and serves as the main electrostatic source. The next longest probe is the low-level probe. As soon as the liquid makes contact with this probe, current flows in the sensor. If this is the only probe active in the sensor or if none of the probes is active, then the system knows the fluid is low. The third probe, the shortest of the three, is the high-level probe. When current flows through this probe, the system knows the fluid has reached the high level and it is time to stop the flow.

With more probes on the sensor, we can set more levels of response. For instance, the second longest probe could be the low-level probe, the next-longest probe could be used to reduce the flow to half the maximum rate, and the last probe could turn the flow off all together. In the past, I have worked with such a sensor that had three electrodes made of all-thread. The great thing about this setup was that we could change the level of the system by simply changing out the all-thread for a different length piece. In this case, process contained a caustic fluid that inevitably destroyed the probes over time, but it was easy to make replacements and get the system going once more without having to change the whole unit.

Pressure sensors give us information about the force generated by the fluid inside a system or container. There are two main divisions in this field of sensing: gauges that provide information in a visual format and switches that interact with contacts. With any fluid system, if the pressure continues to build without release, there is a high probability of catastrophic failure and all the danger that could entail. If you have no idea what could happen in these situations, check out the YouTube video "Mythbusters water heater explosion"—you will readily appreciate the potential for disaster. Even in cases where the system does not explode, the sudden failure of a component or piping can lead to flying debris, machine damage, or injury in the forms of loud noises, high-velocity impact, or coating individuals in hot/dangerous fluids. We will start with the gauge side of things for this area of sensing.

Plunger pressure gauges use a plunger that is exposed to system pressure, a bias spring, a pointer, and a calibrated scale to measure pressure. As the pressure in the system rises, the force on the plunger increases, compressing the bias spring. The pointer either is directly tied to the plunger movement or uses gearing to turn the movement of the plunger into a reading by indicating the system pressure on the calibrated scale (Figure 7-12). The spring is the part that really sets the range of pressure for this gauge; if it begins to wear out, this system becomes inaccurate. For those readers who have used a spring scale, like the kind often encountered in fishing, you will recognize that this gauge uses the same basic principle of operation.

Bourdon tube gauges use a thin-walled, slightly elliptical, cross-sectioned tube bent in a C shape, which is tied directly to the system, to read pressure (Figure 7-13). In 1849, Eugene Bourdon discovered

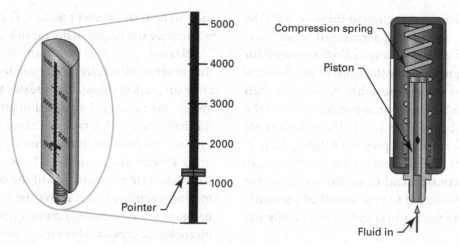

FIGURE 7-12 As the pressure increases, the spring compresses and the pointer moves along the scale.

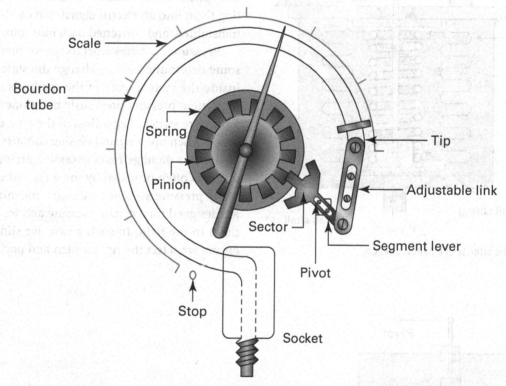

FIGURE 7-13 The interior of a C-style Bourdon tube gauge.

that a curved tube with a free end will try to straighten out or unwind when exposed to pressure. This is the principle behind the Bourdon tube gauge. The moving free end is hooked to a calibrated gearing system that moves an indicator to create a reading. Today spiral designs are available that are based on the same principle, but can work in smaller spaces. These gauges are frequently used in laboratory work due to their precision, and may be filled with fluid to help resist

the force of system pressure. The tube material needs to be somewhat elastic, having the ability to return to its original shape over and over again consistently. The movement of the tube is nonlinear, so careful calibration of the initial system and a tube that responds consistently to pressure are a must.

For lower-pressure applications, we can use bellows, diaphragms, or elastic capsules as the motion device. **Bellows gauges** use thin-walled tubing that can expand or retract in reaction to pressure, with the bellows portion often being augmented by an internal spring (Figure 7-14). These devices are useful for low-pressure readings below the range of the Bourdon tube. **Diaphragm gauges** use a thin membrane, often made of metal in a capsule arrangement, to move the indicator assembly (Figure 7-15). These devices are typically used for absolute pressure readings in low-pressure systems. Single-element diaphragm gauges use a single piece of metal or rubber to create the motion but are reserved for the lowest of pressures. Regardless of the type, a bias spring is typically not

used to help counteract pressure. For higher pressure, we increase the height of the capsule stack.

Piezoelectric pressure transducers use crystals that generate electricity in reaction to pressure to generate an analog signal. The greater the pressure, the greater the generated signal, though it is only in the millivolt range. This type of sensor is a great way to turn system pressure into an electric signal for use by other systems and controllers. Depending on the construction of the system, one of the other mechanical means of sensing pressure may be tied directly to the pressure, with the motion created interacting with the piezoelectric crystals. These transducers also work well with pressures that fluctuate thousands, if not millions, of times per second, which is why they are at the heart of some microphone systems. In this application, piezoelectric pressure transducers turn the air hitting them into an electric signal that can be amplified, transmitted, and converted back into sound.

Pressure switches react to system pressure and, at some determined point, change the state of contacts inside the system. Any of the methods of monitoring system pressure previously mentioned can trigger a pressure switch. Regardless of the type used, a snap action when opening and closing contacts is necessary to prevent damage from excessive arcing; this snap action is often provided by some type of spring setup. Many pressure switches are static, meaning that they are designed for a certain pressure and we cannot alter them in the field. In such a case, we simply need to ensure we select the right switch and pay attention to

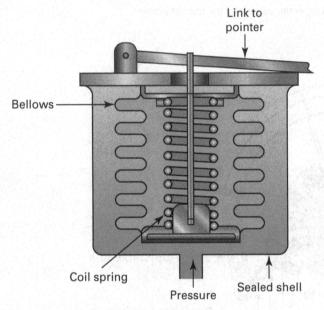

FIGURE 7-14 The interior of a bellows gauge.

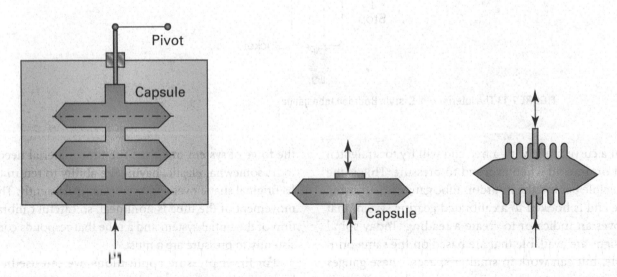

FIGURE 7-15 The capsule arrangement for a diaphragm gauge.

the wiring so we hook the right types of contacts to the right wires.

Pressure switches also come in a differential and adjustable variety. Differential pressure switches are triggered when the difference between the two provided pressures reaches a specific level. These pressures could be the system pressure and the atmospheric pressure, or the switch could be based on the pressure difference between two sections of a hydraulic system, for example. For the adjustable variety, we need to verify the switch covers the range of pressure exerted by the system and then make sure we understand how to set the pressure for the switch. Never assume you set it correctly without doing some kind of verification testing.

Pressure relief valves are valves set to either vent pressure or return fluid to the tank in case the pressure exceeds a set point. While not strictly sensors, these valves can prevent system damage and help to save lives. In some systems, the pressure relief valve is how we limit pressure; in these cases, it tends to be adjustable, such as in a hydraulic system. At other times, the pressure relief valve is the last line of defense for a system that is no longer under control. In that case, it is usually set for a specific pressure, such as in a pneumatic system. For systems where the pressure relief valve is a fail-safe condition, we can put a flow sensor downstream from the relief valve to alert us that the system has started to vent fluid due to excessive pressure.

Other pressure sensors are available, but this discussion is a good starting point if you have not had the chance to work with some of these devices in the past. In many cases we can work with the sensors and switches in the field without knowing all the ins and outs of how they work. Even so, the more you know, the easier it is to diagnose failures and ensure you have the right sensor for the application. Paying careful attention to how the sensor is installed and wired when you change out one of these devices will save you a lot of misery during installation of replacement sensors.

POSITION SENSORS

One of the great benefits of the robot is its ability to perform tasks with precision repeatedly. To do this, the controller has to record and track the position of the robot as points in space with some form of reference point, often known as the origin or zero point. Each axis has its own zero point, and each frame of reference has a zero point from which all points are calculated. Without these zero points, the robot cannot find the designated positions. If the robot forgets where the zero point is or a crash causes a shift in position, we have to reset the zero point so that the robot can run normally and accurately again. Unless we use a precision fixture to zero the robot, there is a high probability that all the programs will need a bit of tweaking or offset to work correctly.

When it comes to determining position, two main types of control systems are used: open loop and closed loop. **Open loop** systems work on the assumption that the control pulse—whatever it may be—activates the motion system and the robot performs as expected. This type of control was common in the early pneumatically powered robots, since their motion depended on valve sequencing, and in stepper motor systems, since the motor moves a precise amount each time voltage is applied. Often open loop systems would include limit switches or prox switches at designated points to check for robot position, but those devices were few and far between, meaning that a large amount of positional error could build up in complex programs with no correction. This chance for error made the open loop control system unpopular with industry, especially when the benefits of the closed loop system became apparent.

Closed loop systems send out the control pulse—whatever it may be—to initiate movement; they then receive a signal back that confirms movement and sometimes identifies the direction and distance of that movement (Figure 7-16). In the robotics field, the preferred feedback sensor for closed loop control is the motor encoder. **Motor encoders** are devices that directly monitor the rotation of a motor shaft and turn that information into a meaningful signal. The robot controller uses this signal to determine what the motor is doing as well as when it has reached the desired point. Keep in mind that if a robot has six axes of movement, it will need six feedback devices to be a fully closed loop system; it will also need six pieces of positional data, one for each axis, to find a specified point in space.

A simple encoder you may encounter uses the Hall-effect sensor (discussed earlier in this chapter). By attaching a disk with a specific number of magnets evenly spaced around the edge where they can activate

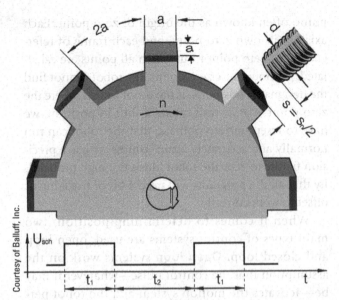

Courtesy of Balluff, Inc.

FIGURE 7-16 In this setup, a prox switch tracks the metal protrusions as they pass. This could be used as a basic form of feedback in a closed loop, identifying changes of state or detecting the protrusion that the controller could count and manipulate into useful data.

a Hall-effect sensor, we can track the rotation of the motor and thus the position of the system. Example 1 explores this kind of setup.

Example 1

For this example, we want to move the axis in question 90°. This robot has the motor coupled directly to the axis. The motor has a Hall-effect encoder with 40 magnets positioned evenly around the edge of a disk mounted on the motor shaft. How many pulses should the Hall-effect sensor register if the motor moves the desired 90°?

To determine how many pulses, we need to determine over which portion of a full rotation (360°) the axis needs to move. To do so, we will divide 90° by 360°:

$$\frac{90}{360} = 0.25 \text{ (one-fourth of a full rotation)}$$

Next, we will take this portion of the rotation, 0.25, and multiply it by the number of pulses or magnets on the sensing disk:

$$40 \times 0.25 = 10$$

Thus, 10 pulses should be sent back to the controller if the motor moves the proper amount to change the axis position by 90°.

The downside of this kind of system is the fact it gives us pulses, but no information on which direction

it moved. Yes, we can verify the system moved the proper distance, but what if it went in the wrong direction? This is where our good friend the optical encoder comes into play.

Optical Encoders

Incremental optical encoders consist of a disk that has either holes for light to pass through or special reflectors to return light, an emitter, a receiver, and some solid-state devices for signal interpretation and transmission. The principle of operation is simple: The transmitter sends out light and the light either passes through the holes or reflects back to the receiver, triggering the electronics of the encoder to send a signal to the controller (Figure 7-17). Because the input consists of light, it is possible to make the divisions on the rotating disk extremely small, allowing for 1024 or more units. In other words, we can take our circle of 360° and divide it by 1024, giving us movements as precise as 0.35° (slightly more than one-third of a degree). The more divisions on the encoder, the finer the positioning options.

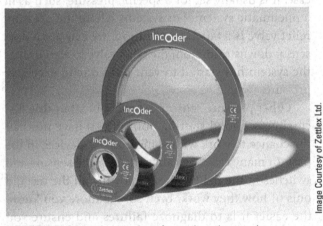

Image Courtesy of Zettlex Ltd.

FIGURE 7-17 Several examples of encoders that can be attached to motors.

"What about the direction of rotation?" you ask. By adding a second row of reflectors or light windows, offset from the pulse count, and another emitter and receiver, we can determine the direction of rotation by comparing the signals from the two rings. The offset creates a unique type of signal for each direction of rotation.

Another encoder option is the addition of a zero pulse (Figure 7-18). This specialized area of the encoder disk is used to establish a zero or home position. We can use this pulse to count the number of full

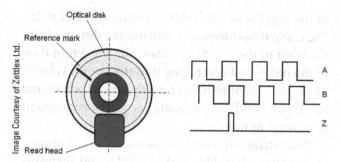

Optical disk

Reference mark

Read head

Image Courtesy of Zettlex Ltd.

A

B

Z

FIGURE 7-18 An incremental encoder setup with the zero pulse as well as the type of signal it generates to the controller.

rotations made as well as to ensure proper positioning of the system should the encoder or motor need adjustment, replacement, or maintenance.

Absolute optical encoders add enough emitters and receivers, usually four or more in total, to give each position of the encoder its own unique binary address (Figure 7-19). A **binary address** is a unique set of 1s and 0s that the controller can understand and use. Absolute optical encoders have more internal parts, cost more, and require more of the controller's processing power to interact with, but they give the system accurate data about the specific position of the motor as well as the direction in which it is moving. When these kinds of encoders are used, the robot knows where the motor is at any given point instead of just how many pulses it moved from the last point and possibly in which direction (depending on the incremental encoder type). This added level of information makes it much easier for the controller to detect when axes are out of position or when the motor is not moving correctly, thereby allowing for a greater degree of control and monitoring.

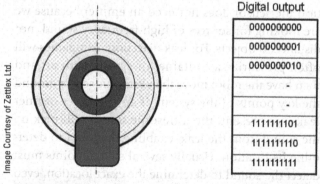

Image Courtesy of Zettlex Ltd.

Digital Output
0000000000
0000000001
0000000010
-
-
1111111101
1111111110
1111111111

FIGURE 7-19 An absolute encoder setup along with the binary data it generates.

Encoders are noncontact sensing methods with a large number of solid-state parts, so they are robust and able to last years. The biggest enemy of encoders

is oil, metal chips, shavings, or any other contaminant that gets inside the unit and either clouds the disk that interacts with the light or damages it. I have removed many encoders and sent them off to the manufacturer for cleaning or rebuilding due to oil in the unit or damage to the disk during machine repairs.

The **global positioning system (GPS)** determines a geographical position based on the time it takes to receive signals from three or four separate satellites in orbit around the Earth. You most likely have used a GPS system if you have gone on a long-distance trip and used an electronic device for navigation. Just as you might use GPS to make sure you are going the right direction and to get turn-by-turn directions, mobile robotic vehicles can use GPS to navigate, providing they can get a signal. For outdoor applications, GPS navigation works well. When indoors, however, most robots need visual references that they can detect with cameras, something a laser can interact with, or devices that broadcast a signal that the robot can interact with for positioning.

There are more ways to track motor position, but these methods are the ones you are most likely to see in the field. Depending on the age of the robot, you may run into resolvers and synchros, which are types of encoders that work on a principle similar to that used in variable transformers. These options were popular for many years but are most likely to be found on older robots. Regardless of the positioning sensor you find in the field, take some time to understand how the system works and which types of signals it generates, as this can save you hours of frustration down the road when you troubleshoot or program the system.

SOUND SENSORS

Hearing—that is, the ability to detect sound—is one of the five major senses. We consider that a disability exists when a person is deaf or can hear only a limited range of sounds. Thus, it should come as no surprise that we have figured out ways to give robots information about sound. As in many other areas, the robot's mechanical nature allows for options in sound detection and manipulation that we mere mortals can only envy. Obviously, not all robots need to deal with sound in any manner to complete their functions, but adding sound sensors opens up new options for operation and environmental response that increases the flexibility of robotic systems.

To add sound detection to a robotic system, all that is needed is a microphone, with the proper hardware, and programming that enables the robot to turn the signal from the microphone into something useful. It truly is that simple. Once we have done this, we can record patterns of sound and use them as triggers for the robot's actions. We can have the robot listen for sounds over a set intensity or decibels, such as what a crash might generate, and use that ability as an added safety feature in the robot. We can also program the robot to recognize a specific set of sounds or voice commands so that the system can start or stop programs based on verbal commands. Once the robot can hear, we have many options to work with.

What if we need to know where a sound originated? With a single microphone installed, it is nearly impossible to determine the source of a sound unless the sound is continuous and the microphone can be moved to different points while taking sound samples. To detect where a sound came from, we add multiple microphones to the robot so the controller can compare signal strengths and determine a probable location for the sound's origin. When robots need to pinpoint or follow sounds, the designers will often add multiple microphones positioned strategically around the robot chassis. Usually designers add microphones in even-numbered pairs, with two being the minimum. Nevertheless, it is not uncommon to have four or six microphones integrated into the system to improve sound detection accuracy.

So far, this discussion has not considered anything especially radical in the world of sound detection, but what if I told you that a robot could use sound to see? Yes, just like bats that chase insects at night or Daredevil (for those readers who are comic book fans), robots can use ultrasonic sensors to detect or "see" objects via the high-frequency sound they reflect back. **Ultrasonic sensors** are similar to the photo eyes or photoelectric prox switches, except that they emit and receive sound instead of light. An ultrasonic sensor consists of a transceiver, a comparator, a detector, and an output device. The transceiver sends and receives the ultrasonic signal. Because it is both a sender and a receiver, it has a "dead" zone located within a few millimeters of the transducer. Signals are sent out in precisely measured pulses at precisely measured intervals, and if a signal should happen to return to the sensor while it is transmitting, the receiver will not detect it. The distance that the dead zone extends from the front of the receiver is determined by the frequency of the signal and the width of the transmitted pulse. The comparator/detector calculates the distance from the front of the sensor by measuring the return time of the pulse, and comparing it to the known velocity of sound. The output can be either a discrete digital signal (on or off) or a digitally encoded measurement of distance or time.

Industrial ultrasonic sensors operate between frequencies of 25 kHz and 500 kHz, and the sensing distance is inversely proportional to frequency. Lower-frequency sensors work at ranges up to 10 m (approximately 33 feet), while higher-frequency sensors have much shorter ranges, typically less than 1 meter at frequencies above 200 kHz. Maximum sensing distance is limited by both the frequency of the sensor signal and the type of material being sensed. Softer sound-absorbing materials are more difficult to detect than denser materials like metal or glass. Liquids can be a problem depending on the temperature of the liquid. Colder liquids are denser, and tend to reflect sound more readily than warmer liquids, but some stand-alone ultrasonic sensors are available for liquid level detection. Notably, the surface temperature of a metallic object can affect sensor range: A hot surface will distort the reflected sound waves, causing problems with the sensor and reducing its effective range. Air temperature and turbulence can also affect an ultrasonic sensor. The velocity of sound decreases at higher temperatures and air currents can cause refraction (bending) of the sound waves.

One use for ultrasonic sensors is the detection of air leaks. Small air leaks emit high-frequency sound that we cannot hear, but the ultrasonic detector can pick them up with ease. Obviously, in this application the sensor does not need an emitter because we are looking for sources of high-frequency sound, not distance to objects. For leak detection, companies will often pressurize a container or system with air and then have the robot move the ultrasonic sensor around the key points of the system. If any pinholes or finer leaks are present, the ultrasonic sensor will pick up the sound from the leak, enabling the robot to determine its location. Usually several spatial points must detect the sound to determine the exact location, even if those points are just before the leak, over it, and just past the leak. This kind of inspection is crucial for tanks and pressure vessels, in which the smallest leak could spell trouble, if not outright disaster.

Many industrial robots work without any type of sound sense, either for detection or for measuring distance. As we look for new ways to make collaborative robots safer, however, we may well see the use of these sensor increase. The realm of voice commands may also expand as the AI abilities of robotic systems improve and users become accustomed to using voice commands in other parts of their lives. If this happens, microphones may become a standard feature on industrial robotic systems.

CONNECTION TO THE ROBOT

When connecting a sensor to the robot, the type of signal generated must be considered. Some DC systems need a specific type of connection, referred to as sinking or sourcing. A **sourcing signal** input provides the positive connection or DC voltage (usually 24 V) to the input module (Figure 7-20). A **sinking signal** input provides the negative connection or ground to the input module (Figure 7-21). The internal switching circuitry of the input unit will provide the other half needed to complete the circuit.

Analog signals are converted to a digital signal called a word, which consists of a number of data bits and a sign bit. The sign indicates if the voltage is positive or negative. The number of data bits determines the accuracy of the conversion and is usually specified by the designer of the controller hardware. The greater the number of data bits, the greater the accuracy, but more data bits also means more cost and more complicated circuitry. Equipment designers usually make a tradeoff between the number of data bits and the desired level of accuracy.

Proper connection is important. Connecting a sinking input unit to a sensor that provides a ground will most likely not cause damage to the unit, but the circuit will not operate. Connecting a sourcing input unit to a sensor that provides 24 V may cause damage to the sensor, the input unit, or both. Even if no damage to the sensor or the input unit occurs, the circuit will not work.

Many input and output cards use optical isolation to protect the internal circuitry from damage in case of improper connection or wrong/excessive power feed. **Optical isolation** works by creating a dead zone where the only connection between the internal electronics and the outside connections is light. An LED lights up when the proper voltage is present, applying light to a photoresistor whose resistance decreases when light hits it. This allows the circuit that includes the resistor to fire, thereby transmitting the signal to the controller. Optical isolation is used primarily for digital signals, but can provide a layer of protection to the main controller, which we can use for digital inputs or outputs.

There are three main things to verify when making connections to or from the robot/controller:

- Make sure it is the right type of voltage, AC or DC.

- Make sure it is the right level of voltage, as too much can cause damage and too little will likely not trigger the input or run the output.

- Make sure it is the right type, digital (on or off) or analog (a range). When the signal is analog, make sure it is the right type of signal for the card it is tied to and the device it works with.

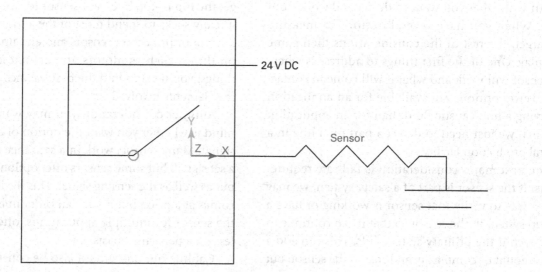

FIGURE 7-20 Wiring of the connection of a sourcing input.

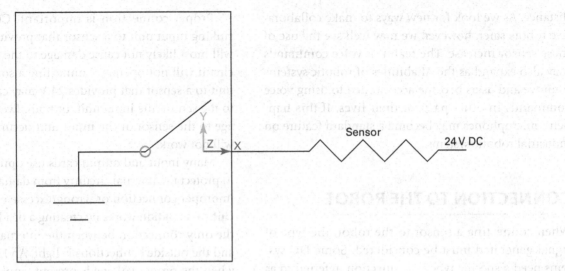

FIGURE 7-21 Hookup of a sinking connection for an input.

If you keep these caveats in mind and pay careful attention, you can avoid most of the major connection issues that occur with sensors.

SENSOR SELECTION CRITERIA

A number of criteria must be considered when selecting a sensor:

- What are you measuring
- Operating environment
- Fail-safe requirements
- Accuracy
- Type of output
- Cost versus performance

Not all sensors work with everything, so it is best to start with what you want to detect and work from there. When you have several options to measure the target, the rest of the considerations then come into play. One of the first things to address is where the sensor will work and what it will come in contact with. Fewer options are available for an application involving a hot, caustic fluid than for an application in which we just need to detect a part on a line in a general production facility.

The next major consideration is fail-safe requirements. If the sensor is part of a safety system, we may need a way to verify that sensor is working or have a backup system in the sensor so that it can continue to work even if the primary system fails. This can add a large amount of complexity and cost to the sensor, but when lives are on the line the cost is worth it.

Next, we should consider accuracy. The better the accuracy of the sensor, the more it typically costs. If we need to answer a basic yes/no type of question, we likely do not need a top-of-the-line, high-end accuracy sensor.

As most sensors offer digital and analog output options, this consideration is fairly low on our list of selection criteria, but still important. Ordering a sensor with the wrong type of output can lead to costly delays, on top of the hassle of having to send the original sensor back to obtain a refund.

The last criterion is the cost of the sensor. Once you have winnowed the list of candidates down based on the other criteria, you will likely have plenty of options in terms of style, manufacturer, and perhaps even type of sensor, so it is time to begin the budget constraints versus "what I really want" battle. While it is nice to get the top-of-the-line, best sensor for the application, it really sucks to spend most of the allotted budget for a repair or project on sensors and end up going cheap on things such as motors and actuators. Like most things, you need to find the best balance among all of the elements involved.

Another consideration you may want to keep in mind is whether you want the option of an adjustable signal. Many sensors work in a set range and generate a set signal, but some sensors offer options for the output as well as the sensing signal. This flexibility usually comes at a price, but if you can order in bulk by using the sensor for multiple applications, often the cost is less on a per-sensor basis.

Cabling options should also be considered. There are two types of sensors in this regard: sensors with quick-disconnect cables and sensors with solid cables (Figure 7-22). Quick-disconnect sensors have a definite advantage if it becomes necessary to replace the sensor. You simply unscrew the connector from the sensor body, replace the sensor, and reattach the connector; there is no need to change any connections on the card or controller. If the sensor has solid cables, it will require extended maintenance time to dismount the sensor, take out all the sensor wiring, and reinstall the sensor and the sensor wiring. Solid-cable sensors have an advantage in hostile environments because there is no mechanical connection to corrode or disintegrate, eliminating a potential source of failure from the system. Because they are sealed units, they are also preferred for applications in which intrinsic safety is a requirement, as the chances of a loose connection are reduced.

As you move into the field and start working with sensors, you will get a feel for what is important and what is simply nice to have in the way of sensor function. Keep in mind that there is always more than one way to measure or quantify whatever it is you are working with. Many times we get into the habit of using one sensor or replacing the old sensor with the exact same one so we can just swap and go, without truly analyzing the process and situation to see if it is

(a)

(b)

FIGURE 7-22 (a) A quick-disconnect connection. (b) A solid-cable system.

the best fit. There is a high probability that the old sensor that finally failed is no longer the best option for that task. You may even find that the old sensor is no longer made, leaving you with the options of digging one up somewhere off a forgotten shelf or updating to a current sensor model. Change in and of itself is not bad, especially when we are thoughtful in how the change is applied.

REVIEW

There are many sensors we did not mention in this chapter, but we have examined some of the devices you are most likely to see in the field. As you continue along your journey of robotic knowledge, do not hesitate to research the new sensors you encounter and learn more about how they work. As the field of robotics continues to mature and grow, you can expect new uses for current sensors as well as new sensing systems to become part of the robotics world. Without sensors, robots are just lumbering creations that have no idea how they affect the world around them. Sensors transform robots from dumb machines into high-tech systems capable of performing astonishing tasks.

During the course of exploring sensors, we covered the following topics:

- **Limit switches.** We looked at one of the basic switches used to sense the presence or absence of objects.

- **Proximity switches.** This section detailed the different types of prox switches available as well as the basics of their operation.

- **Tactile and impact sensors.** We looked at two sides of the force-sensing question along with some of the ways of sensing force.

- **Fluid sensors.** We looked at flow, level, and pressure sensing methods.

- **Position sensors.** We use these sensors to determine where the robot is in space as well as where it is moving to.

- **Sound sensors.** This section covered the standard sound detection sensors as well as ultrasonic sensing.

- **Connection to the robot.** We covered sinking and sourcing signals as well as a few other considerations to keep in mind when hooking up sensors.

- **Sensor selection criteria.** This section had some guidelines to keep in mind when picking out sensors.

KEY TERMS

Absolute optical
  encoder
Bellows gauge
Binary address
Bourdon tube gauge
Calibration
Capacitive level
  sensors
Capacitive proximity
  switch
Capsule gauge
Closed loop
Color detection
Diaphragm gauge
Diffuse
Eddy currents
Float sensors
Flow

Fluid
Global positioning
  system (GPS)
Hall-effect sensors
Impact
Incremental optical
  encoder
Inductive proximity
  switch
Infrared radiation (IR)
  temperature sensors
Level sensing
Light level detection
Limit switch
Magnetic flowmeters
Motor encoders
Negative temperature
  coefficient (NTC)

Noise
Open loop
Optical isolation
Photoelectric
  proximity switch
Piezoelectric pressure
  transducer
Plunger pressure
  gauge
Positive temperature
  coefficient (PTC)
Pressure relief valve
Pressure sensor
Pressure switch
Proximity switch
Receiver
Resistance tempera-
  ture detector (RTD)

Retroreflective
Sail flowmeter
Semiconductor
  temperature
  sensors
Sensing array
Sinking signal
Solid-state device
Sourcing signal
Tactile
Target flowmeters
Thermistor
Thermocouple
Through-beam
Transmitter
Turbine flowmeter
Ultrasonic sensor

REVIEW QUESTIONS

1. If you have a limit switch in the field that has the actuator in the wrong position and a removeable head unit, how would you correct this situation?

2. What do you need to know when selecting a limit switch?

3. What are some of the disadvantages of solid-state devices?

4. What is the sensing range for inductive proximity switches from low power to high power?

5. Describe the operation of an inductive prox switch.

6. Describe the operation of a capacitive prox switch.

7. What is the typical sequence for photoelectric switch operation?

8. Briefly describe how through-beam, retroreflective, and diffuse photoelectric switches operate.

9. What can we use light level photoelectric prox switches to detect?

10. How does a Hall-effect sensor work?

11. Which types of information are we looking for with tactile sensors?

12. Which types of information are we looking for with impact sensors?

13. Which information can we gain from a tactile array that senses 1 or 0?

14. Which kinds of information can a high-end, complex tactile sensor provide to a robot?

15. How do we commonly monitor impact today?

16. What are some things other than an impact that could trigger an alarm when we use current monitoring for impact?

17. What is the benefit of disengaging the axes as opposed to the E-stop method of stopping the robot when impact is detected?

18. What are the differences between thermocouples, RTDs, and thermistors?

19. What are the differences between turbine, target, and magnetic flowmeters?

20. Describe the typical operation of a three-probe capacitive level sensor.

21. What is the difference between a plunger and a Bourdon tube pressure gauge?

22. What is the difference between an open loop control system and a closed loop control system?

23. What are the parts of the optical encoder? How does this device work?

24. How can we determine the direction of rotation with incremental encoders?

25. How does an absolute encoder work?

26. How do ultrasonic sensors work? How do we determine the distance from the sensor to the object detected?

27. What is the difference between a sourcing signal and a sinking signal?

28. What is optical isolation?

29. What are the three main things to verify when making a sensor connection to the robot?

30. What are the six main criteria for selecting a sensor?

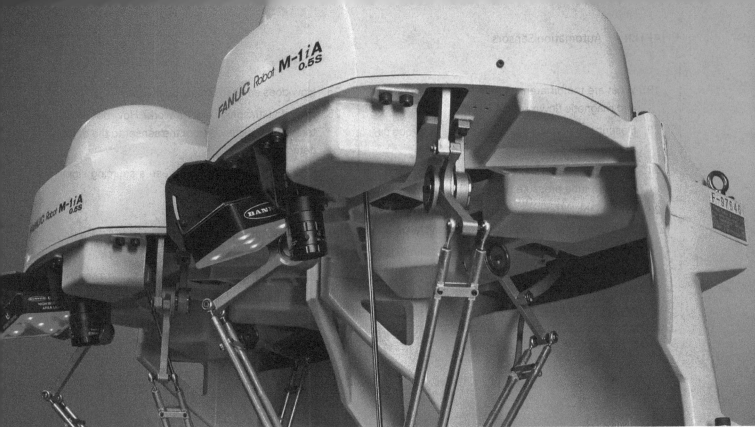

CHAPTER 8

Vision Systems

WHAT YOU WILL LEARN

- Where vision technology began and some of the milestones along its evolution

- The basic parts of a vision system and the role of each part

- How the vision system turns data about light into something useful

- The differences between various types of light sources

- Different placements for illumination and the benefits of each

- How we can use vision systems for measurement

OVERVIEW

Most humans rely heavily on their vision every day to navigate and interact with the world around them. We get up in the morning, grind the gunk out of our eyes, and then spend the rest of the day using them to observe the world around us, find items, identify things, avoid hazards, and so on. To some extent, we utilize visual information in almost every decision we make, which is why the loss of sight is such a detrimental condition. Conversely, many robots work day in and day out with zero visual information about the world around them, instead relying on programming and various sensors to complete their tasks. In this chapter, we focus on what it takes to give sight to a robot and the basics of the hardware involved. During our exploration of robotic vision, we will hit on the following areas:

- History of vision systems
- Components of a vision system
- Image analysis
- Lighting

HISTORY OF VISION SYSTEMS

As with most technology, vision systems did not just suddenly appear in the last decade or two but rather are the result of centuries of human ingenuity and advancement. While it is true that a large number of improvements have occurred over the last few decades, we can track image capturing and enhancement back to a pinhole camera that dates from 500 BCE. The first camera for photography is credited to Loui Daguerre, a French inventor, who created the Daguerreotype camera in 1839. It was followed by the first cathode ray tube scanning device, invented in 1897 by German scientist Karl Ferdinand Braun, who called it the cathode ray oscilloscope. This first **cathode ray tube (CRT)** was a glass envelope that contained an electron gun and a fluorescent screen in a vacuum, and had some means of accelerating and deflecting electrons. Many refinements and improvements of this technology have since been made, but they all work off the basic principle as Braun's CRT.

In 1907, Russian scientist Boris Rosing proposed a method to turn the CRT into a display device that we know and love: the television. Around the same time (1908), but working completely separately, Alan Archibald Campbell-Swinton published a letter on producing television using the CRT; he subsequently expanded on this idea in a lecture in 1911 in London. Swinton's system differed from Rosing's proposal in that the system Swinton described used the CRT to both project *and* record images, thereby laying the groundwork for the first video camera. In 1926, Swinton attempted to prove his theory by projecting an image onto a selenium-coated metal plate that the CRT scanned to. In 1927, Taylor Farnsworth demonstrated the image dissector, the first working video camera that utilized a version of Swinton's experimental system. CRT technology remains in use today, though it is being phased out by other solid-state methods of image manipulation. Many early vision systems utilized CRT technology in the capture process, but the use of these tubes made the cameras fairly large when compared to other image-gathering methods.

Around the same time as the birth of the video camera, from 1924 to 1930, Russian researcher Oleg Vladimirovich Losev published a series of papers that detailed a comprehensive study of the light-emitting diode (LED) and its applications. Losev died at the age of 39 and had no formal training, yet he outlined a solid-state device that has become one of the cornerstones of vision systems and modern lighting.

The earliest machine vision systems date back to the 1940s and 1950s but were primarily used by the military to analyze reconnaissance photograph images and (later) early satellite images. Keep in mind that this was the era of the room-size, relay-based computer, which took days to reprogram in many cases. It was also around this time that the initial research into artificial intelligence (AI) began, though the computers of the day proved ill-suited to this field.

In 1950, the Vidicon tube was introduced by the RCA Corporation; it was developed by P. K. Weimer, S. V. Forgue, and R. R. Goodrich. The target material of this video tube was a photoconductor material, representing a major advancement in the image recording field. Also during the 1950s, James J. Gibson introduced mathematical models that detailed optical flow computations on the pixel level. His models were based on his observations that humans do not analyze the images we receive as raw data but instead compare them to what we know. Gibson's ideas were considered a bit radical at the time, but they laid the foundation for modern machine vision. The 1950s were capped off by Jack Kilby creating the first microchip in 1958. This invention was the basis of our modern computing world and is found in a large variety of devices today. Kilby's discovery sounded the death note for the room-size computer and marked the beginning of the move toward shrinking-size, increased-power processors.

Between 1960 and 1963, Larry Roberts wrote his thesis on how to extract 3D information from 2D views, which would both earn him his PhD and lead many to consider him to be the father of computer vision. Roberts continued to work with computers and advance networking principles for machine communication. His thesis sparked research at MIT's AI laboratory as well as other institutions' research into computer vision of simple objects. Some other notable math came out of the 1960s as well. Jean Serra and Georges Matheron introduced ground-breaking math for dealing with binary images; their work was expanded to grayscale images and led to the creation of a mathematics foundation based on their work in Paris, France. In 1969, Azriel Rozenfield published his book *Picture Processing by Computer*, in which he described many of the vision algorithms still in use today.

On the hardware side, in 1963 C. T. Sah and Frank Wanlass published a conference paper that introduced the CMOS principle to the world—a principle that Wanlass would later patent and RCA Research Laboratories would first produce. CMOS—an acronym for "complementary metal-oxide semiconductor"—is a combination of p-type and n-type metal-oxide semiconductor field-effect transistors (MOSFETs) used to create logic gates and other solid-state circuits. This technology was eventually used to create logic and memory circuits, but it would be decades before vision system would use this technology to replace charge-coupled device (CCD) camera technology, as the early CMOS sensors were inconsistent in their output from image applications.

In 1969, Willard S. Boyle and George E. Smith invented the CCD, which was made up of pixels constructed of metal-oxide semiconductor (MOS) capacitors that converted the light photons falling onto them into electrons. CCDs work similarly to the way a solar panel operates: The photons knock electrons out of the silicon portion of the pixel, and they collect in the pixel's capacitor, creating an analog type of signal. This analog signal is then converted to a digital signal, which the processor can do use. The standard output of a CCD is black and white, but applying a red-, green-, or blue-colored filter to the pixel can turn this output into color information. Only one color can be read by each pixel in this manner, so to get true color the pixels need to be grouped and the information processed carefully.

The 1970s saw several developments related to machine vision. In 1971, William K. Pratt and Harry C. Andrews founded the USC Signal and Image Processing Institute (SIPI), dedicated to advancing image processing. That same year, Thomas B. McCord and James A. Westphal built the first digital camera, which they would patent in 1972. In 1973, Fairchild Semiconductor began commercial production of the CCD, and Fairchild Camera and Instruments Corporation released the first commercially available CCD camera. In 1974, Bryce E. Bayer created the Bayer filter mosaic, which enabled a single CCD or CMOS sensor to capture any color, thereby negating the need for dedicated color pixels. The Bayer filter pattern is 50% green, 25% red, and 25% blue, which mimics the way our eyes perceive color and is the basis of filters we still use today. Fairchild Imaging Systems released the first CCD capable of night vision in 1977.

In 1978, David Marr developed an approach to turn 2D images into 3D images by building them from the bottom up. This method starts with a two-dimensional image, which is then viewed in stereo or from two different locations to create a 2.5D image—that is, an image halfway between the flat 2D image and the full 3D image. The computations and knowledge about shapes are then used to turn the 2.5D image into a 3D image.

The 1980s saw an explosion of devices and maturing of vision technology, as we found better ways to utilize the discoveries made in earlier decades. In 1981, Robert J. Shillman introduced the first industrial optical character recognition system, which gave machines the ability to read. In 1985, the first digital imaging processor was released by Pixar Corporation in California. In 1986, Kodak released the first megapixel CCD, which had a 1.4-megapixel capacity. Of course, the greater the number of pixels, the sharper the image and the greater the amount of data we can collect from an image. In 1987, the first megapixel digital cameras hit store shelves. In 1986, Stanley R. Sternberg introduced his theory on mathematical morphology; it suggested new ways to process grayscale image data, which improves color recognition.

The combination of maturing vision technology, a boom in robotics use by industry, and other advancements in electronics led to machine vision being used widely in industry for robotic and other applications in the 1980s. Advancements in electronics shrank the size of the camera and created single-board image processors, replacing the larger control systems of old (Figure 8-1). The improvements in CCD technology allowed for higher-detail images to be obtained with smaller cameras, increasing the power of the system while reducing the weight and mounting space required. The semiconductor industry was one of the first industrial fields to buy into machine vision, using this technology for inspection and quality control. By no means was it alone in applying this technology.

Unfortunately, these early systems did have their share of problems. The user often had to spend a large amount

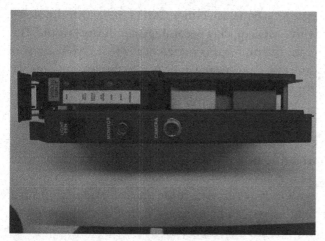

FIGURE 8-1 An example of a machine vision process board—one of the great advancements of the 1980s.

FIGURE 8-2 An older vision system, along with some lenses that it might use.

of time setting up the system, calibrating it, and teaching it what to look for. Lighting was often overlooked in these early applications, which was a sure way to invite failure into the system. Also, as with most emerging technologies, many glitches needed to be worked out. Sometimes the system had trouble processing an image and would have no protocol for response, leaving it stuck in a computing loop. Sometimes the processor simply glitched out, requiring a full reboot or other human intervention. Sometimes the part would be slightly out of position or the lighting would change, causing the system to mistake the image for something else and respond differently than it should. These types of problems would persist well into the current century, causing more than one company to shy away from the technology.

The 1990s was a time of working the bugs out and getting down to business. By the end of the decade, more than 100 companies were selling vision systems, giving consumers a choice in what they used and driving the movement to make the technology more user friendly and reliable (Figure 8-2). We also started to use LEDs for lighting in the 1990s, having learned from the past how lighting can make or break vision processes. In 1994, SanDisk introduced CompactFlash (CF) as a storage medium, which became the go-to technology for storing images, among other kinds of data. Around the same time, personal computers began to increase in power while shrinking in size, creating new options for working with vision systems.

In 1997, the first CMOS digital camera was sold by Sound Vision, based in Boston, Massachusetts. While it took some time before CMOS became a common option for industrial vision, this marked a turning point: CMOS was recognized as a viable imaging technology. During the 1990s, the electronics involved

with vision systems shrank in size, but increased in power. The resolution and number of CCD elements also increased, allowing for better image processing and providing more information to work with.

The new millennium saw a large push toward maturing vision technology and truly utilizing all the computing power and options available. Many manufacturers began offering robots with integrated vision systems, making it easier for users to work with the systems and reducing the problems that can arise from using third-party hardware. In 2005, Sony introduced its first smart camera. This camera housed everything needed to take and process images, and all one needed to do was hook it up to the controller to pass data. By 2008, the technology had improved to the point that we began to work toward 3D image capture, with Microsoft releasing Photosynth, which could create a 3D image from several 2D images.

Today image systems can identify parts, take measurements to gauge quality, create 3D scans, provide their own light and processing, and detect problems that human eyes could never find. The vision systems of today represent a mature technology that is user friendly and rapidly becoming a standard feature of many robotic systems. For instance, the Baxter robot developed by ReThink Robotics relies on a series of cameras that provide a 360° view of its surroundings, allowing Baxter to slow down when it detects people and to find the parts and/or things it is programmed to work with.

If you work with robots, at some point you will likely work with a vision system. When that happens, take a bit of time to learn the specifics of how you set up, monitor, and edit the vision system so that it identifies and responds to objects. The more you know about a vision system's normal operation, the easier it will be to fix any problems that might arise.

COMPONENTS OF A VISION SYSTEM

Now that you have some idea of how the modern vision system evolved, let us take a look at the basic components of the system and what each part does. The specifics of a vision system depend greatly on the manufacturer and the application for which it is designed, but we can break most vision systems down into the following basic components:

- Imaging device
- Lens
- Processor
- Communication
- Light source
- Extras

As we cover the various hardware categories, keep in mind that there is no perfect camera—no camera that fits every application. If there was, then we would all use that camera, and the focus of this chapter would be operation and setup of said system. When selecting hardware for a vision system, it is best to keep in mind what you need the camera to capture and what the working environment of the vision system will be like.

Imaging Device

Imaging devices, or cameras, are the heart of a vision system, where the light is captured and turned into some form of electrical signal that we can transmit (Figure 8-3). Truthfully, you can have the top-of-the-line hardware for the rest of the system and still have a relatively useless system if the camera is junk. In the past, we have used CRT, Vidicon tube, CCD, and CMOS technology to capture pictures, but most of today's

FIGURE 8-3 A close up of a camera used in a work cell to monitor things.

vision systems use either CMOS or CCD technology, with CMOS quickly becoming an industry favorite. The main reason for this preference is that CMOS sensors can embed processing directly into the sensor elements, improving the ability and speed of image processing.

When selecting the camera or imaging device, the first question we need to answer is whether we need color images. In many applications, the answer is no, which makes the system less costly and helps with some color identification issues. Today's vision systems can easily work with color images, and with proper lighting we can highlight or hide features or items to improve the vision process.

The next big question we have to ask is whether we want 1D or 2D scanning. In 1D scanning, also called **line scanning**, the imaging device uses a single line of the CCD or CMOS element to gather the picture. Such a system works well with metals, nonwoven items, paper, and plastics. We can scan larger items by moving the item and continuing to scan the image. This technology is great for unwrapping cylinders for inspection, working in tight spaces where a larger camera might not fit, performing higher-resolution applications, and carrying out continuous-motion inspection.

With 2D scanning, also known as **area scanning**, the imaging device has an array of CCD or CMOS elements, giving the system the ability to take a picture of an entire area or object all at once. The area covered by the image is a function of the size of the array, the type of lens used, and the distance between the camera and the object. The larger the number of elements or pixels in an array, the better the resolution and the more data we can collect about an image, which in turn increases the accuracy of any measurements derived from this data.

Another measure of an array is the number of grayscales it can detect, referred to as the depth of the array. **Grayscale** is any measurement of an element or pixel that is between zero, which is considered a full absence of light, and the maximum output, which is considered full light. Grayscale is how we determine color differences, detect various features, and gather information beyond the basic boundaries of an object. The larger the number of grayscales an array can detect, the finer the details it can pick out and the more closely the captured image will resemble the actual item(s).

With area scanning, we usually take a picture of something static or a moving object that we illuminate with a strobe light to freeze it for image capture. (We discuss lighting later in this chapter.) With 2D scanning, we can capture almost anything we can take a picture of for comparison or analysis.

To create 3D scans, we can use multiple cameras to take pictures of the same object and then process that data through special algorithms. When used this way, the vision system is basically processing visual data in much the same way that we humans interpret the world around us. Each image provides information about the object used to create a 3D representation. We can create 3D images with only one camera, but we have to take multiple pictures of the object from different locations or slowly move the object while using a laser to illuminate it. (We cover the process of laser measurement in the lighting section.) With the multiple-picture method, the system creates the same data as a multiple-camera system but a bit more time is required because the camera or object must be moved for each picture. With the laser method, we create a cloud of points in space that are then stitched together by the system software.

Lens

The **lens** is the part of the camera that focuses the incoming light directly on the CCD or CMOS elements of the camera. Lenses are standard features on vision systems and part of the core cost, but some systems give you the option to change the lens out, allowing for improved image capturing in a variety of situations. To get a good image, we need the right lens, and we need to position the camera at the right distance for the lens installed. To find the approximate focal length for the lens, you can use the following equation:

$$F = \frac{WD \times M}{(1 + M)}$$

F = focal length of the lens

WD = working distance

M = magnification (ratio of the image sensor size, from the camera data sheet, to the required field-of-view size, which is usually application related and describes the size of area to be imaged)

Once we find the focal length, we must ensure that the working distance is greater than this distance, or the lens will not transfer the image to the sensing unit correctly. You may need to adjust either the working distance or the required field-of-view if you plan to use an off-the-shelf lens system. You could perform your calculations and then order a specially made lens, but this will likely add unnecessary time and cost to the vision project.

The power of the lens is a function of the focal length in meters and is measured in units of **diopters**. One diopter is equal to 1/1 meter of focal length. We can use the following formula to find diopters or, more importantly, to find the focal length if we know the diopter value of the lens:

$$P = \frac{1}{F}$$

P = power of the lens in diopters

F = focal length in meters

Other formulas take into account the lens focal length, magnification, refraction, and curvature. They are based on the thickness and curvature of the lens (for those readers who would like to delve deeper into lenses).

If the vision system came already installed from the robot manufacturer or integrator, there is a good chance that the vendor did all the necessary calculations and initial setup, so all you need to do as the end user is to train the system and go. Some integrators can even setup the system so all you have to do is press the go button.

If you decide to use multiple lenses but do not want to worry about the math, you can use any of the free online calculators to help with these calculations, or you can contact a sales representative and let him or her worry about it. If you can move the camera setup or sacrifice some of your field-of-view, then it may be easier to just try several lenses until you find the one that gives the best picture or the desired effect. Just make sure that your camera system has the option of swapping lenses, and that the lens you have in mind will attach to the camera without the need for major system alterations.

Processor

The **imaging system processor** is the portion of the vision system that takes the raw data from the sensors, usually in the form of voltage or current, and converts it into a digital signal that other controllers can use one frame or picture at a time. In the early days of machine vision, this was a separate computer, controller, or a processing card (like the one shown in Figure 8-1). In simple systems, the processor turns the analog signal from the imaging device into a set of binary numbers for processing by another system. Many of today's vision systems utilize a processor that converts the analog signal into binary data, runs image analysis algorithms, and perhaps compares the processed data to parameters or images supplied by the user to determine which information should be sent out. The result could be a set of offsets that enable the robot to pick up a part, an image on a display screen, or a command to initiate a fault sequence for bad parts.

When it comes to the processor, the more it does and the smaller it is, the more you can expect to pay for it. The high-end smart cameras house the processor in the body of the camera assembly and just require a connection to transmit the data to the appropriate controller. This is a great space saver, but it increases the cost of the system in many cases. It may also add weight to the camera assembly, which may cause payload problems in applications in which the camera is mounted on the robot.

Often vision systems supplied by the robot manufacturer have a dedicated card that fits into the robot controller for processing the vision application. These cards tend to be smaller in size than the vision system cards of old and often work with several models of cameras and sometimes even with multiple cameras. Because they are designed to work with the controller, problems with communication rarely arise with these systems. Nevertheless, you may have to set **dip switches** (small on/off-type switches used to set options or parameters) or add **jumpers** (connections between two or more settings pins) to ensure proper operation.

When dealing with third-party vision systems, communication issues are more likely to occur. If the integration of the vision system with the robot is happening in-house, the possibility of problems seems to greatly increase. Often the cost of having a specialist come in and set up the system is well worth the savings in down time and frustration. Also, talking with representatives of the company that sells the system and working with them to find the best vision system for your particular application can prevent a large number of issues before the project starts. When working with third-party systems, the least expensive system is not likely to be the best system for your application: You may very well get what you pay for—a cheap, junk, vision system.

Communication

The **communications** portion of the vision system is responsible for transmitting the refined data from the processor to the robot controller or other control system so it can respond appropriately. In earlier days, this component consisted of an RS232 communication standard or some proprietary pin out cable that required specific wiring on the opposite controller side. Around the turn of the century, the FireWire protocol became popular because it could transmit a large amount of data quickly. Today USB 3.0, coaxial cable, and wireless connections are popular, but by no means the only ways of transmitting vision data. Utilizing the CoaXPress protocol, we can reach transmission speeds of 6 Gbps

(gigabits/second) over coaxial cable. This is followed in speed by USB 3.0 (good for up to 5 Gbps) and then by wireless connections, which top out at 1 Gbps on average.

For the communications portion of the vision system to work correctly, three important things must happen. First, the information must be formatted for transmission in a way that the receiving system can understand. This is typically the job of the software run by the main processor for the vision system.

Second, there has to be a transmission device that works for both the sending and receiving systems. When using multi-pin connection cables, we need to make sure the pins-out connect to the correct pins-in at the receiving system. Many modern systems no longer offer RS232-type connections or support some of the other older connection styles, requiring the use of adaptors or special cables to get the information across when this technology is employed. When using wireless communication, we have to ensure both systems are on the same network and are utilizing the proper protocols for communication. (We will look at networking in Chapter 9.)

> ## Tech Note
> Tightening, removing, or wiggling communication connections under power can cause damage to the system and, more importantly, unexpected startup or rapid robot motion! It is best to adjust, tighten, or add system communications with the power off.

Third, all of the data has to reach the receiving unit in an intact state for storage and/or use. If it does not the system may fault out, behave erratically, crash, or simply stop working due to faulty or partial communication.

Light Source

The **light source** provides a specific type of light that strikes the object in a specific manner to facilitate taking an image. Some cameras have a built-in lighting system, but most vision systems leave the lighting up to the user, as it is best to match it to the application. When choosing a lighting option, the user has to consider several key points:

- *What is the environment like around the vision system?* Is there a lot of dust? Could anything create bright flashes that might interfere with the image capture? Which type of lighting is in use for workers in the area? Could anything else interact with the system?

- *Which type of part are we working with?* Does it have a shiny, matte, or combination finish? Is any part of the object transparent or especially reflective? What color(s) is the part? Do we need to image the whole object or just a small part of it? If the latter, where will that be? What are we looking for specifically with the object? Could anything else cause problems?

  - *What are the specifics of taking the image?* Is the part sitting still or is it moving? If it is moving, how fast is it moving? Is there a chance that what we need to see might be out of view because of the part moving? Could anything else cause a problem?

Once all the data has been gathered, we can begin the process of choosing the right type of light as well as the lighting application. When I was learning about vision at FANUC, one of the instructors told me that most vision problems had nothing to do with the hardware or the software; instead, they could be traced to the lighting. Whole textbooks have been dedicated to this subject, and it might be worth your time to learn more about this topic, especially if you find yourself designing vision system applications. Because of the importance of this topic, we will devote the last section of this chapter to lighting.

Extras

Extras for vision systems include the various options and add-ons that might come with a system—spare or different cables, mounting brackets, display screen(s), storage device(s), network adaptors, automation sensors, filters that alter the light coming in, thermal imaging equipment, or anything else we might add into a vision system. Sometimes to compete with the higher-end vision systems, a manufacturer offers a basic model that takes images and transmits the data, allowing users to add in the options and features they want and avoid paying for things they do not need. In these situations, the wise course of action is to add only the options you truly need, rather than things that just sound interesting. For instance, if the application is to take a picture of a part to verify dimensional quality, which is a measurement function, it would be frivolous to add in **thermography** (thermal imaging capability).

As with any other system involved in robotics, the specifics of the extras for the vision system depend on

the manufacturer and the application for which the system was designed. Given the large number of companies that sell vision systems, it can be a bit intimidating to narrow their offerings down to the best ones for your particular situation. I usually start with the manufacturer of the robot to see if it offers a vision system designed to work specifically with its robots, and then move to the application at hand. If the manufacturer does not offer a vision system or the one it offers does not work well with the application, I then undertake a broader search based on the application. I look for big names like Cognex, Sony, Panasonic, and other companies that have a history in image recording, and start comparing from there. If I run across a company that is new to me, but its system sounds like a good fit, I start looking at online reviews of its equipment and reading up on how long the company has been in business and what its operational philosophy is. If that all seems good, I then make contact and see how easy the company is to work with and how much information it can provide. I can assure you that a company that is hard to deal with before the sale will most likely completely ignore you after the check clears—which is often when you need tech support the most.

IMAGE ANALYSIS

The world of machine vision has seen and benefited from many innovations when it comes to image processing algorithms and techniques. These advancements allow vision systems to take measurements from a two-dimensional image without the need to move the camera, create three-dimensional images from multiple two-dimensional images or groups of data points, and work with colors and light spectra beyond human vision. The good news here is that we can use vision systems without understanding the math going on in the processor to convert the raw data from the sensors into useful information. In this section, we focus on two approaches for turning the raw data into something workable but present a broad overview of them rather than studying the specific equations involved.

Edge detection is the data sorting process in which the processor looks for sharp differences in the light values between elements or pixels and then uses the elements nearby to confirm that it has, indeed, found an edge. This process works best with lighting situations in which there is a stark contrast between the part and the area around the part. The processor compares elements that are beside each other for a difference in value that meets a set threshold amount. Once it finds a set of elements that meet the designated condition, it examines the pixels nearby to see if more of these stark differences are in evidence; it uses a mathematic formula to assign a probability value to those pixels. If enough of the surrounding elements meet the threshold value, the area is considered an edge. If not, the initial reading is considered a false reading or noise and ignored. The greater the number of elements sampled around the first set, the greater the probability of an accurate reading, although this comes at the cost of increased processor usage and time.

Clustering, sometimes known as region growing, sorts data by finding elements that have a similar value and growing the clusters outward from there. One way to do this is to first determine all the pixels that have a different grayscale value than the background. From there, the system uses math to pick a starting point, called the **seed pixel**. Once it has created some kind of mathematical profile for the seed pixel, the system begins looking for pixels around it that share similar properties from a mathematical standpoint, filtered by the specifics of the algorithm. The clusters continue to grow until the processor has mathematically processed all the raw image data.

Once the processor converts the raw data to an image, the next step is to do something with that data. Often this step involves comparing the freshly processed image to either a template or data about the part so the system receiving the vision data can respond accordingly. **Template matching** involves comparing the processed data to the data stored from a previous image. Ideally, the template is a picture of the optimal object in the most probable position, making it easier to compare. If a threshold value, often set by the user, is reached, then the new image is considered to be the right object and the system proceeds as programmed. If the user sets the threshold too low, the wrong parts or bad parts may be missed and get processed. If the user sets the threshold too high, the part quality will increase, but there is also a greater chance that good parts will be rejected because of imaging issues. The template method also tends to have problems when the new image is at a different scale or the object's orientation varies from the template.

Edge and region statistics uses data about unique features of the object, usually selected by the user, for comparison to the newly processed image to determine if it is indeed the object and how it is oriented. This approach helps to overcome some of the problems encountered with template matching and facilitates measurable math comparisons. Features used in this method include the following:

- Center reference point from which all other points are described
- Major part axis
- Minor part axis
- Number of holes
- Angular feature relationships
- Perimeter squared divided by the area, which creates a unique mathematic value that is not affected by scale
- Object texture data

With the edge and region statistics approach, the system again has a threshold value for confirming it has found the desired feature or features, and from there it proceeds as programmed. Ideally the system would check two or more unique features to avoid false positives and negatives.

The **algorithm method** uses a specific mathematical formula to analyze the sample part and then compare newly processed image data to this initial calculation result. Several algorithms are utilized with vision systems, but they work similarly to the edge and region statistics method; that is, they turn the desired image into something mathematically measurable. When an algorithm is used, the user may not need to specify which features to look at or in other ways detail the system's operation.

A great example of the algorithm method is seen with the Baxter robot from ReThink. When the operator teaches the robot what to pick up, the robot uses the built-in cameras to take a picture, which it then runs through internal software to process and store. During operation, the vision system scans the work area. When it finds data matching the stored information, the robot typically responds by offsetting the initial grab point to pick up the item and then continues on as programmed.

Regardless of how the vision system processes data, your responsibilities as the user should be outlined in the literature that comes with the system as well as any training you receive. The more you know about how the system works and what it is looking for, the easier it will be to determine the proper method of lighting and otherwise improve the chances of successful image processing. Knowing the specifics of the vision system in your facility can save you hours or days of frustration in many cases.

LIGHTING

Light is a form of electromagnetic radiation that is broken into three separate ranges. The lowest range is infrared, which is below the visible spectrum and is frequently used as a heat source. The infrared part of the electromagnetic spectrum covers the range from roughly 1 mm to 750 nm. It can be divided into three ranges:

- Far-infrared: 10 μm to 1 mm
- Mid-infrared: 2.5 to 10 μm
- Near-infrared: 750 to 2500 nm

Light in the mid-infrared and near-infrared ranges is most commonly used in vision and photographic applications.

The next range is the visible spectrum of light, from red at the lowest wavelengths to violet at the high end of the spectrum. (You may remember the colors of light in this spectrum from science class as ROY G. BIV.) This range of light is the most critical in machine vision systems and is the part of the spectrum to which the human eye is most sensitive. Visible light wavelengths fall between 380 nm and 760 nm. Just as our eye encodes light information and sends it to our brains, so the camera encodes this light and sends it to the processor, where it becomes usable data. The color wavelengths can be broken out as follows:

- Red: 620 to 750 nm
- Orange: 590 to 620 nm
- Yellow: 570 to 590 nm
- Green: 495 to 570 nm
- Blue: 450 to 495 nm
- Indigo: 440 to 450 nm
- Violet: 380 to 440 nm

Above the range of visible light is the ultraviolet range, which is not commonly used in vision systems. Ultraviolet radiation from the sun is what causes sunburns in the summer. It resides above visible light on the electromagnetic scale but below X-rays and gamma rays (which have their own unique dangers). UV radiation is broken into three ranges:

- Near UV: 315 to 400 nm
- Middle UV: 280 to 315 nm
- Far UV: 180 to 280 nm

The lighting system, including the selection of the light source, is perhaps the most critical part of any machine vision system. It must highlight the features of interest and obscure or minimize any features that are not of interest, while producing a high contrast between the object and its background. It must minimize the effects of natural lighting, work area lighting, and lighting from process sources. Failure in any of the areas can cause huge issues with image acquisition.

Types of Lighting

The following types of lighting are often used with vision systems:

- Incandescent
- Halogen
- High-intensity discharge
- Xenon
- Fluorescent
- Light-emitting diodes
- Laser

We will take a quick look at each type so you can have an idea of their strengths and weaknesses (Figure 8-4).

SOURCE	Strobe	Cost	Intensity	Life	Heat
Incandescent	Poor	Good	Fair	Poor	Poor
Halogen	Poor	Fair	Good	Poor	Poor
HID	Poor	Good	Good	Good	Poor
Xenon	Good	Poor	Good	Fair	Poor
Fluorescent	Poor	Good	Poor	Good	Fair
LED	Good	Good	Good	Good	Good
LASER	Good	Poor	Good	Good	Good

FIGURE 8-4 The various types of lighting covered in this section.

Incandescent lights pass electricity through a tungsten filament or wire, which in turn produces heat and light. By varying the electrical properties of the bulb or the supplied voltage, we can control the amount of light produced. Although once the go-to source for many lighting needs, this type of bulb has fallen out of favor due to its inefficiencies and is giving way to LED lighting. Due to the response time between power applied and light produced, these bulbs do not strobe well. They also tend to degrade or produce lower light levels over time with continued use. This kind of change in the lighting could cause imaging problems and/or require adjustments to the imaging process.

Halogen lights are the evolved form of incandescent lights that utilize a tungsten filament inside a pressurized, gas-filled bulb made of high-silica glass, fused quartz, or aluminosilicate. The high-pressure gas inside the bulb increases the light output considerably, making halogen light a long-time favorite in industry and among filmmakers. On the down side, they produce more heat than a standard incandescent bulb, enough to burn if touched; they have a slow response like incandescent lights, which makes them a poor choice for strobing; and the light produced decays over time. On top of all this, halogen lights can explode! If you touch a halogen bulb, the oils left behind can cause an imbalance in the surface of the bulb, leading to a rupture that can send hot shards flying. Halogen lights are giving way to LED, fluorescent, and HID lighting.

High-intensity discharge (HID) lights use an arc tube filled with gas to vaporize mercury, sodium, metal salts, or other substances. When current flows between two tungsten electrodes, it ionizes the gas and creates heat and pressure inside the tube. As the heat and pressure build, more of the nongas substance vaporizes. This produces an intense light, with metal halide lamps producing up to 1500 watts. HID lights usually have an inner arc tube and an outer protective bulb that helps to stop debris if the arc tube fails and protects the arc tube from the outside world. In some applications, the outer bulb has a coating that helps to improve the color output of the light. Because of the need to establish an arc, these lights use an electronic or magnetic ballast with a capacitor to start the arc and then regulate the current of the system to prevent arc tube damage.

HID lights have a good duration of use, but the light generated depends on the materials used. Many of these lights require several minutes to reach full output, making it impossible to strobe those that require a warm-up time. Also, when you turn off a HID light, it may have to cool for several minutes before you can restart the lighting process. As this kind of light nears the end of its life, the light output may decay and the arc tube may pull excessive current, causing the ballast system to turn off the light, wait for it to cool down, and then restart. HID lights have been used to illuminate industrial settings for years, so you may have to deal with them as part of the ambient light.

Metal halide lamps are specifically used in photography applications. These HID lamps produce the greatest amount of light in the HID family, with their output being close to daylight frequencies. The downside of these lamps is that they are so bright that they can produce too much light, causing problems with image applications where they provide the ambient lighting and we want to create a specific light setting. These lamps also take 5 to 10 minutes to warm up.

Xenon lights also use tungsten electrodes and an arc tube filled with xenon gas to produce intense light. The application dictates the system components from there. When the xenon light is used as a flash lamp, a control circuit dumps charge from a capacitor across the electrodes, creating light that is 10 to 100 times brighter than an incandescent bulb and that lasts from 0.001 to 0.2 second. This flash is often timed with the same pulse that captures the image, effectively freezing fast-moving parts for imaging or simply providing needed light. In the early 1990s, this technology was developed into a HID lamp, used in vehicle headlamps and other applications. When it is used as an HID application, the gap between the electrodes is kept very short, around 4 mm (0.2 inches), and a ballast system is needed to start the arc and control current. Unlike other HID lamps, xenon lamps do not require a warm-up time. The downside to xenon lights is that they tend to have a bluish tint and will degrade over time. Also, the capacitor in the flash circuit must recharge between flashes, potentially adding time to the process.

Fluorescent lights produce light by using high-voltage electricity to excite gases inside a glass tube coated with a phosphor material, which gives off light when struck by the ultraviolet radiation released by the excited gases. Fluorescent lights come in various shapes and sizes. They can provide a diffuse light in colors ranging from cool whites to warm whites, which results in minimal glare from shiny objects. The bulbs are fairly cheap and last a long time, which helps on the cost side of things. We cannot strobe these lights, however, and dimming them is difficult at best. They also degrade over time. The ballast used to excite the gas in the tubes can cause radio interference and even fires when poorly installed or maintained. Of all the disadvantages associated with fluorescent lights, however, their flickering is the most problematic. Flickering can cause headaches, eye strain, a greenish or yellow hue to images, and inconsistent lighting, with all the problems therein.

Light-emitting diodes (LEDs) are a special type of semiconductor with a p-n junction that emits light when current is applied, via energy released as photons when electrons from the n-type material cross the boundary and fill holes at a lower energy level in the p-type material. In recent years, the high-brightness (HB) white LED has opened new avenues and applications for LED lights. These LEDs use semiconductor materials that produce blue or ultraviolet photons; phosphors on the lens are excited by these photons. Some of the blue photons escape and the phosphors emit yellow photons when excited, producing a bright white light. LEDs can create light from the infrared spectrum, through the visible spectrum, and into the ultraviolet spectrum.

LEDs are inexpensive to create, emit specific colors of light, and consume very little power during operation. They run cooler than HID, incandescent, and halogen lighting systems and can be strobed LEDs if needed. The HB-type LEDs have opened up a whole new realm of lighting options and increased the use of LEDs all the more. The control circuit for LEDs keeps the voltage at safe levels and ensures that current flows in one direction only (they are diodes, after all), but the control circuit rarely causes any problems outside the lighting system. On the downside, LEDs are less dependable at temperature extremes. They can generate enough heat to damage their semiconductor material and may require heat sinks or fans to cool them; if the cooling system fails, there is a high probability of LED damage. Moreover, their light will degrade over time. Despite these disadvantages, the LED remains one of the mainstay sources for light in photography, industry, and business.

Laser lights are based on the principle introduced by Gordon Gould in 1957. The word *laser* is actually an acronym for the working principle of the system: light amplification by stimulated emission of radiation (Figure 8-5). Lasers intensify light by circulating light between two mirrors—one curved mirror and one flat, partially reflective mirror (known as the laser resonator). They then amplify the light by use of a gain medium. If we just circulated the light between the two mirrors, it would eventually fade out due to losses each time it is reflected. To prevent these losses,

the laser feeds light or electricity into the gain medium (depending on its source material), which is placed between the two mirrors. The gain medium amplifies the light passing through it, intensifying it to the point that the light can break free of the mirror (the output coupler), which is only partially reflective, and exit the system to do whatever it is the laser does.

Modern diode lasers provide a highly collimated light beam that is easily strobed and can be used to provide three-dimensional images (discussed in detail later in this section). On the downside, lasers can be expensive, they are somewhat fragile, and they tend to produce images with a grainy texture. Safety is also a major issue in any industrial application of lasers, as the intense light from these devices can cause blindness. When working around lasers, you must wear special safety glasses, not just tinted or regular safety glasses.

Lighting Techniques

Now that you have learned about some of the different types of lights used with vision systems, we will focus on the different uses of light, including placement and type of light used to enhance certain features of the part. Be aware that ambient light or the lighting around the vision application may interact with the specific lighting in use and cause unwanted changes. The ideal setup for any vision system is a consistent amount and type of light, even if that light is not placed optimally or the best choice for the application. It is better to have consistent light that is subpar than to have perfect light part of the time and fluctuations the rest of the time. To help shed some light on the subject, we will look at the following lighting techniques:

- Bright field lighting
- Axial diffuse lighting
- Diffuse dome lighting
- Dark field lighting
- Backlighting
- Structured lighting

With **bright field lighting**, the lighting system is applied such that a large portion—if not almost all—of the light from the illumination source is reflected from the object back into the camera (Figure 8-6). The most common application of this technique is the **partial bright field**, in which one or two light sources are set at a greater than 45° angle to the object so that most of the light transmitted is reflected into the camera lens

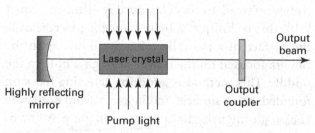

FIGURE 8-5 While oversimplified, this diagram shows the basic operating principle of a laser.

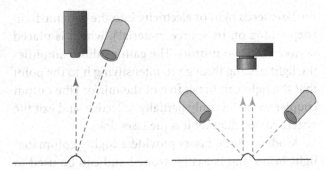

FIGURE 8-6 Two common configurations used for partial bright field illumination.

(Figure 8-7). This setup provides great contrast, illuminates details, and can enhance topographical details. There may be glare from shiny surfaces, however, and feature-obscuring shadows may occur with certain geometries. To address the glare problem, we can either adjust the angle of the light to reduce the amount reflected or increase the working distance between the light and the object. To minimize shadows, we can add a second light. For moving parts, we often strobe the light to prevent ghost images or smearing.

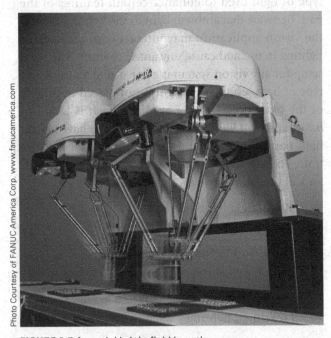

FIGURE 8-7 A partial bright field in action.

To get **full bright field lighting**, we use flat array lights (Figure 8-8), axial diffuse lighting, or diffuse dome lighting to rain light down on the object and return the maximum amount of light to the camera. This technique is commonly used with reflective objects that may have mirror-like qualities or objects

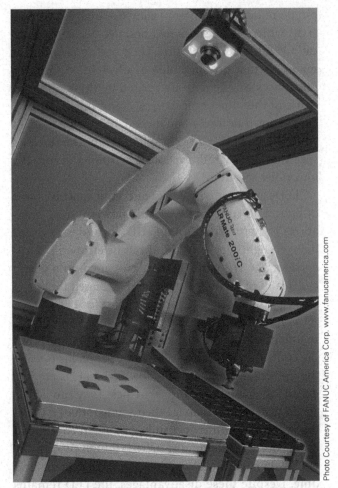

FIGURE 8-8 If you look closely, you can see the camera in the middle of a flat light plate, which is designed to provide a full bright field illumination.

for which even but multidirectional light is needed for their illumination.

Axial diffuse lighting (Figure 8-9) uses a light source set at 90° to the part; a mirror in line with the camera reflects the light straight down to the part, and then allows the returning light to pass through to the camera. This technique is good for detecting flaws in flat, shiny surfaces and illuminating small cavities, but the mirror can produce a double image if it is too thick.

Diffuse dome lighting (Figure 8-10) is sometimes referred to as "cloudy day illumination." With this technique, a light source is placed inside the surface of a partially reflective dome, with the camera focused on the object through a hole in the middle. This method is good for detecting flaws on rounded, shiny surfaces and helps to avoid **hot spots** (i.e., high intensity light spots) and glare. A downside of diffuse dome lighting is the dead spot of light where the camera hole is. In addition, this method

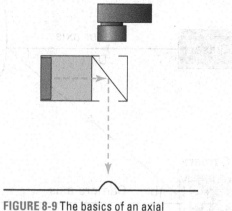

FIGURE 8-9 The basics of an axial diffuse lighting system.

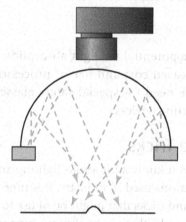

FIGURE 8-10 The basics of a diffuse dome lighting system.

does not deliver the same intensity of light as the other methods. With all three full bright field methods, the lighting source needs to be close to the part to get the full effect.

In **dark field lighting**, most of the light from the light source is reflected away from the camera, with the lighting usually at an angle of less than 45° or focused off the object (Figure 8-11). With this lighting style, defects and certain features of the object catch the light

and send it to the camera, making them show up in an otherwise dark background. This method, which avoids issues with hot spots, is very good for detecting edges and mapping the topography of an object. It can also easily differentiate between a hole or other such feature and a dark-colored spot on the object—a task that is sometimes difficult with bright field methods. Just as with the bright field techniques, some features may block the light and prevent reflection from areas of the part.

Backlighting techniques provide the greatest contrast, as the light source is placed opposite to the camera, with the object between the two (Figure 8-12). This type of lighting is a great choice for edge detection. It creates a silhouette of the object without most of the surface details, other than a bit of texturing at the edges. This method is good for inspecting and measuring certain object dimensions, determining the presence or absence of features that can be backlit (such as holes), and finding an object's location in the illumination field. Backlighting is used for specific applications, rather than being a general-purpose lighting application.

Structured lighting is a variation of partial front lighting that typically relies on a laser. Specifically, a fiber-optic line light is used as the light source to capture object features or measure specific points. Structured lights cover a specific, targetable area instead of creating a wide area or large spot effect like most other lighting methods, allowing the user to illuminate only the part or area of the object that is of interest. Some fiber-optic lighting systems are flexible, allowing the user to move and adjust the light as needed without having to modify light fixtures or other components. Fiber-optic light is fast and essentially heat free, but the fiber-optic material is made of glass. This glass can crack or break when exposed to pressure or vibration, disrupting the flow of light.

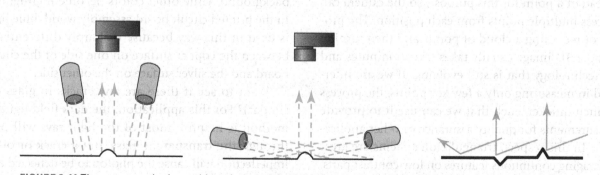

FIGURE 8-11 These two methods would produce dark field lighting. Notice how the light is reflected to the camera.

FIGURE 8-12 This backlighting light source is built into and under the primary work area for a FANUC delta-style robot. The camera is at the top of the unit, just underneath the FANUC logo.

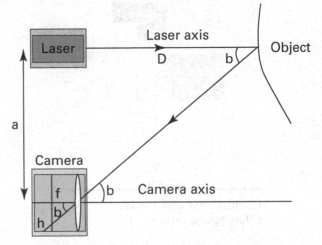

FIGURE 8-13 The setup to take measurements using a laser and camera.

If we use a laser and set up a right triangle in which the laser, camera, and object are the three corners, we can triangulate the distance of the recorded light for measurement or as part of a point cloud used to create a three-dimensional image (Figure 8-13). To create multiple points, we have to move the camera and laser setup or move the part to reach and map all the various points. Some systems use a laser line instead of a point for this purpose, so the camera can process multiple points from each position. The process of recording a cloud of points and then turning it into a 3D image usually takes several minutes and is a technology that is still evolving. If we are interested in measuring only a few key points, the process is much quicker, such that we can use it to provide measurements for quality assurance or other applications. In this respect, triangulation systems are great for gauging continuous features on low-contrast parts. The disadvantages with using laser light in this way

include the potential for light absorption by certain parts, increased cost and use of processor capability, and the need for special safety glasses or other light-protecting devices.

Tips and Tricks

Now that you know a bit about lighting and some of the applications used in industry, it is time to provide a few tips and tricks that might be of use to you in the field. Obviously, there is no way to cover every possible problem you might encounter or every way to deal with these problems, but we will hit some high points here.

Need to look for missing material? A good solution is to use axial diffuse bright field illumination. The missing or rough spots on a part will appear as dark spots in the image with this lighting method, as long as you set it up correctly.

Looking for a specific colored feature? By using a specific wavelength of light, or color, you can hide or accent features. For example, if you bathe the work area in red light, anything red will likely fade into the background while other colors become highlighted. In the printed circuit board assembly world, blue light is used in this way because it sharply differentiates between the copper surface on one side of the circuit board and the silver surface on the other side.

Want to see if there are any cracks in glass (or similar)? For this application, the dark field lighting method is helpful. Most of the light rays will pass through the transparent glass, but a crack or other imperfection will cause the photon to be deflected and reach the camera.

Need to look inside blister or other clear packaging? The dome diffuse lighting technique would be a good fit here. The light coming from multiple angles eliminates reflections and shadows that could otherwise cause issues.

Have a highly reflective part? Infrared light eliminates reflections. It is a different grayscale value than the ambient light, so the system ignores the reflecting ambient light in the image. This method also works well when you have multiple colored objects and you want to reduce them to a more uniform shade.

Looking for fluorescent material? Certain types of ink, glues, and other organic material will emit fluorescent light when exposed to ultraviolet light. When using ultraviolet lighting, filters should be added to the camera to block ultraviolet light and allow only the fluorescent wavelengths into the camera.

Ultimately, if the method or lighting system you are using is not producing the desired results, you need to do something different. There are plenty of options out there and many vendors that would gladly help you find a solution. The Internet has a wealth of information about lighting and lighting techniques—all you have to do is start searching. You may even find help in the form of another department or manufacturing facility that can offer some sage advice. The point here is help and information are out there, but you have to look for them or ask for them.

REVIEW

Vision systems have grown in popularity to the point that many robot manufacturers provide them as a standard feature. While this chapter has covered the basics of these systems, each system has its own strengths, weaknesses, and requirements for proper operation. Taking the time to learn the specifics of the system(s) you work with is time well spent and will save you from the depths of frustration later on. Be aware that many imaging problems relate to the lighting, more than the camera, and that stable, consistent lighting is a necessity. When set up and used correctly, vision systems are a great benefit. Conversely, when they are applied incorrectly or when only minimal thought has gone into the process, they can become a nightmare that never works right.

During our exploration of vision, we covered the following topics:

- **History of vision systems.** This section gave some brief background information on the people and inventions that led to the modern vision system.

- **Components of a vision system.** We looked at the various pieces of a vision system and discovered what each part does.

- **Image analysis.** We looked at the basics of image analysis and explored a few ways to process images.

- **Lighting.** This section covered both light sources and techniques for lighting parts. It serves as a starting point for exploring the complexities of lighting.

KEY TERMS

Algorithm method	Charge-coupled device (CCD)	Diopters	Grayscale
Area scanning		Dip switches	Halogen lights
Axial diffuse lighting	Clustering	Edge and region statistics	High-intensity discharge (HID) light
Backlighting	CMOS		
Bright field lighting	Communication	Edge detection	Hot spots
Cathode ray tube (CRT)	Dark field lighting	Fluorescent lights	Imaging device
	Diffuse dome lighting	Full bright field lighting	

KEY TERMS

Imaging system processor	Lens	Metal halide lamp	Template matching
Incandescent lights	Light-emitting diode (LED)	Partial bright field	Thermography
Jumpers	Light source	Seed pixel	Xenon lights
Laser light	Line scanning	Structured lighting	

REVIEW QUESTIONS

1. Who invented the first camera? The first CRT? When were these discoveries made?

2. What is Oleg Vladimirovich Losev's claim to fame?

3. Who invented the complementary metal-oxide semiconductor?

4. Who invented the charge-coupled device?

5. What does the Bayer filter do?

6. What are the main components of a vision system?

7. Which kinds of technology have been used in the past in vision systems to capture images? Which kinds of technology are used today?

8. What is the difference between line scanning and area scanning?

9. How can vision systems create 3D images?

10. What is the importance of the focal length of a lens?

11. What does the imaging system processor do?

12. What is the difference between edge detection and clustering?

13. What are some of the features we might look for with edge and region statistical matching?

14. Which of the lights covered in this chapter can be strobed?

15. Which of the lights covered in the chapter generate large amounts of heat?

16. Which type of light is the go-to light for many vision applications? What are the pros and cons of this type of light?

17. What does the term *laser* stand for? How do lasers operate?

18. What is the difference between bright field lighting and dark field lighting?

19. How does backlighting work, and what is it good for?

20. How can we take measurements using a laser with a vision system?

21. What can we do to improve a vision system's chance of finding a specific color?

22. Which lighting method would be best to look inside of clear packaging to detect items?

23. What are two problems that infrared light can help to eliminate?

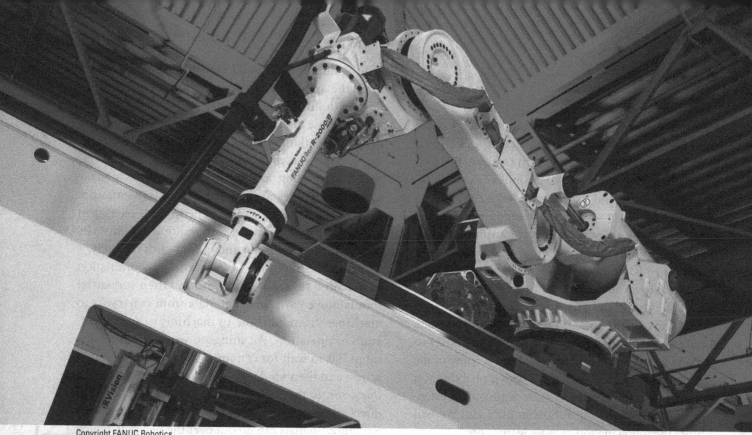

CHAPTER 9

Integration and Networking

WHAT YOU WILL LEARN

- What an industrial network is
- Why we use networks
- The common wiring topologies
- The basics of common industrial recommended standards
- The basics of several common industrial communication protocols
- Some things to keep in mind when integrating the various components of a network

OVERVIEW

Many attribute the saying "Knowledge is power" to British philosopher Francis Bacon, who supposedly made this statement in 1597, although Shakespeare used a similar quote and the Old Testament expresses this idea as well (Titelman, 2000). Over the millennia, our ability to gather, infer, and transfer knowledge has led us to where we are today—a modern world in which many devices have the ability to store and transmit information about the equipment itself as well as the world around it in some cases. Entire professions are dedicated to the transmission of information and ensuring all the data flows as it should when individuals run cables, using specialized equipment to route data transfers, and creating rules for how the data moves so nothing is lost. In this chapter, we look at some of the basics of data transfer, as there is a high probability you will work with this process to some degree, if not extensively, as you work with robots in the field. We will explore the following topics as we learn more:

- What is a network?
- Types of networks
- Communication protocols
- Integration

WHAT IS A NETWORK?

When you hear the word "network," what comes to mind? Perhaps a bunch of interconnected lines or a web? Maybe a group of people who know one another or work in the same profession? Any of those ideas is correct in terms of the word's definition, but this chapter focuses on the computer and data side of networks. For the sake of this chapter, a **network** is the connection of two or more devices for the purpose of information sharing. This simple definition often encompasses days of setup, hours of troubleshooting when something goes wrong, and a full-time job in most facilities. Some networks use cables with communication modules, whereas others work wirelessly via either built-in technology or an added wireless card. (We will get into the specifics in the next section.) Regardless of the physical makeup of the network, the goal is to create a path for the flow of information to maximize operation, data gathering, and troubleshooting.

Now that you have some idea what a network is, you may be asking yourself, what is it good for? One great feature of a network is the ability to share resources. One computer that holds the master set

of programming can send new programs or updates to all the appropriate equipment on the production floor without the need to physically connect to each machine one at a time—which was our only option before networking became available. This is great way to make sure everything is running the correct and current programming, as there is only one place to make whatever changes are needed before transmission. Another example of the value of networks is seen when one processor or controller runs multiple machines by connecting to their input and output cards remotely.

Another common use of networks is coordination of machine function and motion. Often industrial robots move parts and materials from conveyors to machines, from machine to machine, or from operation to operation. For things to work smoothly, the robot has to wait for certain signals before initiating action and then send a go signal to equipment when it is done. In the past, this required a large amount of direct wiring between inputs and outputs. Today, however, those signals can go out over the network, which usually requires one wired connection at most. With highly automated facilities, we can go a step further and actually start and monitor production runs remotely. One engineer I know automated a facility that made industrial cleaners to the point that he started production runs from his house with his cell phone.

Of course, another key aspect of networks is the transmission of data. How many parts has a machine produced, how long did it take, how many are left to fill the order, and just about any other pertinent data can be shared via the network. Management loves this ability to see the metrics of a manufacturing process in real time in a format that is easy to save, manipulate, and transfer. From the technician's standpoint, a newer trend helps greatly with the troubleshooting process— namely, the ability to send error codes and alarm histories. In the traditional model of operation, something goes wrong with the equipment, the operator notices/ identifies the problem, and a call goes out to the maintenance person, who in turn visits the machine to fix it. This process can take anywhere from a few minutes to hours, depending on what is involved. With proper networking and machine setup, at the same time the machine lights the alarm light or displays an alarm, a message goes out via the network to the appropriate person, alerting that person to the problem with the system. Sometimes this leads to the maintenance person or engineering staff showing up to fix things before the operator even knows there is a problem. The message going out may be as simple as the alarm code or a simple message containing the machine ID and

"I'm broke," or it may be complex enough to include a snapshot of where the machine was in the program, what it was doing, the state of inputs and outputs, and/or any other history data of importance.

Another new trend thanks to networking is remote maintenance. Sometimes equipment problems stump the in-house repair staff, and they need help from the manufacturer. In the past, the only options at this point were to make a phone call with the hope that step would be enough or to hire someone from the company to come to the facility at great cost. Through a network connection, someone halfway across the country or the world can gain access to the machine and look at the program, data tables, and other machine information in an effort to fix things or direct repairs. Obviously, the network must tie to the Internet for this approach to work, but many facilities already have this capability built in. The downside of connecting the plant network to the Internet is the danger of someone hacking the system and doing damage or some form of computer virus wreaking havoc.

The specifics of the network connections and the data flowing along those paths are primarily based on the application, the user's desired data, and the experience of the technician(s) involved in the setup and maintenance. The greater the amount of data required and the faster the response, the more complex the system and operating rules. If the persons installing the network are more comfortable with one type of network over another, they will naturally gravitate to that style, which often sets the network style used in the facility from that point onward. We can mix different types of networks that work under different sets of communication rules but that often leads to unique problems, lots of downtime, and frustration. Sometimes we have to mix and match because older equipment can communicate only with certain styles of networks and specific communication hardware; in those cases, the frustration is just a cost of doing business. One thing is for certain: Networks have carved out a place in industry and there is no sign of them going away anytime in the near future.

TYPES OF NETWORKS

When it comes to networks, there are five primary configurations for the system between nodes. A **node** is anything tied into the network, such as controllers, smart sensors with a processor, communication modules, smart switches that help with information transfer, computers, programmable logic controllers (PLCs), or anything else that can send and receive data.

The primary configurations or topologies for nodes are bus, ring, star (also known as hub-and-spoke), tree, and mesh. Each has its own strengths and weaknesses, which we will look at in greater detail. The data sent from node to node, referred to as a **packet**, includes the address of the sending node, the address of the receiving node, and the transmitted data.

Network Topology

A **bus topology** network, also called a linear bus, has one main cable called a **trunk cable**, with terminating resistors attached at each end (Figure 9-1). The trunk cable carries data and power, with physical and electrical connections called a **tap** being found at various points along the trunk cable. The tap is connected to a **drop cable**. The drop cable may, in turn, have several taps along its length, allowing for node connections. In this type of network, data is sent along the trunk, and all nodes will receive the information. Specific address data is coded into the packet, so that only the node with the matching address can actually read and use the data. The advantages of a bus-type network are the ease of adding or removing nodes, the easy setup, and the use of only small amounts of wire. The disadvantages are that a short or opening in the wire will cause part or all of the network to become inoperative; as you add either more nodes or length to the bus, the data transmission speed will slow down; and such a system is easy to crash.

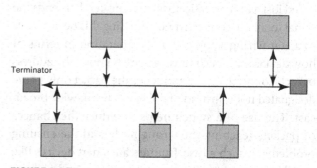

FIGURE 9-1 The basic configuration of nodes in a bus topology.

A **ring topology** network has all the nodes connected in series to form a continuous loop or ring (Figure 9-2). This topology often uses a **token**, which is a bit of code that allows a node to transmit information. The token is passed from one node to the next along the ring. If the node with the token does not have data to send, then it just passes the token along. This prevents multiple transmission and garbling of data. The standard packet is used when data is transmitted. This system has the advantage that information can travel in both directions, thereby reaching the intended node even if there is a break in the line.

The downside to this topology is slow transmission of data, as the token cycles through the network, and the requirement for a large amount of cabling.

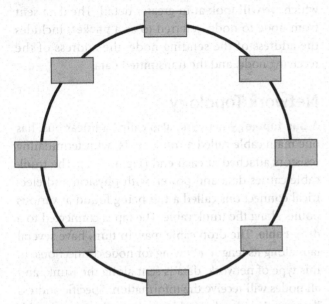

FIGURE 9-2 The basic layout of a ring topology. While in the field, keep in mind that the ring-style connection simply means a cable in and a cable out of each node.

A **star topology** network uses a central connection point known as the hub node, which has connections (often hard-wired) leading to each separate device on the network (Figure 9-3). For the hub node, we can use a network hub, switch, or router. A network hub, usually called just a **hub**, is basically just a repeater that sends the signal to all the connected nodes. Using this device leads to a star wiring setup but a bus operation in terms of how data is sent. In contrast, a **switch** knows the address of each connected node and sends the packet only to the designated node, instead of doing a network-wide broadcast. The use of a switch helps to reduce the chances of packets reaching the wrong node and the ensuing problems in such a case. **Routers** are smart devices like switches; their connections can be either wired or wireless, depending on the application. When working in a wired system, routers can send information to specific nodes. When they are used wirelessly, the only sorting that occurs is to make sure the information is specific to the network of devices connected to the router. The advantages of the star topology include easy setup, easy troubleshooting, and the fact that one bad cable does not crash the network. On the downside, this method, when cabled, uses a lot of wire. Moreover, if we lose the hub node, that entire network section goes down.

A **tree topology** network is the combination of a bus network and multiple star networks, usually comprising three levels of devices (Figure 9-4). The first

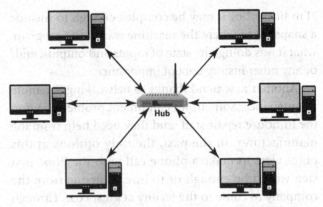

FIGURE 9-3 The basic idea behind the star topology. While in the field, remember that each connected device uses one cable for communication when wired.

level or lowest level is the devices connected to the hub nodes; the second level or next level up is the hub nodes; and the third level or top level is the main hub that controls the flow of data in the network. This type of network configuration is easy to expand and supports more devices than a pure bus, which is limited by transmission traffic, or a pure star, which is limited by the number of connections a hub can manage. The disadvantages of this topology include the initial setup cost, the large amounts of cabling, and the fact that a tier three node failure kills the whole system.

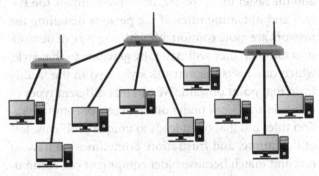

FIGURE 9-4 The lower two levels of a tree topology. It is missing the usual third tier, which would be the main hub.

A **mesh topology** network consists of multiple connections, where each node is connected to all or several nodes, with the ability to share information along those connections (Figure 9-5). If the topology is a full mesh setup, each node connects with every other node. By comparison, a partial mesh topology has nodes that connect to only three or four other nodes instead of all the other nodes. Although wired mesh networks do exist, this has become the preferred method of setting up wireless networks. The beauty of this topology is that signals can take multiple routes, making the network self-healing when a connection fails. Most often, setup simply requires getting the device hooked into

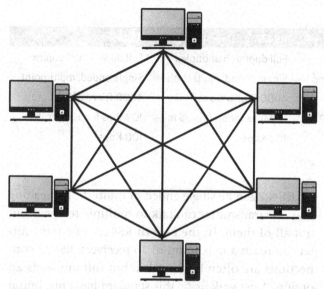

FIGURE 9-5 A full mesh topology configuration.

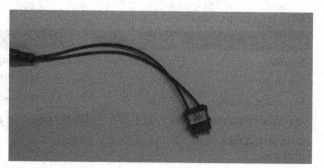

FIGURE 9-6 A fiber-optic cable. If you shine a light on one end, you can see the light through the pins on the other end if the cable is undamaged.

the **local area network (LAN)**. (A LAN covers a small area such as a business or manufacturing facility.) This is also a great way to bring Internet connectivity to areas where it would otherwise be unavailable, as the information is passed along the network from node to node as needed. Local information can pass directly from node to node without the need to route it through unnecessary hubs or nodes, speeding up communication. On the downside, if this system is hard-wired, there is a huge amount of cabling to run. In fact, sometimes the initial setup can get complicated regardless of whether the network is wired or wireless.

Network Connections

While we are on the topic of networks, this is a good time to talk a bit about how we connect nodes to the network. In a wireless network, each node connects using built-in wireless technology, a wireless network card, or physical connection to a device that is connected to the wireless network. For wired connections, the first broad choice is either copper wires or fiber-optic cabling. Fiber-optic cabling provides for faster data transmission, as it uses light instead of electricity to transmit data, but copper wires are more common because they are cheaper and easier to repair.

Fiber-optic cables have a central core made of glass or plastic (Figure 9-6). Data is transmitted in the form of a laser light signal through the length of the cable, with equipment on the sending end that converts data into light and similar equipment on the receiving end that converts the light back into digital data. The cable consists of a single fiber-optic core surrounded by a cladding protective material, which in turn is

surrounded by an insulating jacket. Fiber-optic cables are more expensive than copper conductors and harder to work with, but they can transmit large volumes of data over long distances, they are immune to electromagnetic noise, and signals move at the speed of light.

When working with fiber-optic cables, extreme caution should be used, as a bend radius that is too small may result in a **kink** in the cable—that is, a sharp bend that causes damage to the fiber-optic material. Another concern is any impact, such as a part falling on the cable, or excessive pressure, which occurs when we step on them; this can damage the cable and prevent light transmission. Damaged cables can be repaired, but it often requires specialized equipment, a skilled technician, and a fair amount of time. Often a damaged cable is simply replaced, which incurs the cost of a new cable and the amount of time needed to swap out the bad cable.

Data Transmission
Recommended Standards

As mentioned previously, copper cables are the most common method of wiring nodes into the network, and there are several options and configurations of cable to choose from. To pick the right cable, we need to know how many wires are needed, how they connect to the node and hub or tap, and how far the data has to travel. In an effort to make things easier, several **Recommended Standards (RS)** have been developed that designate specific pin connections and connector types to aid in the transmission of data. Because these are only recommendations, there is no governing body to make sure everyone uses the standard as it is written and there are no repercussions for changing the pin connections. In other words, just because we have a cable with the right end connectors, that does not mean we will have proper communications. The common Recommended Standards found in industry are RS232, RS422, RS423, and RS485 (Table 9-1), with the RS485 being the more current and popular of the group.

TABLE 9-1 Four Recommended Standards

Standard	RS232	RS423	RS422	RS485
Duplex	Full duplex	Full duplex, half duplex	Full duplex, half duplex	Full duplex, half duplex
Topology	Single ended	Single ended, multi-drop	Single ended, multi-drop	Single ended, multi-point
Maximum distance	50 ft or 15 m	4000 ft or 1200 m	4000 ft or 1200 m	4000 ft or 1200 m
Maximum data rate	19.2 Kbps	100 Kbps for 40 ft or 12 m	10 Mbps for 50 ft or 15 m	10 Mbps for 50 ft or 15 m
Data rate at maximum distance	19.2 Kbps	1 Kbps	100 Kbps	100 Kbps

How the system transmits signals also has an impact on the cable, as some methods require more wire than others. **Simplex** transmissions flow in only one direction, with no way to verify whether the data was received, and are rarely used in networks. **Duplex** transmissions flow in both directions, overcoming the problems of simplex transmission, which is why this is the preferred transmission method for networks. Duplex transmissions are further broken down into **full duplex**, in which data is transmitted and received by devices simultaneously, or **half duplex**, in which the system transmits or receives data but does not do both at once. In simple systems, there is one wire per connection for half duplex transmission and two for full duplex transmission, but systems that use reference voltage to help with electromagnetic noise double the number of wires to two for half duplex and four for full duplex.

Noise—that is, voltage induced into the line by magnetic fields or radio waves—can create false signals and alter data transmissions. One way to minimize noise is to use **twisted pairs**, which is simply spiraling two wires together. With this arrangement, the induced voltage happens in both wires at the same time, but in opposite directions of positive and negative, so it cancels out. If twisted pairs do not provide enough protection or if only one wire is used, we can put **shielding** around the wires in the form of wire mesh or metal foil that is connected to ground. This creates a low-resistance path for the inductive force to travel down, thereby preventing it from passing through the wire and causing problems.

RS232 was introduced in 1962 and can still be found in industry today, though it is no longer the most popular method of data transmission. The full duplex communication method can send data up to 50 feet (15 m) at a maximum speed of 19.2 Kbps (kilobits per second). It is **single ended**, meaning that one transmitter sends signals to one receiver. **RS423** is similar to RS232 but adds half duplex operation and transmission speeds up to 100 Kbps out to 40 feet (12 m), which dwindles to 1 Kbps at the maximum range of 4000 feet (1200 m). RS423 is rarely used in industrial applications.

RS422 can be single ended or **multi-drop**, meaning that one transmitter can talk to multiple receivers but not all of them. In the case of RS422, one transmitter can reach a maximum of 10 receivers. RS422 connections are often half duplex, but full duplex is an option. A network using this standard has a maximum transmission rate of 10 Mbps (megabits per second) for 50 feet (15 m), which decreases to 100 Kbps at 4000 feet (1200 m). RS422 has been used as an extension cord to transmit RS232 communications over long distance, with the use of inline converters, and is an industrial standard still in use.

RS485 is the current favorite and is the next step up from RS422. It can handle true **multi-point** communication, in which a transmitter can talk to any receiver. A total of 32 transmitters and receivers may be present on the same cable, and every transmitter has access to all the receivers on the cable. RS485 is often half duplex in operation, although full duplex communication is an option. This standard provides the same transmission speeds as with RS422.

Cabling

For the various recommended standards, we can use special cables with the correct number of wires or Cat5 (Category 5) and higher cable. Both options use twisted pairs, come in shielded and unshielded varieties, and have various options for number of wires. You can buy them as rolls or as ready-to-go cables with connectors already attached. Generally speaking, the higher-end cables specifically designed for RS485 connections tend to have 120 ohms of resistance on the wires, whereas the Cat family tends to run around 100 ohms. Thus, when we use Cat cable, which is typically used for Ethernet communications, the amperage transmitted over the wires is slightly higher and the voltage drop on the wires is slightly less, which in turn means a slightly higher voltage reaches the receiver. This can lead to a false signal in certain conditions but often is easy to adjust on the device through a setting. There is a good chance that using Cat cable instead of a specialty type of cable will not cause any problems, as the difference in the two is very slight. Cat cables generally have 8 or 10 connectors, which may be too few for an RS485 application.

Other types of wiring you may encounter in industry include coaxial cable and ribbon cable. **Coaxial cable** consists of a single copper wire conductor surrounded by a layer of insulation, which in turn is surrounded by a braided metal shield that is connected to ground, which in turn is covered by an outer jacket (Figure 9-7). The center wire is the electrical conductor through which all network data is transmitted, and the braided shield prevents electric noise from corrupting the data. These cables are often used with slower Ethernet protocols or camera-type applications.

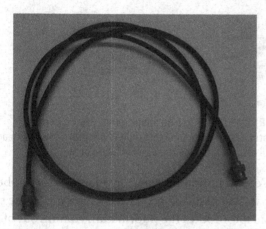

FIGURE 9-7 A coaxial cable. You may have a similar cable bringing satellite or cable TV services into your TV.

A **ribbon cable** consists of a series of conductors laid parallel to each other in a common insulator, with one edge of the cable marked with a red stripe that usually connects to pin 1 on the connector (Figure 9-8). Since the conductors are parallel to each other and the cable is flat, it resembles a ribbon—hence the name. Ribbon cables are used widely in computer systems, but because the data transmission speed in a ribbon cable

FIGURE 9-8 Ribbon cables. Notice how the one on the left has the red marker for reference.

is relatively low, typically less than 500 Mbps, they are restricted to fairly short runs of a few feet or less.

Connectors

Of course, a cable is worthless to the network if it cannot connect to the devices at both ends. In some applications, we simply tighten the ends of the wires under screws and go about setting up the network, but this method takes a bit of time and creates the potential for improper connections. To combat this and make our lives easier, we can use **connectors**—devices that are designed to fasten together and make electrical connections through cable wires being attached to their numbered internal pins. These connectors are built into many devices. The male connectors (sometimes called plugs) have pins that stick out, and the female connectors (sometimes called jacks) receive the pins, giving us an easy way to connect the device to the network. While certain standards tend to use specific connector types, the chaotic jumble of old and new equipment in production facilities has blurred the lines to the point that you can get almost any type of connector on either end of a cable to make communications easier.

The DB or D-sub style of connector was very popular for many years. These connectors come as DB9, HD15, and DB25 options, with the number designating how many pins are in the connector (Figure 9-9). Mini DIN connectors use a circular-style connector, usually with a guide pin or grove to prevent pins from being placed in the wrong spot. The 9-pin is most commonly used for networking; it is similar to the S-video connection in Figure 9-10.

FIGURE 9-9 A male and female DB9, which is a common RS232 cable.

We often use RJ-style connectors with Cat cables and Ethernet-type connectivity. The RJ-45 connector is the most commonly used member of this group (see the Ethernet/RJ-45 connection in Figure 9-10), but other options include the 4-pin RJ-11, 6-pin RJ-12, 8-pin RJ-45, 10-pin RJ-45, and 8-pin shielded RJ-48, among others.

The USB family includes the A-type connector, which is found in most computers and on flash drives;

the B-type connector, which is the square-looking one found on some printers; the micro family, which is often used to charge and connect cell phones; and a few others not often used (Figure 9-11).

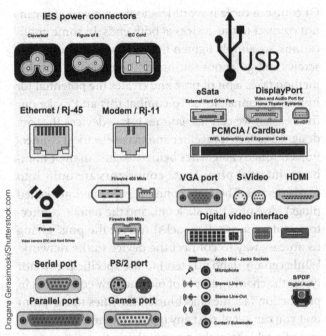

FIGURE 9-10 Different styles of connectors used in networking.

FIGURE 9-11 USB connections used in industry. Types A and B are widely used.

A few different configurations of fiber-optic connections are possible, but they typically carry one or two light paths, with designations such as ST, SC, FDDI or MIC, MTP, LC, and MTRJ. Since there is a good

chance you will order these cables premade instead of creating them in-house, just make sure you have the right cable number or type of termination at the end.

Adaptors

When you cannot find a cable with the correct connectors or you need to patch new equipment into an existing cable, you can use an adaptor. **Adaptors**, as the name implies, change the style of the pinout so that communication or power can flow. Among the commonly used adaptors today are those that can convert a USB to a DB or Mini DIN connection (Figure 9-12).

FIGURE 9-12 A USB to DB9 adaptor that you may find in industry. Most of the time this type of adaptor comes with an install disk that has the needed software on it.

Keep in mind that using an adaptor may not be as easy as "just plug and go." It is important to make sure the adaptor routes the information coming out of the system to the correct pins on the cable or disaster can follow. (I learned this lesson the hard way when I used an incorrect RJ-45 style connector that shorted power to the wrong place on an older PLC, causing insulation damage in the system.) Some adaptors need a driver or software loaded into the device to define the pinout for correct operation. The key here is to verify the internal connections of the adaptor or ensure the software has the pins defined correctly. If the adaptor is specifically designed for the application you have in mind, you are likely safe. If it is just a generic USB to DB9 connector or something similar, however, be prepared for some hassle and frustration while hoping for the best.

Obviously, it is impossible to cover all the ins and outs of network hardware in one section of a chapter, but you now have a good basis on which to build your future learning. As you work with networks in the field, the missing pieces will fall into place as you learn the specifics of your network system. If you find yourself intrigued by how all the hardware works together, numerous classes, seminars, training courses, and videos are available to teach you more about how all the hardware works together.

COMMUNICATION PROTOCOLS

It takes more than just a bunch of hardware to make a network—there has to be a set of rules governing how data is sent, which voltages and currents are involved, how the data is read, and how packets are addressed, among other things. We call these rules **protocols**.

Early industrial networks were often developed by equipment manufacturers that developed their own hardware and software for use specifically with their equipment. In these **proprietary networks**, the specifics of the protocols were typically a closely guarded secret, and the systems were often incompatible with instrumentation and devices built by other companies. This incompatibility meant that if you had a certain vendor's network, you had to buy your equipment from that vendor.

In an attempt to create common networking standards, several nonprofit consortia were formed to establish standard communication protocols that would allow devices from multiple vendors to work together. A partial list of such organizations includes the following:

- Foundation Fieldbus
- HART (Highway Addressable Remote Transducer)
- Open DeviceNet Association
- Profibus
- Modbus
- Netlinx

Fieldbus is a generic term for a number of industrial network protocols designed to provide real-time control of various devices and work with all the common network topologies. Although the technology has been around for many years, in the late 1990s a committee was formed in Europe to create a standard document that would cover all the various protocols. This document became the IEC 61158 standard, which encompasses the following network protocols:

- Foundation Fieldbus H1
- ControlNet
- Profibus
- ProfiNet
- Foundation Fieldbus HSE (High-Speed Ethernet)
- World FIP
- Interbus

The original intent of the IEC 61158 specification was to provide a common unified communication protocol, but due to the variety of applications and control structures in use, this goal was not realized. In the United States, the Allen-Bradley division of Rockwell Automation developed a group of standards that eventually became DeviceNet, ControlNet, and Ethernet/IP. In Europe, the Siemens Corporation developed a group of standards, which evolved into Profibus and later ProfiNet.

DeviceNet

DeviceNet is based on a Controller Area Network technology that uses a bus topology. The Controller Area Network was originally developed by the German automotive supplier Robert Bosch for automotive applications, with the goal of making automobiles safer and more reliable while reducing the amount of wiring used in them. Because of its early success, this protocol has gained widespread popularity in industrial automation applications around the world.

DeviceNet uses a trunk and drop line system with a 121-ohm terminator resistor at each end, and all cables surrounded by a grounded shield (Figure 9-13). In the four-wire flat cable version, two wires provide power at 24-V DC and the other two wires transmit data using the RS485 standard. The round thick and thin cables have five wires; the extra wire is used as a drain and is bare of insulation, similar to a ground wire. If additional power is required, multiple power supplies can be incorporated into the network.

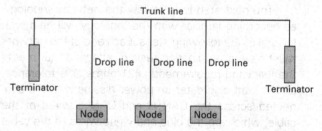

FIGURE 9-13 The basic layout of a network using DeviceNet.

DeviceNet networks support 64 addressable nodes and a maximum of 2048 field devices. They are designed to control sensors, switches, valves, and other such devices and are often controlled by a PLC. Each device has its own unique address known as a **media access control (MAC) address**. The main controller, which uses one node, is called the master scanner. Data can be sent on a DeviceNet network at one of three baud rates—125 Kbps, 250 Kbps, or 500 Kbps; the speed depends on the overall length of the trunk and the drop lines. Generally, drop lines are limited to a maximum length of 20 feet (6 m) from the trunk. For thin cable trunk lines, the length is limited to 328 feet (100 m) for all baud rates. For thick cable trunk lines, the length is limited to 1640 feet (500 m) at 125 Kbps, 820 feet (250 m) at 250 Kbps, and 328 feet (100 m) at 500 Kbps. For flat cable trunk lines, the length is limited to 1378 feet (420 m) at 125 Kbps, 656 feet (200 m) at 250 Kbps, and 246 feet (75 m) at 500 Kbps. When determining the data transmission speed for a DeviceNet network, the speed of the slowest device determines the overall speed of the entire network. Two fairly common problems encountered when setting up a DeviceNet network are inadvertent duplication of node addresses and use of multiple baud rates on one trunk line.

A DeviceNet scanner converts serial data on the network bus to a format more usable by the controller processor but can handle only a limited amount of input/output (I/O) data. The cumulative I/O size of the field devices cannot exceed the capability of the scanner. DeviceNet supports master/slave relationships, peer-to-peer relationships, and multicasting networks. DeviceNet is used to communicate between discrete I/O devices, robots, other machine controllers, variable-speed drives, and similar devices in industry.

This open network standard is maintained by the Open DeviceNet Vendor Association, which is currently supported by more than 250 active members. For more information on Open DeviceNet Vendor Association, visit its website at www.odva.org.

 FOOD FOR THOUGHT

9-1 Setting Up a DeviceNet Network

The first step in setting up a DeviceNet network is to ensure that the controller has the proper hardware and software interfaces installed. Each individual robot manufacturer has its own protocols for setting up DeviceNet networks on its respective controllers, and some of these protocols may change when software versions change. The robot manufacturer's technical documentation should be consulted for specific information on connecting DeviceNet networks to its controller.

The next step is to review the network topology, as becoming familiar with the topology will help with several of the following steps. Each end of the network must have a terminating resistor attached that meets the following requirements: 121 ohms, 1% tolerance, and ¼ watt or greater on power dissipation. It is connected across the CAN_H and CAN_L wires of the cable, which are the blue and white wires of the cable respectively.

Selection of the appropriate cable is influenced by a number of factors. To help with the selection, make a list of all devices on the network, including the electrical characteristics of the device, the power requirements, the baud rate, and the MAC address of the device. The power requirements will assist in selecting the proper power supply (supplies) for the network. Verifying the MAC address ensures that no duplicate MAC addresses are assigned, preventing later problems. The baud rate of the device is critical, as the speed of the

entire network is determined by two factors: the speed of the slowest device on the network, which sets the maximum possible speed for the overall network, and the length of the trunk and drop lines. The longer the lines, the slower the overall speed.

The maximum cable length influences two aspects of the DeviceNet network: the speed of the network and the size of the cable. As previously stated, longer network lines will cause a slower network speed. Longer lines may also require larger cables to minimize power losses. Refer to ODVA Publication 27, which can be downloaded from the Open DeviceNet Association website, for additional guidance on cable media selection.

Selection of the proper power supply is essential. If a single power supply is used, it must be capable of providing adequate power to all devices on the network. If two power supplies are used, each should be capable of supplying the power to the devices on its portion of the network. Add the power requirements of each device to determine the total power requirements for the system, and then add in a minimum safety factor. For example, if a network has 16 devices, the total power requirements of all the devices are 5.6 A, and a minimum safety factor of 2 is used, then the power supply should be capable of supplying a minimum of 11.2 A: $5.6 \times 2 = 11.2$.

Another critical concern is the environment in which the system will be operating. Some environments, such as food processing, require special considerations such as wash down and chemical exposure during cleaning operations. Both the devices and the cables must be capable of withstanding these considerations, and waterproofing is a must in this application. Operating temperatures can also be a factor, if extreme high or low temperatures are a part of the network environment—for example, if part of the system runs through a freezer or near a boiler.

Grounding the network is also important. The network should be grounded at only one location, as improper grounding can cause ground loops. If shielded round cable is used, the V-conductor, shield, and drain wire should be grounded at only one place. For flat cable, ground the V-conductor at only one point. Consult ODVA Publication 27 for additional guidance if needed.

The physical installation of the network can pose some difficulties. Proper bend radii for cables can sometimes be difficult to achieve, but make sure you do not bend a cable to a radius smaller than the manufacturer's recommended minimum bend radius. Doing so may cause cable damage down the road and lost data. The location of taps for drops can also be a problem. Avoid exposing taps and the terminating resistors to environmental hazards as much as possible, as connections can work loose and expose the network to damage. The terminating resistor at the beginning of the network is often located near the robot controller, and sometimes inside the controller cabinet itself, whereas the resistor at the end of the network will usually be somewhere in the work cell.

ControlNet

ControlNet is another industrial networking protocol that supports multiple controllers, remote I/O racks, communication with DeviceNet sub networks, and **human–machine interface (HMI)** devices, which are touch screens that display information via interactive graphics that are often tied into the system controller to cause actions. One of the major advantages of ControlNet is its higher data transmission speed—approximately 5 Mbps at its maximum. Additionally, all devices on the network are capable of communicating with each other. ControlNet data is primarily transmitted over coaxial cables or fiber-optic cables, as both of these cables offer higher noise immunity, which is extremely desirable for high-speed data transmission. Since these cables do not carry voltage, electrical power must be provided to the field devices by external wiring. ControlNet works with the standard network topologies, and each ControlNet network supports up to 99 nodes.

ControlNet has a few unique features relative to the other protocols. First, the network is **deterministic**, which means we can set the minimum rate of data transfer within given parameters and the system will keep the timeline we set. While we can set the minimum transmission times with other systems, these settings are approximate and not truly deterministic. Second, ControlNet offers faster response times than traditional master/slave networks. In **master/slave communication**, the controller (the master) asks for data from nodes (the slaves), and the nodes respond only when directed to do so. This method ensures clean communication and allows for precise timing of messages.

ControlNet also has built-in redundancy, as multiple masters can control the same I/O, and constant transmission times that are unaffected by devices leaving or entering the network. Because redundancy is built into the ControlNet hardware, communication is easier and cheaper than with most of the other protocols. Another advantage of ControlNet is its ability to transmit over Ethernet, FireWire, and USB, which opens up several options that some protocols (e.g., DeviceNet) cannot accommodate. The downside of ControlNet is limited support on the vendor side of things and the expensive Rockwell Automation hardware required to use this protocol.

The maximum distance of a ControlNet network depends on the number of nodes that are connected and the number of repeaters in the line. Each segment has a maximum length of 3280 feet (1000 m) with a maximum of two taps, and every segment must have a terminating resistor at each end. Every extra tap we add in past the first two decreases the segment distance via the following formula:

$$3280 \text{ ft} - (53.4 \text{ ft} \times [\text{number of taps} - 2])$$

or

$$1000 \text{ m} - (16.3 \text{ m} \times [\text{number of taps} - 2])$$

Note: This formula is based on standard Allen-Bradley RG-6 cable. If RG-6 high-flex cable or some other brand of cable is used, the calculations are different and need to be researched.

The maximum number of taps allowed on one segment is 48, with a maximum length of 820 feet (250 m). To increase the cable distance, we can create a new segment by adding in a **repeater**, which is a device that takes the signal from a segment and amplifies it for the new segment. One tap must be added in for the repeater, leaving a maximum of 47 nodes for the segment; however, this tap does not count as a node because these devices are not addressed with the controller. Keep in mind that we must have two terminating resisters per segment and the controller can handle 99 nodes per ControlNet network. One of those nodes could be the controller for a DeviceNet network, which would seem to increase the total nodes of the system, but only the controller for the subnet is communicated with; thus, it counts as one. To get around the 99 nodes limit, some controllers are capable of managing multiple ControlNet networks, thereby increasing the overall node capacity.

Ethernet/IP

Ethernet/IP is a network that uses the standard Ethernet protocol and Transmission Control Protocol/Internet Protocol (TCP/IP). One significant advantage of using TCP/IP is that data packets are routed the best way to their destination, so as to avoid data bottlenecks at either 10 Mbps or 100 Mbps. To minimize data collisions, switches are used at the hubs of the network, and the industrial networks can be isolated from other networks in the plant area.

The development of Ethernet/IP was inspired by the need to gather more data within the plant and process it. This protocol offers several distinct advantages over both DeviceNet and ControlNet:

- Transfer of large data files
- Network diagnostics
- The ability to perform controller operations
- Device configuration and programming

Ethernet/IP can easily accommodate new products and devices that do not work well with other protocols due to its flexibility. In many cases, the most challenging aspect of adding new devices is assigning the device an IP address that works well with the network.

Ethernet/IP incorporates the open communication architecture known as **Common Industrial Protocol (CIP)** inside the standard network protocols used by the Internet. CIP is an information transmission method that creates a common language among protocols so information can flow between them without the hassle and loss of time associated with translating everything into a different format. CIP is used by Ethernet/

IP, DeviceNet, CompoNet, and ControlNet, allowing systems based on these protocols to share information easily without extra translation hardware or software.

Ethernet/IP complies with the IEEE 802.3/TCP/UDP/IP standard, so much of the hardware is readily available. Switches are generally connected using Cat5 or Cat5e unshielded twisted-pair cables with RJ-45 connectors. In environments characterized by high levels of noise, however, special shielded cables are preferred. The maximum distance between a switch and a node when Cat5 cables are used is limited to 328 feet (100 m). The maximum distance between the switch and a node may vary when using other wired cables or fiber-optic cables, so make sure to verify the distance for the cabling used in your network.

Profibus

Profibus is a network protocol created in Germany in 1989 by a consortium of factory automation suppliers working with the German government. Siemens, one of the world's largest makers of PLCs and other industrial controls, has adopted Profibus in all its forms and designs its equipment to work seamlessly with this protocol. Profibus is the go-to protocol in Europe, but you will also find it in North America, South America, parts of Asia, and portions of Africa. Siemens, with its use of Profibus, is competing with Allen-Bradley, with its use of DeviceNet, ControlNet, and Ethernet/IP, to become accepted as the standard for plant automation in North America. As a consequence, there is a good chance you will run into both systems and types of networks on the production floor. Adaptors are available that talk between the two systems, but whenever you mix protocols that are competitors instead of allies, you can expect trouble from time to time.

Profibus uses the multi-drop type of network, with multiple devices on a single cable. It is capable of sending process values, diagnostic information, device parameters, performance data, and other pertinent information across the network. The devices are extensively specified, which means that the hardware performs well regardless of manufacturer and is designed to be backward compatible, thereby avoiding problems when new equipment is placed on older networks. Profibus also comes with software tools for troubleshooting and engineering applications that can save time and frustration. Profibus works with intelligent field devices (slaves) such as remote I/O devices with control centers, control valves, variable-speed drives, and various measuring devices, as well as with robots, CNC machines, safety systems, and other complex systems. Profibus includes several products: Profibus DP, Profibus PA, Profibus FMS, and ProfiNet.

Profibus works with a bus or tree topology and allows a maximum of 32 stations or nodes per segment; we can use a repeater to add segments. The recommendation is to include no more than 4 repeaters, creating a maximum number of 127 nodes per network. Maximum cable distances range from 328 feet (100 m) for a single segment with copper cable to 78,740 feet (24 km) with repeaters and fiber-optic cable. Transmission rates range from 9.6 Kbps to 12 Mbps, with 244 bytes of data being sent per message. Data is controlled by polling in Profibus DP and Profibus PA and sent peer-to-peer in Profibus FMS. **Polling** controls communication by having the master (requesting unit) send a signal asking the slave (sending unit) if it has any data to send. If the answer is yes, the data is transmitted from the sending/slave unit. If the answer is no, the master/requesting unit moves on to the next node. In **peer-to-peer (P2P)** communication, each node can talk directly to the others without having to go through a third node, such as the main controller, and can act as sender or receiver of data.

Profibus uses a **general station description (GSD)** file, which provides detailed information about the device that is readable by the configuration software. In short, the GSD file resides on the device and is tied to its unique identification number (similar to a MAC address). It allows for easy integration because the configuration tool has all the data needed to add the device to the network for communication. Another advantage of Profibus is that the different versions of this protocol work well with the majority of manufacturing processes and needs.

Like all protocols, Profibus has some downsides. Each message contains 12 bytes of overhead control data as part of the maximum 244 bytes of message data, so the overhead is relatively heavy. Also, all versions of Profibus do not support sending power over the network, so separate power is sometimes a must, and the equipment is a bit costlier than some of the other options.

Profibus DP

Profibus DP (Decentralized Periphery) is a device-level bus that supports both analog and digital signals and is generally designed for controlling a specific machine

from inside the control cabinet. This bus, which was originally developed to replace conventional I/O systems that required separate wires between a controller and the field devices, is designed to handle time-critical communication. Profibus DP communicates over RS-485 media at speeds between 9.6 Kbps and 12 Mbps, at distances from 328 feet (100 m) at the low end to 3937 feet (1200 m) with fiber-optic cable and repeaters. Power is supplied to the field devices separately from the communication bus. For Cat5e or better cable, this system uses DB9 or M12 connectors, which are 3- to 12-pin round connectors, similar to the Mini DIN connectors. When fiber-optic cable is used, the connector of choice is an ST style with a single round, light conductor connector. Multiple master nodes are possible with Profibus DP, but each slave is assigned a specific master node. As a consequence, multiple nodes can read from the slave device, but only the assigned master node can write to a slave node.

Profibus DP has V1 and V2 extensions that add flexibility to this version of Profibus. V1 extensions integrate the functionality of Profibus DP and Profibus FMS so they work seamlessly together, instead of as two separate systems. This allows Profibus DP to work with newer, high-complexity equipment. V2 extensions add a synchronization feature so multiple motion axes or multiple devices can work and coordinate using the same clock function. It also adds **publisher/subscriber messaging**, which lets nodes talk on a one-to-one or one-to-many basis; these capabilities are helpful for coordinating multiple-axis motion.

Profibus PA

Profibus PA (Process Automation) was introduced in the late 1990s to support process automation applications and manage smart process instrumentation. This protocol was designed specifically for the process industry, with the goal of replacing 4–20 mA transmissions using a two-wire connection carrying power and data. Whereas Profibus DP is designed to work in the machine controller, Profibus PA is designed to work outside a specific machine; it is mounted in a safe location where it can manage multiple processes at once. Data is transmitted at 31.25 Kbps with a maximum length of 6233 feet (1900 m) on the Manchester Bus Powered (MBP) cable connected with M12 or gland screw connectors. The power on the Profibus PA network is kept to an intrinsically safe level, so no sparks are generated; only low, safe levels of power are released that will *not* ignite dangerous atmospheres or hurt people. Communication between Profibus DP and Profibus PA systems require a barrier converter to ensure the power on the PA network stays at safe levels.

All devices on the Profibus PA network must conform to a mandatory profile, which is part of the intrinsically safe operation. Another aspect of this profile is the origination of data, which includes standard floating-point format for analog information, standard digital format for discrete data, and a status value that denotes whether the quality of the data is good, bad, usable, or some other designation. The profile also sets how acyclic, or non-time, data use occurs so that various software tools can do their jobs without causing issues on the network. Collectively, these properties create a very stable network with built-in redundancy that can work with dangerous processes such as gasoline production or hazardous waste management.

Profibus FMS

Profibus FMS (Fieldbus Message Specification) is used for communicating at the cell level, including with PLCs and industrial-grade personal computers (PCs), although it can also communicate at the device level. (A **cell**, or work cell, is a logical grouping of machines that aid in the work flow required to turn raw parts into finished goods.) This protocol uses an **object dictionary** that holds all accessible values; the index, name, and type of each variable in the system; and descriptions, structures, data types, and relationships for device addresses in a manner that it can broadcast over the network as needed. Profibus FMS uses peer-to-peer communication, which allows all nodes to designate themselves as a master if desired, but comes at a higher overhead in terms of byte price.

Profibus FMS uses the RS485 transmission standard over copper cable, with the maximum cable length set by the baud rate of transmission. At 12 Mbps, the maximum distance per segment is 328 feet (100 m); at 1.5 Mbps, the maximum distance per segment is 656 feet (200 m); at 187.5 Kbps, the maximum distance per segment is 3280 feet (1000 m); and from 93.75 Kbps to 9.6 Kbps, the maximum distance per segment is 3937 feet (1200 m). Thus, the slower the transmission, the greater the distance the system can cover. Ideally, we would like to keep all cabling as short as possible, but in refineries and other operations with miles of pipes, conveyors, and other transmission media, short-distance runs may not be an option.

ProfiNet

ProfiNet works seamlessly with standard IEEE802.3 Ethernet protocols and hardware and is an open industrial Ethernet standard. Transmission speeds can exceed 100 Mbps on cables, and the protocol can work with gigabit hardware in a wireless setup to achieve even faster transmission speeds. The TCP/IP channel is used for non-real-time communication, such as diagnostics, device management, and downloading parameters. A special channel is used for real-time transmission of data related to critical events, such as process data, alarms, critical messages, and monitoring of communications in hazardous situations. ProfiNet works with all members of the Profibus family and uses software to create proxy connections to DeviceNet, ControlNet, and other protocols to enhance functionality.

The maximum transmission distance for ProfiNet depends on the method used. Twisted-pair copper cables have a maximum network length of 16,404 feet (5000 m), with a maximum segment length of 328 feet (100 m). Fiber-optic cables can reach a maximum network length of 492,125 feet (150 km) in single mode or 9842 feet (3000 m) in multi-mode, with a maximum segment length of 64 feet (50 m) for plastic fiber-optic cable and 328 feet (100 m) for glass fiber-optic cable. For wireless transmission, the maximum coverage area is usually limited to 3280 feet (1000 m), and segments are limited to 98 feet (30 m) between nodes indoors and 328 feet (100 m) between nodes outside.

One of the things that makes ProfiNet faster than Profibus is the former's use of a **producer/consumer** communication method, in which a device "produces" or sends information with specific identifiers attached and other devices "consume" the data that matches the identifiers they are looking for. This method can allow multiple devices to consume a single produced transmission. Another benefit of ProfiNet is that because of the Ethernet standard use, there is no specific node limit. If we keep adding nodes, we will eventually reach a point where the network is too large to manage effectively, but this node limit depends more on the hardware used and the network arrangement than how many connections are on the network. With that said, for wired portions of a ProfiNet setup, I would recommend using the basic guidelines outlined for Profibus and creating multiple sub networks that connect into the main network wirelessly.

ProfiNet uses Ethernet protocols but is more than a standard Ethernet network. This means that the plant IT person(s) could run the network, but there is a good chance the full benefits would not be realized. The real-time channel and proxy functions are different from those in standard Ethernet and deliver part of the power of this protocol. I would recommend sending some of the technical folks who work with the network or, at the very least, IT personnel out for specific ProfiNet training to ensure that you get the full benefits of ProfiNet. With training, the use of smart switches/hardware, and proper use of the built-in software tools, it should be fairly easy to set up a working, stable network within a reasonable time frame with minimal frustration and headache.

Foundation Fieldbus

Foundation Fieldbus, often just called Fieldbus, is another protocol that you may encounter. It was developed by the Fieldbus Foundation, which formed in 1994 when the international Interoperable Systems Project (ISP) and the French Factory Instrumentation Protocol (FIP) joined forces. Their goal was to create a single, international standard for hazardous environments, and their biggest competitor is Profibus PA. While Profibus PA holds sway in Europe, Fieldbus is concentrated in America and Asia with hopes of nibbling away at Profibus's hold in Europe. In fact, if you look into Fieldbus, you are likely to run across some direct comparisons between the two competitors that were made in an effort to sway your decision.

In January 2015, the Fieldbus Foundation joined with the HART Communication Foundation; the merged entity, which is known the FieldComm Group, is seeking to increase collaborative efforts between open digital standard developers and increase market share in the digital field devices market. This merger—along with the understanding between FieldComm and Profibus PA that neither standard is going away, so they would be better served by finding ways to cooperate—opened the door for Fieldbus, Profibus PA, HART, and OPC communication protocols to work with one another in industry with minimal problems.

Fieldbus is a digital communication system that use multi-point peer-to-peer communication. Transmission size is 128 **octets** (8-bit units of data), and Fieldbus supports power over wired portions of the system. This protocol is also designed for intrinsic safety, so it is safe to use in applications where flammable or explosive

atmospheres might be present. Real-time data transmission is possible as well as time-synchronized events, usually based on local time settings that are updated/verified by the main controller on the network through a data pulse with current time information sent out at specific intervals. This protocol uses function blocks for control, with 21 types of standard blocks for data transfer being available that users can modify with their specific parameters. Each device has a resource block, similar to the GSD (covered earlier), that contains all the specifics of the device, such as name, manufacturer, serial number, and other important information about the device for network communication and operation. Function blocks are joined together to form operational systems in a graphical way in the software.

When a new device is added to the network, it joins at one of four special default addresses specifically reserved for this purpose. The configuration tool assigns a physical device tag that holds the permanent address assigned in the next step. Using system management services, an unused address is assigned to the device. The tag is then stored in nonvolatile device memory, which means the information will be maintained even when power is lost. This process is repeated for everything new on the network until all the devices have a tag with permanent address. This approach makes it easy to add new devices to the network as well as to reestablish communication with devices returning to the network. The tags in the system are searchable, making maintenance easier as well as helping host systems acquire needed information.

The wired version of Fieldbus, known as Fieldbus H1, is capable of data transmission speeds of 31.25 Kbps to a maximum wired distance of 6232 feet (1900 m) when #18 AWG individually shielded, twisted-pair wire is used. This distance decreases to 656 feet (200 m) when two untwisted, non-shielded #16 AWG wires are used. Each segment has a maximum of 32 devices without repeaters or a maximum of 240 devices with repeaters, with 4 to 16 devices per segment being the norm. Since this is a power-over-wire system, each device must draw 8 mA from the segment at minimum, with four-wire devices typically pulling 8.5 mA and most two-wire devices pulling 15 mA to 25 mA, and segment maximum amperage being limited to 400 mA. Voltage over the wires can range from 9-V to 32-V DC, but 24-V DC is typical.

Each spur off the segment is limited in length based on the number of devices on the segment and the spur. Spurs typically have only one to three devices, with each device on the spur significantly reducing the maximum length of the spur. For example, if there are 12 total devices on the segment and only 1 device on the spur, the maximum cable length for the spur is 393 feet (120 m). If there are 18 total devices on the segment and 3 devices on the spur, the maximum length drops to 3 feet (1 m).

Fieldbus HSE is the Ethernet portion of the Fieldbus protocol and is an enhanced version of the IEEE 802.3 Ethernet standard. In a typical Fieldbus system, the H1 portion works directly with the devices and the HSE portion links all the H1 networks together, creating a very versatile plant communication system. Fieldbus HSE transmits data at 100 Mbps and works with all the standard Ethernet equipment you would expect. When wired, the maximum range depends on the type of cable used as well as the Ethernet base. The greatest distance possible is 6561 feet (2000 m) when using twin fiber-optic cable with 10BaseF Ethernet transmission; the worst is 328 feet (100 m) when using twin cable to a central hub and the 10BaseT transmission method. HSE offers redundancy that is not possible with standard Ethernet solutions, creating hundreds or even thousands of loops for information to get where it needs to go. This prevents a loss of communication that could be disastrous in many industrial applications.

You may encounter other protocols than those profiled in this section, and there is much more to learn about each of the protocols covered here. Most facilities pick a protocol or two to work with and try to stick with them, so make sure you learn all you can about the ones you have to work with when you get out in industry. Networking is a living field, so new developments are continually emerging that may impact your current networks, especially when new equipment is involved. Also, some of the protocols are at odds with one another, and it may be hard to integrate them directly, so keep in mind most protocols have some way to get on the plant's Ethernet network these days. There is a wealth of information on the Internet as well as resources such as vendor-supplied training and on-the-job experience to help you learn more about the networks in use.

INTEGRATION

Integration is the process of combining equipment and design elements to create a functioning system or work cell. A **work cell** is a logical grouping of equipment in which the processes flow seamlessly from one operation to the next, with the goal of improving quality and reducing per-part production times. Integration is a focal point of the field in industry referred to as **mechatronics**—that is, the combination of mechanical, electrical, and computer engineering fields into one multi-skilled discipline. The concept of mechatronics has been around since 1969, when Japanese engineer Ko Kikuchi coined the term, but it has recently become a hot topic in industry, driven largely by the advancements in networks and technology. As we continue to create new and exciting ways to use and share information, the desire to link everything together only seems to grow in industry.

Since the birth of the factory, we have looked for better and cheaper ways to produce parts and to improve quality as well as profit margins. Older technology for control included punch cards, peg drums, and timing cams, but none of these systems worked well for controlling a large number of machines at the same time. While we could still organize and create work cell arrangements of machines, often each machine was its own island, so to speak, connected by conveyors, robots, operator action, or something else to feed materials in and out. When we started to switch from mechanical-based controls to computer-type controls, we began to see ways to pass information as well as parts. That realization led to the birth of networking as we know it today. In many modern factories, every piece of equipment is tied to the plant network, allowing maintenance, engineering, and management personnel to obtain the data they are interested in at a moment's notice. We also use these networks to time motions, pass parts between operations, and create processes that would be nearly impossible via the mechanical methods of old.

When we integrate systems, we have to take several things into consideration:

- The process involved
- The equipment involved
- The protocol(s) used
- The data tracked
- The work environment

Having a firm grasp of what each part of the process entails makes it easier to figure out an order of operations as well as to determine which functions should be paired together in the work cell. For example, it makes sense to perform all the raw-material processing operations before trying to put the finish polish on any portion of the part, as reversing these steps could mar the surface.

Once we know the *how*, it is time to look at which machines we will use to complete the tasks. The physical constraints of the machine play a large part in layout, placement, and work flow of the cell. Some industrial machines are so large that they are placed inside the facility before the building is completed, and the rest of the structure then built around the machine after its installation. In these cases, there is a good chance that moving that equipment is out of the question, requiring everything else in the work cell to be staged around or somehow connected to the larger machine. Sometimes several machines can be moved, but there is not enough free space to move them all to the same area. At times, we can move several machines around and make space, but other times we either have to come up with a new plan or add on to the facility.

Another concern is the function of the equipment. For instance, if a 200-ton press shakes a large area every time it cycles, it would be a bad idea to put a fine finish process next to it, where the vibration could cause unwanted tool movement.

Once we have the machine placement figured out, it is time to get everything talking back and forth. From the previous section, you should have an idea of the variety of ways we do this as well as some of the difficulties that can arise when we mix and match protocols. Most likely you can find some kind of adaptor to get any two protocols to talk back and forth, but it may be simpler to use that protocol's equipment to get the machines onto an Ethernet network and then use standardized equipment from there. Special adaptors seem prone to failing at 1 o'clock in the morning when no one working has any idea how the network was set up or where to look for more information. Some of the protocols play nicely with one another, making it easier to mix and match the machines as needed. In a perfect world, all machines would use the same protocol, but in the real world we often end up with a weird mix, as equipment manufacturers tend to put what they know in their machines.

Once we have the machine on the network, we then need to make sure we are sending and receiving the right data. Getting something on the network does not magically ensure that data flows to the right places at the right time. The network is just a data highway—we still need something to drive the data transfer. The *how* may be a master/slave, multi-drop, multi-point, peer-to-peer, or other method to make sure data is not corrupted or lost. The *when* may be cyclic, such that data transfer happens at set intervals, or acyclic, such that it happens only when specifically requested, often as a one-time event. Cyclic data transfer is often controlled by some kind of operational software, whereas acyclic data transfer may be software or human initiated. Often the same software that controls the machine handles the network communications and all internal data transfers. When we want a specific piece of data, a personal computer that can communicate with the network is the typical tool.

Another consideration is the work environment of the systems we plan to integrate. Wires that carry large amounts of amperage, transformers, solenoids, motors, and other components used in industry can generate magnetic fields large enough to cause problems with low-voltage network signals. As lines of magnetic flux pass through the signal wires, a small voltage is induced and, in some cases, this may be mistaken for a signal or data. Extremes of temperature, sunlight, and chemicals can damage wire insulation and create shorts and other such problems. Physical damage to wiring can occur in a number of ways—rubbing against something due to wire/machine movement, falling parts, people stepping on them, forklifts rolling over them, or anything else involving contact and pressure. With wireless systems, walls, metal in the roof, and other workplace features may dampen or completely block signals.

You may not always notice an environment problem immediately, as it may take time to develop. If you do run into a work environment issue with the system, be prepared to rework the system as needed to prevent it from happling again.

The ultimate goal of integration is to get multiple systems working together efficiently. The particulars of making this happen are as diverse as the equipment, protocols, and environments we work with. Even with careful planning there is a high probability that something unexpected will occur, requiring a change in plans or equipment to get everything working correctly. Some systems become so cobbled together as new equipment is added one machine at a time that it is better to start over and convert everything to one or two protocols. While I cannot accurately predict the specifics you will encounter in the field, I can tell you that over the last 10 years I have had an ever-increasing number of technicians tell me that networking and machine communication is increasingly something they have to set up, maintain, and work with.

FOOD FOR THOUGHT

9-2 Integration Example

FIGURE 9-14 A work cell.

Figure 9-14 shows a student-created work cell that integrates several pieces of equipment:

- A robot
- A programmable logic controller (PLC)
- Two conveyors
- Two motor controllers
- A light curtain
- End-of-arm tooling
- A work piece holder
- A pneumatic drill
- An operator panel

The work cell was set up as a classroom exercise to teach the students the process of building, testing, and running a production work cell designed to drill a hole in a piece of plastic pipe. The part is placed on the in-feed conveyor and the operator starts the conveyor. The part

travels to the end of the conveyor, where a photo eye detects the part and sends a signal to the PLC to stop the conveyor. It also sends a signal to the robot to initiate the robot program, which picks up the part from the in-feed conveyor and places it in the work piece holder (Figure 9-15). The work piece holder secures the part and the robot moves away; then a pneumatic drill drills a hole in the piece of pipe. When the drill returns to its resting position, a signal goes to the robot to remove the part from the work piece holder and place it on the out-feed conveyor.

A light curtain was integrated into the cell to stop the conveyors when the operator's hands entered the work cell. A gate was incorporated on the side of the cell to allow entry and egress to work on the robot, tooling, and work piece holder, along with a switch to stop the robot and conveyors when the gate was opened. An emergency shutoff for the air supply to the robot's end-of-arm tooling, work piece holder, and pneumatic drill was built into the operator's station. Emergency stops were incorporated at the main control cabinet and operator panel in addition to the emergency stops installed on the robot control panel and teach pendant. The operator panel also incorporated manual controls for the conveyor as well as anti-tie-down logic in the form of two buttons that had to be pressed and held simultaneously to start automatic operation.

Integration included construction of the cell, building and wiring the operator's station, wiring the main

FIGURE 9-15 A view of the robot's end-of-arm tooling, work piece holder, and pneumatic drill.

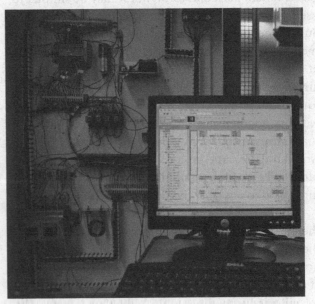

FIGURE 9-16 The control cabinet as well as some of the PLC program driving the system.

control cabinet, and wiring the valves that controlled the work piece holder and pneumatic drill (Figure 9-16). On the programming side, the robot and PLC programs needed tested and verified before the system could run in automatic mode. As is typically the case with a work cell, a few minor modifications had to be made during the construction and testing of the cell.

One question that had to be answered during the design and integration of the cell was whether the PLC would have total control or whether it would share control with the robot. In this case, the decision was made to share control. The advantage of the PLC having total control is that with proper design of the interface, it brings control to a central point; the downside is that it complicates the wiring. By sharing control, the wiring was simplified, but the operator had to go back and forth between the operator's station and the main control cabinet to start production. In this case, due to the small number of input and output signals (eight inputs and outputs), simple point-to-point wiring was used without the need for a networking protocol.

REVIEW

Networking has become a crucial part of modern manufacturing and is a field that continues to advance and evolve. With each new generation of protocol options and device functionality, we discover new and exciting ways to share data among equipment and create functionality options that were previously beyond the equipment's capabilities. While the specifics of your involvement with networks will depend on which job you choose to pursue and which company or facility you work for, it is almost guaranteed that at some point you will find yourself interacting with various networks in the workplace. While this chapter has laid a good foundation of knowledge, a wealth of information is available from other sources on each of the networks covered in this chapter and many others. If you are looking for an elective to pick up that would work well with a technical degree, you may want to take a networking class or two, as most protocols have some way of working with Ethernet.

Over the course of this chapter we hit on the following topics:

- **What is a network?** We defined a network and identified the common tasks that networks perform in industry.

- **Types of networks.** We explored the common wiring arrangements for networks as well as the more common recommended standards in use by industry.

- **Communication protocols.** This section was all about the different ways we send data as well as some of the distance limitations involved with sending messages.

- **Integration.** Integration focuses on getting everything to work together; we saw some of the things to consider along the path to an integrated system.

KEY TERMS

Acyclic	**Connectors**	**General station**	**Local area network**
Adaptors	**Cyclic**	**description (GSD)**	**(LAN)**
Bus topology	**Deterministic**	**Hub**	**Master/slave**
Cell	**Drop cable**	**Human–machine**	**communication**
Coaxial cable	**Duplex**	**interface (HMI)**	**Media access control**
Common Industrial	**Fiber-optic cable**	**Integration**	**(MAC) address**
Protocol (CIP)	**Fieldbus**	**Kink**	**Mesh topology**

KEY TERMS

Multi-drop	Polling	Ring topology	Star topology
Multi-point	Producer/consumer	Router	Switch
Network	Proprietary network	RS232	Tap
Node	Protocol	RS422	Token
Noise	Publisher/subscriber	RS423	Tree topology
Object dictionary	messaging	RS485	Trunk cable
Octets	Recommended	Shielding	Work cell
Packet	Standard (RS)	Simplex	
peer-to-peer (P2P)	Repeater	Single ended	

REVIEW QUESTIONS

1. What is an industrial network?

2. What are three things we use networks for?

3. What is the difference between a node and a packet?

4. What are the advantages and disadvantages of a bus topology?

5. How is a ring topology setup, and what is the common method used to manage data transmission on this topology?

6. How does a star topology work?

7. What is the difference between a hub and a switch?

8. What is a tree topology, and what are its three levels?

9. How is a mesh topology set up?

10. Which problems do you need to avoid when working with fiber-optic cables?

11. Briefly describe the functionality proscribed by the RS232, RS423, RS422, and RS485 standards.

12. Which topology does DeviceNet use? How many nodes and field devices will it support?

13. What are the advantages of ControlNet? Which cables are used by this system?

14. How does master/slave communication work?

15. How does Ethernet/IP work, and what are some of its advantages?

16. Which type of topology does Profibus use? What is the maximum number of nodes per segment and the maximum number of repeaters and total nodes per network?

17. How does polling data transfer work?

18. How does peer-to-peer communication work?

19. What is the difference between Profibus DP and Profibus PA?

20. How does producer/consumer communication work?

21. How are new devices added to Foundation Fieldbus protocol systems?

22. What is the maximum transmission speed and distance of Fieldbus H1?

23. Which of the protocols discussed in this chapter work with the IEEE 802.3 Ethernet standard?

24. What are the five main things to keep in mind when working to integrate systems?

Reference

1. Titelman, G. (2000). *Random House Dictionary of America's Popular Proverbs and Sayings.* New York: Random House.

CHAPTER 10

Programmable Logic Controllers (PLCs) and Human-Machine Interfaces (HMIs)

WHAT YOU WILL LEARN

- What PLCs are and why we use them
- How the modern PLC evolved
- The basic components of PLCs and the basics of what they do

- How the PLC scan cycle works
- Some of the common instructions used in PLC programming
- What HMIs are and why we use them

OVERVIEW

In the modern factory, the process of turning raw materials into finished goods looks very similar to a complex and well-choreographed dance. Machines take in raw parts or materials and alter these materials, often at high speeds with great precision, to produce an improved part that will eventually go out to the customer as a finished item. Robots often dance in and out of machines as a part of this process, feeding in raw parts and extracting the finished ones, which in turn may go to a conveyor, another machine, or the operator for inspection. At times, you may see a robot working on the parts or performing some task before it passes the part on. In the grand dance of industry, there are miles of conveyors and chains moving parts, millions of containers to hold and orient parts, countless part tracking and/or identification steps, and a near-infinite number of ways to get the job done. On top of this, we humans are making changes, initiating actions, loading new programs, and requesting data through various interface methods. To keep it all moving smoothly, we often turn to programmable logic controllers (PLCs) and other industry-hardened controllers to deal with all the data coming in, trigger actions at the appropriate time based on that data, and send the requested information out.

In this chapter, we dig into the operation of the PLC as well as some of the methods used by humans to send data to and get data from machines. This chapter covers the following key topics:

- What is a PLC?
- A brief history of PLCs
- Basic components of the PLC
- Operation of the PLC
- HMIs

WHAT IS A PLC?

Prior to Joseph Henry's invention of the electromechanical relay in 1835, all intricate machine control involved gears, cams, and other such devices that created a rigid mechanical sequence. This sequence was difficult to change or might require extremely intricate configurations to create options of operation. In any event, technicians had to do a lot of maintenance to keep things running right. When something went wrong, the potential for damage to equipment was fairly high. In the worst cases, where something bound up the power or timing transmission systems, the result might be shattered pieces of gears, sprockets, shafts, and other such items flying out with deadly force. (Early instances of these kinds of disasters led, in part, to the machine guarding standards of today.) Although we might have some control options before the relay, they required a lot of mechanical planning and component interaction. The *Writer* created by Pierre Jaquet-Droz in 1772 is a great example of what can be done with complex gearing and some ingenuity. An industrial example is the Jacquard loom, which worked with punch cards; invented in 1804, its introduction predated the relay by more than 30 years.

Although the relay was invented in 1835, it was only in the early 1900s that relay logic began its rise to prominence. First relays were strung together to create a sequence of actions, then patch cords were added to the systems so operation could be altered, and finally punch cards were introduced as the control element, creating a veritable wealth of options for operation. (See Food for Thought 2-1 in Chapter 2 if you need a quick refresher on relay logic.) Relay logic was a stable, effective method of control that dominated the realm of controls until PLCs and CNCs came along.

Relay logic, though versatile, came at a cost. First, there was the monetary cost of the components. Many relay logic systems required a large number of relays to work, ranging from tens to hundreds of individual relays complete with bases. In turn, a large surface area was needed somewhere for mounting purposes, often inside the control cabinet on the door or along the back. Once installed, someone had to wire the system up, which in most cases took days of careful effort. After installation, the technician had to make sure everything worked correctly, as one crossed wire could mean the difference between normal operation and machine-damaging failures. Another consideration was the rather large amount of heat that each relay generated while activated. While one or two relays may not have a large effect on temperature, a hundred relays in a cabinet can raise the temperature considerably, which in turn can cause operational problems with other components in the same area. At some point when a wire worked loose or a relay failed, someone had to go in and figure out which relay, out of the entire multitude, was the source of the problem. I have heard horror stories of hours or days spent trying to find the problem, along with all the frustration that goes with this kind of repair.

It is no surprise, then, that programmable logic controllers (PLCs) and other industrial controllers using microchips began to replace this technology as soon as their major bugs were worked out. While it may take days to reprogram a relay logic system, PLCs allow for major changes of operation in minutes by simply changing the program. Granted, it may take someone a fair amount of time (likely days) to reprogram complex systems or make changes in their operation, but the amount of time during which the machine itself is down can be limited to minutes if the program has already been proven good. With the PLC, there is no requirement to change wiring, there is no need to lockout the equipment, and we can save the operational program on a backup device somewhere. In short, the PLC was specifically designed to take the place of relay logic with a hardened computer-type system.

PLC systems are specifically designed to withstand the harsh environments found in industry that would rapidly destroy a standard computer or laptop. In fact, PLCs are so durable that you can find a large number of them for sale on eBay; those PLCs may have outlived the piece of equipment they controlled or been pulled out of service, still working correctly, when the equipment was upgraded to a newer model.

PLCs have one specific function: They filter the information coming in through the program created by the programmer and either create data or trigger output connections as directed. That's it. They are not screen savers, there is no update checker crawling the net, and they do not simultaneously run multiple programs, each of which requires a portion of the processing power. All PLCs do is run the PLC program and respond as directed, and they do so very efficiently. In fact, each scan from start to finish usually takes less than 50 ms (0.05 second), with 10 ms to 20 ms being a common range for scan cycle time. This fast scan cycle creates a response time that the relay logic of old cannot match, which is another reason for the rise of the PLC.

Another thing that the PLC can do and the relay has only ever dreamed of is communicate on a network. In Chapter 9, you learned about several different protocols and some of the network communications involved with them. To send and receive data on a network, some kind of processor must be present in the device, which the standard relay lacks. In today's industrial world, the ability to network equipment has become a standard expectation in many, if not most, facilities. We are addicted to information in the modern age. We want to know where parts are, how many have been produced, how uptime compares to downtime, how many parts we can produce before the tool wears out, how long the machine was down for maintenance—and the list goes on. Our desire for data is so great that technicians tell tales of changing

out PLCs and other hardware just to make it easier to data mine all the equipment in the facility. You might think this kind of demand would be found only in newer facilities, but you would be mistaken. One company I work with has been in business for more than 100 years, making wine and whiskey barrels. It has the ability to track each individual stave as it goes through the process from coming in as a part of a tree to leaving as part of a barrel. (Staves are the boards that make up a barrel.) When a customer buys one of its barrels, the company can tell that customer the geographical area where the tree used to make the stave came from.

Another feature of the PLC is the ability to monitor system conditions for troubleshooting purposes. You can get online with PLCs and monitor in relative real time what is occurring in the system—that is, the data state of things—and see if any problems are occurring with the program. (I say "relative real time" because there is always a communication delay between the PLC and the computer used for monitoring purposes, which can hide high-speed issues that occur between updates.) Indicator lights on the PLC can give a quick look at which data is flowing in, which components should have power going out, and what the state of the processor is. A savvy technician can spend a few minutes looking at the state of the PLC and digging deeper, as needed, to quickly narrow down the potential sources of the problem. With relay logic, this process often involved a meter and a large amount of time trying to figure out which relay or wire was the offender.

As you begin to understand what a PLC is and what it does, you might ask, "Do we even use relays anymore?" The answer is yes. While the PLC can take the place of relay logic, where relays would otherwise be used to make decisions, the PLC is not suited to situations with high-amperage needs. Typically, most PLCs have a maximum amperage tolerance of 2 A or less for each output. That means we cannot use the PLC to fire up a single-phase motor directly, nor can we use it to power most heating elements, or anything that requires more than a couple of amps to run. Moreover, some outputs can handle only a half amp, which even further restricts our options. To get around these constraints, we use the PLC to fire the coil on motor starters, contactors, and relays, and then let these other devices handle the load.

Contactors are similar to relays, but they are generally rated for higher amperage and tend to have one to three main contacts and one or more lower amperage contacts called auxiliary contacts (Figure 10-1). The main contacts control devices such as heating elements or lights. In contrast, the **auxiliary contacts** are used to

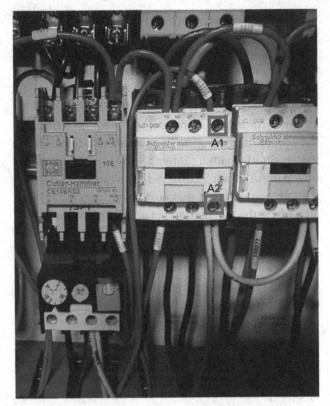

FIGURE 10-1 On the right and in the middle, you can see a couple of contactors. The black part is the base where the three main wired connections are made; the white part is where the coil connections and auxiliary contacts are. A1 and A2 are the coil connections. On the left is a Cutler-Hammer motor starter. The black module below the white contactor is the overload; it is detachable and can be set by changing the dial in the lower-left corner.

latch circuits and maintain power after a pushbutton is released, power other devices at the same time as the main device, or provide voltage signals as needed. Just like relays, contactors change state when power is applied to a coil of copper, which in turn creates a magnetic field, which then moves a plunger or other mechanical device to change contact states. Contactors are normally open or not passing power; they close when the coil is energized.

Motor starters are contactors with an overload module that are added to start motors. **Overload modules**, often just called overloads, are designed to allow the large inrush of current needed to start a motor but change state if the flowing amperage exceeds their set value for more than a few seconds.

Some solid-state relays do not have a control coil. With solid-state relays, contactors, and motor starters, the PLC provides a trigger pulse that causes current flow. There are no physical contacts to change position, so these devices can easily last for years, but they are susceptible to damage from vibration and voltage surges.

Smart relays have built-in mini-processors (Figure 10-2). These relays can often perform multiple timing functions and may offer the ability to control how outputs act, almost like limited PLC systems. These devices are the relay industry's attempt to hold on to market share in the modern industrial world. There are times when a small amount of control is needed, but not a full PLC system, and the smart relay nicely fills this niche.

 FOOD FOR THOUGHT

10-1

When motors are first powered up, they draw a large amount of current to get things spinning. This currency draw reflects the fact that two magnetic fields are present in the motor—one in the stationary part and another in the rotating part. Before the motor starts spinning, there is no counter magnetic resistance, so the only limit to current is the resistance of the wire and a little bit of resistance from the magnetic field generated in that wire. As the motor begins to spin, the interaction of the two magnetic fields creates a resistive force that lowers the amount of current coming in. At top speed, this resistive force is at its maximum and the motor pulls the least current. As the motor slows due to load, the magnetic fields fall further out of sync, the resistive magnetic force lessens, and the motor pulls more current. If the motor continues to slow, the shaft eventually stops spinning, the magnetic resistive force drops significantly,

the current draw increases greatly, and the copper wires heat up. Ultimately, the resin coating on the wires melts, shorts develop, and there is a high chance of fire.

That is a very simplified overview of how motors work, but it serves our purposes here. You may be asking yourself, "How can a motor pull enough current to catch fire and *not* blow the fuse?" In fact, the fuse protects the motor circuit only from dead shorts, in which the voltage jumps to very high levels. Because of the large amount of inrush current when a motor starts, we have to fuse the system high enough to allow for this high voltage at start-up, even though it should last for only a few cycles of the sine wave.

Example

As an example, suppose a three-phase, 150 HP (horsepower) motor running at 460 V has a full load running amperage rating of 163 A. The locked-rotor current rating for this motor is given as *G*. When we look up this value in a chart in the National Electrical Code (NEC), we find that this motor has kVA per HP of 5.6 to 6.3. It is always best to use the highest number to be safe. The calculations are as follows:

$$150 \text{ HP} \times 6.3 = 945 \text{ kVA}$$

$$I_{line} = \frac{945 \text{ kVA}}{\sqrt{3} \times 460 \text{ V}} = 1186 \text{ A}$$

(See the NEC code book, an Ugly's reference book, electrical manual, or a good motor text for more information on these formulas.)

In this case, the inrush current is 728% greater than the running current, so we have to oversize the fusing to handle this inrush. In reality, the fuse cannot allow the inrush and also protect the motor when running normally. The overload's job is to allow the motor to run above the rated amperage, plus a small variance factor, for a short period of time. This lets the inrush through to the motor without tripping the fuse as well as handles short-term current spikes used to deal with added or changing load on the motor. If the current draw exceeds the set limit for a period of time, usually no more than 30 seconds maximum, the overload will open and stop the motor.

Some overloads work via heat generated by the current. Often these devices have two dissimilar metals joined together that expand at different rates. As the heat increases, the metal bends and eventually triggers contact change. The greater the amperage, the higher the heat, and the faster the metal bends. Another way to accomplish this task is through magnetic fields. With this method, coils of wire are used to generate a magnetic field that attracts a metal piece to change contact state. The higher the current, the stronger the magnetic force.

In both methods, a spring or some other device is present so that the contacts change abruptly and not slowly—a gradual change would burn the contacts because an arc of electricity is created when contacts open and close. With the heating method, ambient temperature can be a problem for overloads. Hot environments can cause false trips, while cold ones can prevent contact change and prolong the duration of high current to the motor. In contrast, the magnetic method is unaffected by ambient temperature.

Most overloads have two or more contacts of both the "normally open" and "normally closed" varieties. Make sure you use the right set when wiring them into your system. The standard practice is to wire the normally closed contact in line with the motor contactor control coil so that the overload acts like a stop button and kills the motor circuit when it trips. With PLCs, you could use the voltage from the overload as an input signal, but this is a bit dangerous. If someone messes with the program or does not set up the system correctly, the motor would then have no protection. I recommend always having the overload wired in as a part of the control coil circuit, with the option to add an input signal to the PLC, rather than replace it.

FIGURE 10-2 In the upper-right hand corner, notice the smart relays; they have some of the options found in a PLC.

Maximumm/Shutterstock.com

Another question that may have come to mind is, "What is the difference between a PLC and other controllers?" The controller for an ABB, FANUC, Baxter, or other robot was designed by the robot manufacturer and is a proprietary unit that runs specialized code to which you may not have any access. Sure, you can change the robot program, but anything beyond that typically requires the vendor's involvement.

Computer numerical control (CNC) controllers are similar to PLCs in that they run a specialized type of program and you can change the operation fairly easily. The key difference is that CNC controllers focus on axis-based motion control. CNC controllers use **G-code**, which is a specific programming method that has proved very effective for motion control and is relatively easy to learn. I have programmed G-code, simple stuff only, and found it more or less intuitive once I learned the specific commands it uses.

Where the PLC differs from the rest of the control options is its versatility. Just as the robot can paint, weld, inspect, load, or perform a variety of other functions, so the PLC can control robots, conveyors, safety systems, machines, processes, and almost about anything else found in industry. PLCs are designed to be versatile controllers: Thus, once you learn the specifics of programming the brand of your choice, you can use the PLC to control almost anything and/or everything in your facility. Just as we can change the operation of the robot with tooling changes, so we can change the operation of the PLC with card changes. (We will cover this topic in greater detail in the components section.)

In summary, the PLC is a highly flexible, specialized, industrial control system hardened to survive the extremes of temperature, contaminants, and magnetic fields found in industry. It was invented/designed to take the place of relay logic, but its application shave continued to expand to the point that many systems now offer most of the programming options found in the computer realm. The PLC of today accepts multiple programming methods and provides an instruction set to make the job of information sorting easier for those writing the programs. PLCs range from simple units with a few inputs and outputs to units that hold a large number of cards, creating a multitude of options for operation and offering expansion options to meet any need found in industry. It is this versatility and stability in operation that has led to their widespread use.

A BRIEF HISTORY OF PLCs

To find the birth of the PLC, we have to travel back to New Year's Day, 1968, when General Motors (GM) consulted with Dick Morley and his company Bedford and Associates. GM was looking for a solution to the relay logic problem—some kind of computerized controller that could take the place of all those relays, and the hassle involved with them. Some of the requirements included:

- A solid-state system that was flexible like a computer but priced competitively, when compared to the relay logic system it would replace

- Easily maintained and programmed in line with the accepted relay ladder logic way of doing things

- Able to work in an industrial environment, with all its dirt, moisture, electromagnetism, and vibration

- Modular in form to allow for easy exchange of components and expandability (Automation Direct, 2017)

The first attempt at creating a system that would meet these needs failed, as it had too little memory and was too slow. In 1969, though, the Modicon 084 was released. Modicon is short for MOdular Digital CONtroller, and the 084 designation came from its being the 84th project of Bedford Associates. The Modicon 084 is considered the first PLC and was the first sold on the market. In 1973, Bedford Associates unveiled the Modicon 184, the brain child of Michael Greenberg and Lee Rousseau, which was an upgraded version of the 084 with more bells, whistles, and memory. This model prompted the company to add 90 employees, and Bedford Associates sold $5 million of the PLCs as industry adopted the technology. In 1975, it released the 284, which had a microprocessor, and the 384, which boasted the first digitized process algorithms.

In 1994, the Modicon brand released the Quantum range as a truly open approach to automation control designed for complex processes and the integration of multiple networks. It utilizes **programmable automation controllers (PACs)**, which are a combination of PLC operation and PC-based control systems. In other words, a PAC is an industrial environment–hardened PC that can interact with various components and run PLC software in a similar fashion to a dedicated PLC.

Before looking at some of the other big names to come along in the PLC world, let us take a look at how ladder logic—one of the most common ways to program PLCs—came to be. The point of the PLC was to replace ladder logic but to do it in a way that was friendly to those involved with the changeover and day-to-day operations of the systems. Because relay logic was based on electrical components, it was often drawn in a schematically method (Figure 10-3). With this method, we start by drawing two vertical lines on the page, one on the left side that is the main power in and one on the right that is the return path in a 120-V or DC system or the second power line in a 220-V system. These are called the **power rails**. Next, we draw a line from the left-hand side to the right, placing various components in the electrical sequence of the path, thereby mapping out the logic of the circuit and showing what has to happen to get power flow from left to right. We call these **rungs**. **Ladder logic** is the PLC equivalent of this electrical schematic, in which the physical components of a relay logic schematic are replaced with symbols that represent instructions for data sorting in the PLC. It is *crucial* to understand the distinction here: The relay logic ladder-type schematic is an *electrical path* drawing and the PLC ladder logic structure is a *data path* drawing. We will dig into this issue in greater detail later on.

Another major player in the PLC world, Allen-Bradley, introduced its first PLC in 1970, called the PDQ-II; this PLC turned out to be too large, too complex, and too hard to program, so Allen-Bradley had to go back to the drawing board. In 1971, it rolled out the PMC (Programmable Matrix Controller), which still didn't really fit the PLC bill. In 1974, Allen Bradley filed a patent for the first PLC with parallel processing and introduced the first CRT-based programming panel for PLCs. The programming panel might not sound

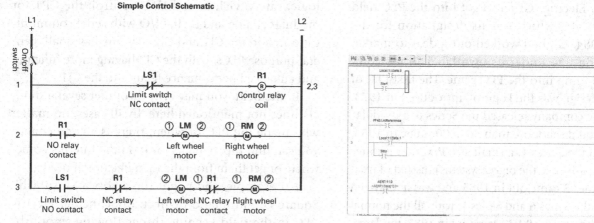

FIGURE 10-3 (left) The control schematic for a coaster bot I built, drawn in the relay logic schematic format. (right) An example of an Allen-Bradley 5000 PLC program, utilizing the same basic structure but containing instructions instead of electrical components.

like much, but it was a major improvement for working with PLC code and systems. In 1976, Allen-Bradley introduced remote I/O, which meant user could have control without needing a processor for the various input and output cards. (We will dig deeper into this concept later.) In 1978, the company released the PLC-2 controller, followed by the PLC-3 in 1981.

In 1986, Allen-Bradley rolled out the PLC-5, which is still found in industry today, and linked the PLC to the personal computer to make monitoring and programming much simpler. Use of the PLC-5 system is waning now, as it is no longer supported by Allen-Bradley, making it hard to get parts when something breaks. The SLC 500 replaced the PLC-5 as the flagship for Allen-Bradley PLCs in 1991 and was still supported at the writing of this book. Allen-Bradley has stopped supporting the early processors in this line, however, and is focusing all of its energy on the ControlLogix (launched in 1997) and CompactLogix (launched in the early 2000s) lines. In 1994, Allen-Bradley released the MicroLogix line, which was a reduced-capability system that used the same software as the SLC 500 systems. You can still find MicroLogix systems in use by industry, but the only models still supported at the writing of this book were the 1100, 1200, and 1400 models. Given Allen-Bradley's focus on the ControlLogix

and CompactLogix lines, it is reasonable to suspect that the company will continue to phase out the SLC and MicroLogix lines over the coming years (Figure 10-4).

Another big name in the PLC industry is Siemens, which holds most of the market share in Europe and a good chunk of the U.S. market. In 1974, Siemens debuted the Simatic S4 as the first soft programmable controller, beginning the rise of the PAC system. The Simatic S5 was introduced in 1979 and became one of the most widely installed PLCs in the world. In 1980, Siemens added its fail-safe controls technology to the S5, paving the way for the safety-oriented PLCs of today, which offer multiple levels of built-in redundancy. In 1994, Siemens released the Simatic S7 and Step 7 PLCs, with the Step 7 being a stripped-down, cheaper system for less complex applications. In 1997, the company released the PCS7 controller for the S7 system, which combined PLC function with a **distributed control system (DCS)**—that is, a system in which autonomous controllers are distributed throughout the equipment, but there is a central operator supervisory controller. The S7 remains Siemens' flagship PLC, though it has been upgraded many times since its introduction in 1994 to the point that it can handle anything industry might ask of a PLC these days.

FIGURE 10-4 (left) A MicroLogix 1100. (middle) An SLC system. (right) A ControlLogix system.

General Electric (GE) also got into the PLC field with the PC-45, which was its designation for the Modicon 084. GE had worked out a deal to market the Modicon 084 under its own label so the company could jump into the PLC game. The first PLC of GE's own design was the Logitrol, introduced in 1973. In 1981, the company released the Series 6, a modular PLC that would sell more than 500,000 units. In 1983, GE released the Series 1, a small-size PLC that features removable cards (like the bigger systems) instead of fixed I/O. The Series 3 came out in 1985 and was in between the sizes of the Series 6 and Series 1, with all the normal bells and whistles. The 90-70 came out in 1987; this large-size PLC is designed for complex applications requiring large numbers of I/O points and a large amount of memory usage. In 1988, GE released the Series 5, designed for mid-range PLC applications. In 1990, it launched the 90-30 model, which is smaller and sleeker, has more than 100 different I/O modules, and is comparable to the Allen-Bradley SLC systems in size and capability.

These historical systems have since been moth-balled, but GE does have some current models in the field. The VersaMax, released in 1997, is designed to work with remote I/O, a stand-alone PLC unit, or distributed control for as many as 256 I/O points with a processor boasting 12K of memory. Although this sounds puny, keep in mind many of the early processors ran on 4K of memory with no problem, as the files and data saved in the PLC consume data in the bits and bytes range. In 2002, GE introduced the Series RX7i, which is a PAC-type system designed specifically to replace the 90-70. The RX7i accepts 90-70 modules, making changeover to the new PLC easy; is designed to provide application portability across multiple hardware platforms; sports 10 MB of memory; and is built for mid- to high-end applications. In 2003, GE launched the Series RX3i PA-type system, which is smaller in form and designed to work well with networks, human–machine interfaces (HMI), and process control, among other processes, with system performance and flexibility being a focus.

Another player in the PLC field is Omron. In 1972, this company developed the SYSMAC PLC for use in Japanese automation, which had developed into a standard line with microcomputers by 1978. In 1983, Omron released the C-Series, an upgrade of the SYSMAC line. The C-Series boasts more than 40 models, some of which are still in production. Omron's production line at the writing of this book included the NX, NX1P, and NJ, which are designed for motion, logic, safety, vision, and HMI control; the CP1 for modular analog and digital I/O with serial communications; and the CJ1 and CJ2, designed as small, general-purpose PLCs, with the CJ2 having more function and enhanced performance relative to the CJ1.

In the field, you may also encounter several models/lines not mentioned here. In all cases, no matter who made the PLC or when, there is a good chance you can find the manuals for it on the Internet somewhere or get them from the manufacturer if needed.

For example, you might run across a Mitsubishi, Square D, Automation Direct, or Schneider Electric PLC in the field keeping things flowing smoothly. Some of these units might be obsolete, while others are relative newcomers, hoping to carve out their own share of the PLC market. Automation Direct is a great example of the latter, with its low-cost PLCs and free programming software. While the products may not be as tough as Allen-Bradley or Siemens equipment, a company can buy several sets of replacement hardware over the years of use and still come in under the initial cost of other equipment. Many maintenance technicians encounter these PLCs when they come in as a part of new equipment purchased by their company. In the future, you should expect the field of PLCs and other such controllers to continue to change and evolve as we continue to find new and better ways to control things.

BASIC COMPONENTS OF THE PLC

Now that you have a basic understanding of what a PLC is and how PLCs came to be, this is a good time to dig into the mechanical side of the things. While the specifics of the construction vary from PLC to PLC, they all have the following components:

- Inputs that provide information about the process
- Outputs that send power in one form or another to devices in the field
- Processor(s) to sort, store, and manipulate data
- A backplane or something similar to allow the various components to communicate

As we look at each of these areas, we will explore how each performs its function to make the PLC a cohesive system. Keep in mind that just as we cannot take a teach pendant from a FANUC robot and use it to control an ABB robot, we cannot take the card from an Allen-Bradley PLC and expect it to work in a Siemens model.

Inputs

The input portion of the PLC responds to voltage coming in from field devices and creates data in the processor accordingly. If the input is digital, where the only states are 1 or 0, on or off, then usually the incoming voltage must exceed some threshold value to be considered at a 1 state. There will also be a lower limit, below which the data is changed to 0. The voltage coming in can be either AC or DC—but be careful not to mix the two or send excessive voltage to the contact on the input card lest it damage that connection. Devices that send digital inputs typically include proximity (prox) switches, pushbuttons, photo eyes, light curtains, limit switches, simple level sensors, and anything else that is designed to answer a yes/no type of question.

If the input is analog, where the incoming value varies and that variance has a meaning, then the value stored in the process will be a binary representation of the variable value coming in, usually taking a word or more of data. Analog inputs can monitor voltage or current, and some input cards allow the user to decide what the contact monitors. Common amperage ranges are 0 to 20 mA or 4 to 20 mA; common voltage ranges include 0 to 1 VDC, 0 to 5 VDC, 0 to 10 VDC, 1 to 5 VDC, −5 to +5 VDC, and −10 to +10 VDC. While any of these may be used in industry, the 4 to 20 mA and 0 to 10 VDC or −10 to +10 VDC ranges are the most common. With the 4 to 20 mA range, the 4 at the low end helps to avoid transmission of a false signal when something happens to the connection and power is lost, as 0 is below the accepted range. An alarm can instead be programmed to trigger specifically with power loss. Analog sensors are used to measure temperature, distance, volume, flow rates, and anything else that has a varying value that we want to track. The input module has an analog-to-digital converter that turns the variance in voltage or current into a digital signal that the processor can understand.

You may have noticed that analog inputs typically use a word of data, but you may not know exactly what a "word" is in this context. Each 1 or 0 in the processor's memory is called a **bit**. A group of 4 bits is called a **nibble**; a group of 8 bits is called a **byte**. **Words** of data consist of multiple bytes, with the number depending on the design of the system. Many of the older PLCs used 16-bit or 2-byte words. With the advancement of technology, we have developed systems that use 32-bit or 64-bit words, which PLC companies can build into their own systems. Having more bits in each word enables larger numbers or more information to be stored in memory, thereby expanding the system's abilities and increasing the overall number of bits the system can use. Comparing a 4K memory, 16-bit system and a 4K memory, 32-bit system, the 32-bit system has twice the memory capacity of the 16-bit system, as 4K of memory means 4000 words of memory. Because of the way computers are structured, 1K of memory actually means 1024 words of memory (not 1000), so 4K of memory is actually 4096 words of memory.

The input portion of the PLC creates data based on the voltage or current present at the terminal screws. There is no capacity, no ability, and no way for the input portion of the PLC to check the field condition of the device to which it is connected. There is no camera. It cannot get up and walk over to the device. It has no voice to ask questions and no ears to hear answers. The only connection between the part of the input that creates data and the outside world is the terminal screw to which the wires are attached. That's it. If the field device is working correctly, but the connection is lost between it and the input portion of the PLC, then the information sent to the processor will be a 0. If someone changes two wires around, the input module or connection has no way to know this and will create data based on the current or voltage present.

Some PLCs use **modules**, which are removable devices designed to serve a specific purpose or function. In PLCs that are not designed to accept modules, the inputs are limited to a set number and type, with the terminals being mounted in the main body of the PLC unit. In contrast, in modules (also called **cards**), the electronics inside reside on a flat platform that connects with the system in some fashion for communication and power draw. Some modules feature a removable connection pad that the field wiring attaches to, which allows you to swap out the module without having to rewire everything. You can get modules with 8, 16, 32, or even 62 connections—but the more connections per module, the less current they can withstand. Generally speaking, more connections also mean that the screws are smaller and you have less space for each wire.

The types of input modules range from simple digital modules, to various types of analog modules, to specialty cards such as high-speed counters that have small processors on them to store input data that might otherwise be lost due to the length of the PLC's scan cycle. Modules are available that cover all the common industrial needs as well as a few special situations, so there is a good chance you can find whatever you need with a bit of searching.

To keep the processor safe from power spikes, misconnections, and other such electrical mishaps, optical isolation is used. With **optical isolation**, the power coming in lights up a light-emitting diode (LED) or similar light source that interacts with a photo resistor or other light-sensitive device. When the light from the LED hits the resistor, the resistance value drops until voltage can flow through the circuit, signaling the processor to put data into memory. Several different components can be used for this process, but the underlying principle is the same—the only thing linking the physical connection in the world to the processor is light. With optical isolation, if someone makes an improper connection or a device shorts out, the only thing damaged is the light source, and we lose only one connection on the card instead of the high-dollar main processor/controller.

The same circuit, if not the same LED, that provides optical isolation also illuminates either a specific LED on the front of the PLC module or often a specific number on the input module. A savvy technician can watch the input module for the proper LED lighting up when activating an input to prove proper operation. If you trigger the input and the proper LED changes state, then you have just proven the connection between the field device and the PLC is working as it should.

Keep in mind that just seeing any LED change could lead to some flawed thinking. If any field input was triggered at that moment, you could be seeing that input, rather than the one you desire. Also, the input you are monitoring could be connected to the wrong terminal but working as directed. Finally, if the field device is wired "normally closed," the light should go off when the device is activated. If it is wired "normally open," the light should come on when it is activated. Some devices, such as pushbuttons, may have both types of wiring or multiples of both, so you may see several lights change. In these cases, make sure you fully understand the field device so you can accurately verify operation.

Outputs

The output portion of the PLC responds to data in specific locations in the PLC and opens or closes connections between terminals on the output and common connections. The connections are made by standard relays or solid-state devices like triacs for AC or diodes for DC. **Diodes** are solid-state devices that allow current to flow in only one direction. **Triacs** are solid-state devices designed like two diodes in antiparallel orientation, with a trigger connection to start current flowing. When the connection is triggered, whatever power is supplied to the common connection for that terminal is passed, if possible, to the output terminal screw and then onto whatever device.

The primary difference between an input and an output is that an input creates data in the PLC processor while an output acts on data from the PLC processor. Outputs also use optical isolation, but the light is not tied to the processor and the light-reactive portion is between the common and the terminal connection on the output. When the data for the output address is set to 1, the light comes on and the light-reactive device triggers the connection for the output. For analog outputs, a digital-to-analog converter takes the digital signal from the processor and converts it into the proper voltage or amperage signal to flow out to the field device.

The LED on the output card is illuminated when the connection between common and terminal is triggered, and it is triggered either by or a part of the optical isolation circuit. While a light on the input card generally verifies field device operation, the light on the output card simply confirms the common-to-terminal connection on the output card. In other words, you can have the LED on for an output and yet not have a single volt flowing from it. You will need to use a meter to verify operation once the output is triggered.

This leads us to another thing to watch out for: A triac output can show 120 VAC when it is not triggered and there is no actual flow. Because the triac is a solid-state device, a small amount of current continually leaks through it, though not enough to power a device. Meters are low-resistance devices and, as a consequence, may give false readings of full voltage even though it is not really present. To fix this, you can use a resistive load adaptor for the meter, which puts a bit of load on the circuit and shows that power is not really flowing.

The other concern with triacs is that leakage current can build up enough to give you a decent shock. Even if the output is "off" in the PLC and the light is not on, if you touch the terminal while the module has power, it can give you a good zot! While the amperage should be low enough that it is not life-threatening, your reaction to the pain could well cause you to fall, hit yourself on something sharp, or jerk your hand into another live circuit with more juice. If the module has a removable wiring bracket on the front, you can pull that out and then make your wiring changes; otherwise, you should power the system down before making the changes.

Output cards come in the standard digital and analog formats, but a large number of specialty modules are also available. Some are designed to control stepper motors, numerous axes of motion, high-speed operations, and other devices that need a bit more than just a standard digital or analog signal. Just as we have a wide range of input cards to cover the typical industrial situations, you can also find a range of output cards to cover most, if not all, of the tasks needed in your facility.

Specialty Modules

Some modules do not truly fit into either the input or the output category because they do a bit of both. Communication modules and networking modules are great examples. They can bring data into the processor as well as transmit data out to other systems about the state of things in the PLC. Ethernet modules are a widely used type of communication module in today's PLCs, especially for those that do not come with an Ethernet connection or those in which that connection is used by a specific piece of equipment that needs dedicated communications. You can also find modules designed to work with specific network protocols such as DeviceNet or ControlNet.

Another specialty module you might encounter is the PID module. **PID**, which stands for **proportional integral derivative**, is a mathematical process used to vary an output to maintain a specific set point. The math behind this is fairly complex, and the setup of these systems can result in perpetually unstable results if you do not know how all the pieces work together. Nevertheless, PIDs are popular control devices in industry for maintaining level, flow, temperature, and a multitude of other variables. The good news is that with self-adjusting modules, all you have to do is set where you want the system to be, make sure all the connections to the card are correct, and give the PID module enough time to figure out the system. If the variable being monitored takes a long time to change, you may have to adjust a time setting for the module to reflect this to prevent erratic behavior.

Another type of specialty module you might encounter are safety modules. Whether for input or output, these modules have redundant connections so that if one connected path is lost, there is a backup and a good chance of continued operation. Safety modules also have internal error reporting and testing functions that send data to the processor and can trigger alarms as well. The internal testing portion is the part that usually does not strictly meet the solely input or solely output module description.

Other specialty modules include ASCII modules that transmit and receive data using a method similar to the way your keyboard sends data. Some PLCs let you add memory to the system in this manner, using the module like a spare hard drive. BASIC modules are used to interface with bar-code readers, scales, modems, robots, and other such systems. Binary-coded decimal (BCD) modules work well and offer seven segment displays and thumb wheels. You can also get combination modules in which half of the terminals are inputs and the other half are outputs, but otherwise the modules work like regular I/O cards. As you explore the world of PLCs, you are sure to run across options not covered in this section, so make sure you take the time to learn the specifics of the modules in your PLC and save yourself some grief down the road.

Processor

The processor is the heart of the PLC; it is where all the data is manipulated based on the program. In some PLCs, the processor is built into the unit and there is no real option to change it out for another one. This is often the case with cheaper PLC units or those that have a set number of inputs and outputs. In the rack-style PLCs, which allow us to add or remove modules, the processor is often just another module. Because the processor has to communicate with all the inputs, outputs, and other various modules, some PLC brands require the processor module to reside in a specific location. If you put the processor module in the wrong slot, generally all that happens is that it does not communicate with the rest of the I/O and often will fault out. The specifics of what happens vary between manufacturers based on how they monitor things.

Most PLCs have LEDs that provide important information about the state of the processor (Figure 10-5). The RUN light indicates that the PLC is in run mode and is capable of scanning the program and altering data. If the RUN light is solidly lit, the processor

FIGURE 10-5 An Allen-Bradley CompactLogix controller. Notice the LED section on the upper-left side. While the specific location and lights may differ from brand to brand, they all should have some means to give technicians information about the system at a glance.

is scanning the program and manipulating data as directed. If it is flashing, it may be transferring memory or faulted out. PLC processors traditionally have four modes of operation that are selected with three different key or switch positions. In **run mode**, which is one of the extreme switch or key positions, the processor scans the program and manipulates data. In this mode, we can pull the program from the PLC to our interface device, generally a laptop, which we call an **upload**, but we cannot send a program from the laptop to the PLC in a **download**. (Note: In the world of PLCs, we use the terms "upload" and "download" in the opposite way that we often use these terms to refer to data manipulation with the Internet.) The middle position of the key or switch is the **remote mode** (Rem); in this mode, we can use the laptop to upload or download data. The third position, the other extreme of travel, is **program mode**. In this mode, we can upload or download data, but the processor does not scan the program and the I/O does not update with changes.

The difference between remote mode and program mode is that we can change the mode of operation with the laptop or another communication device when the switch is in the remote position. When the key or switch is in program mode, we have to physically change it before the program can start scanning. The same is true when the key or switch is in the run mode: We have to change to remote or program mode to make any changes to the code of the PLC. Two modes are possible when the key or switch is in the

remote position: remote run or remote program. These modes work like the standard run or program modes, respectively, but we can switch between the two via commands from the software on the laptop. When using the switch or key, the remote mode that the processor enters depends on the mode it was in before going to remote. If it was in run mode, it switches to remote run; if it was in program mode, it changes to remote program. This is how we get four modes out of a three-position key or switch.

Many companies use run mode as a kind of fail-safeguard to keep employees from changing the code/operation of the PLC during normal operations. While it is true that you cannot download to the PLC and thus make major changes to the program during run mode, a technician can make changes that have dire results. The value of certain instructions can be changed with the processor in run mode, leading to disaster.

At one of my previous jobs, we had an overhead chain system for moving parts from the work cells to a wash station in another part of the department. Each cell had a loading station that could hold up to three racks in reserve for parts loading, to prevent parts from stacking up. Occasionally something would happen and the count would be off, requiring someone to go in and change the count in the PLC. Sometimes we would change the count, thinking it was off, only to discover the missing rack(s) were in transit but had not received their final destination yet. The result of this error was too many racks trying to park off the main chain line, with the last rack having the front part in the parking zone and the other half being pulled along by the main line. When this happened, the chain system would bind up and eventually stop when the load on the motors for the chain drive became too high. We would then have to climb up on a catwalk and take a sledgehammer to hit the trolley release for the rack, taking the pressure off the chain. This would cause the chain to jump forward as much as 10 feet. At times, other racks on the chain would lurch forward as well, making life a bit interesting on the catwalk, to say the least. This was all because we changed the count on an instruction in the PLC program while the system was in run mode.

Another LED that is fairly common is the OK or fault signal. If an OK light is used, if the processor is running normally, the LED for this state is solidly lit. A flashing LED typically indicates some type of error condition that has stopped the scanning process. If it is completely off, then something serious is going on—for example, no communication. The fault LED generally works in the opposite way: Off means everything is working well, whereas flashing means a major error is preventing scanning of the program and full on means no communication.

Force is another LED found on many processors and something pretty serious in its own right. **Force** is a condition in which we set the state of data for an input or output, ignoring the actual field condition for inputs and the data result of the program for outputs. Forces are a great tool for troubleshooting purposes, but they are dangerous because you can make a motor come on with no regard for machine position or operator safety. If they are used incorrectly, you could kill someone! If the force light is blinking, there are forces in the system, but they are not turned on at the moment. If the light is solidly lit, that means forces are active and something is no longer under normal control. If the light is off, that means there are no forces in the PLC and none are active.

Some PLCs use a battery to maintain the program and other volatile or non-stored memory when the system loses power. If the light is on, that means the battery is low; if it is off, that means everything is good to go. Usually a capacitor in the PLC can maintain power for a short period of time so you can power down the system to change out the battery. My recommendation, however, is to change out the battery with the power on if you can do so safely. If the battery is dead and you power down the system, there is nothing to preserve the program and other such data, and they will be lost. You can download the program to the processor once more after changing the battery and getting everything going again, but this can lead to a fair amount of downtime if you have trouble finding the right program, do not have a copy of the current program, or have trouble connecting to the PLC.

Other LEDs may indicate the status of various communication paths or other important items specific to the processor in question. With communications lights, often they blink when data is being transmitted back and forth and then go dark when nothing is happening. Many processors have multiple communication methods, meaning that they can connect to other devices and still have a path for technicians to use. A word of caution: If you tell the processor to communicate on one of the other paths, make sure the right cables and such are in place so that the processor can actually use that path. The Allen-Bradley SLC series

processors typically had at least two paths, but you could use only one for communication with the laptop. If you changed the default communication path but did not have the right cables or such, you could no longer make changes to the program, though the current program would run normally. To fix this problem, the technician would have to either get the right cables and setup or reset the processor card to the factory default, wiping all the volatile memory in the process.

The memory for the system often lives on the processor module, or in the same area as the processor for fixed units. In older PLCs, the only way to get more memory was to upgrade to a processor module with more memory or add in a memory storage module, which was not an option with all brands. Some of the newer PLCs have taken advantage of flash memory and often use an SD card of some type to store programs and other important data. In these systems, you can add memory up to a specific size as well as change out the cards as needed, such as during changeovers to run a different product line and operating program. With this style of memory storage, you may find that the battery is no longer needed, so there is no battery light and no expensive battery to change out.

The specifics of the processor depend on the brand (manufacturer) and the model as well. Sometimes there are pins to reset the processor to the factory settings, or to wipe the memory should someone set a password and then leave the company. The type of processor and its running speed also vary, but all processors in PLCs are hardened to deal with heat, vibration, humidity, magnetic fields, and anything else industry has to throw at them. The specifics of connecting your laptop to the processor also vary by brand, but there will be a method to do so and a manual or two to help you figure everything out. **Safety PLCs** contain two processors that are identical in nature; they are set up so that if the primary processor fails, the secondary processor picks up where it left off without so much as a hiccup. These PLCs are designed to keep functioning normally even when something breaks, with extra circuits to facilitate this continuing operation as well as dedicated circuits that notify the operator when something has failed. Safety PLCs are used in industries such as oil and gas where a shutdown or blip could result in an explosion, fire, loss of life, or damage to the environment.

Just as you need to learn how all the various modules or I/O functions of your PLC work, so you should take some time to learn the basics of the processor. Find out what your communication options are and how they work. If you do not have the right cables or software, get what you need. Find out whether the unit uses a battery or flash memory. Check the forces light to determine if forces are in use or lying in wait like a landmine. If you encounter an unfamiliar LED signal, read up on the system so you know what it tells you. The idea here is to learn all you can *before* something goes wrong so you can be ready. This would also be a good time to learn where the master programs for each machine are stored and how to access them should you need one.

Backplane

There is one more crucial component of the PLC system we need to look at—the backplane. The backplane provides a path for power and communication between all the various modules or parts of the PLC. For some PLCs, it is part of a rigid structure known as a rack—a metallic shell that the various modules fit into, with guide rails to keep everything straight (Figure 10-6). The backplane is found at the rear of the rack, and is where the cards in the modules plug into. This configuration makes it very easy to swap out modules and not damage the small pins used to make power and data connections. A rack also provides physical protection to the modules with its metal frame and is a great place to attach the ground for everything. Often there are vent holes at the top and there can be metal burs on these holes, so be careful where you put your hands when adding or removing modules.

The construction of the backplane determines how the user interacts with the system and can designate

FIGURE 10-6 A technician removing a module from the rack. Notice how it comes out straight because of the guide rails built into the rack as well as the ventilated metal on top.

specific locations for certain modules. In the SLC 500 and PLC 5 family of Allen-Bradley PLCs, the control module had to be in the slot closest to the power supply; known as slot 0, this slot is always to the right of the power supply. The SLC 500 and PLC 5 devices required this slot due to the way the backplane is constructed, as these processors required some special connections that the other modules did not get. In the ControlLogix and CompactLogix family of Allen-Bradley PLCs, the processor can be in different locations. For the CompactLogix family, the processor has to be within three slots of the power supply due to power concerns. I have not heard of any specific requirement for the ControlLogix family. With that said, most users still put the processor next to the power supply to keep everything consistent. The CompactLogix family goes a step further and allows the user to put the module on the left side of the power supply as well as on the right side (the traditional position).

You might wonder why the older PLCs are so rigid regarding their placement, whereas the newer ones seem to be a free-for-all. This difference is due to the way the backplane is constructed. The older backplane design was very much like that found in older computers, where you had to put communication cards, hard drives, sound cards, and so on in specific places. The newer backplanes are designed to work like a network cable, providing the same communication and power to all the connected items. This design has some distinct advantages. First, the processor is not chained to slot 0. Second, it is easier to get information about all the modules in a PLC from remote locations. Third, this design facilitates communication among PLCs, modules, controllers, and anything else that is connected to a backplane and also communicating with the plant network (Figure 10-7). With these new designs, the backplane becomes an extension of the network instead of a stop-off node, so that the items connected to the backplane are actually network nodes. With the older backplanes, we had to use special communication modules and run most communication through the processor module. With the newer ones, we can simply use the Ethernet connection on the processor module or add an Ethernet module to the rack; everything is then on the network for us to work with.

Some PLCs use modules but do not have a rigid rack structure they fit in. With this type, a ribbon cable and connector setup or similar adds the new module to the group. Often these connections are

Xmentoys/Shutterstock.com

FIGURE 10-7 Two PLCs, one with quite a few modules in it. Notice the modules with Ethernet ports on the front, which make it easy to get these PLCs on the plant's network.

made at the side of the module, near the back, and locking tabs are used to hold everything secure. With this more flexible arrangement, we depend on options such as DIN rail to mount the system and provide rigidity. **DIN rail** is a mounting system that consists of a specially designed metal (aluminum) rail, often 35 mm in width, to which we can clip components to provide stability (Figure 10-8). DIN stands for Deutsches Institute fur Normung, which translates to German Institute for Standardization. In the 1970s, German companies started using a 35-mm standardized rail to mount their equipment, and over time it has become an industry standard worldwide.

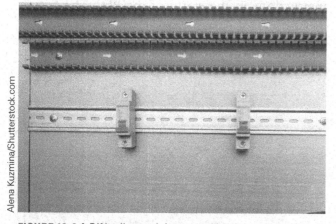

Alena Kuzmina/Shutterstock.com

FIGURE 10-8 A DIN rail containing some circuit breakers. The gray sections above it are plastic wire ways.

When dealing with all-in-one PLCs, like the MicroLogix 1100 from Allen-Bradley, you will find that there is no external access to the backplane other than the main communication ports provided. This can be limiting when you want to add these devices to

the network or connect them with a touch screen that requires a communication link. The devices still have communication and power connections internally, but they are hard-wired to the provided components and accessible only via the provided connections or by dismantling the device. The loss of flexibility is one of the tradeoffs when you select a lower-cost PLC. There are many applications where an all-in-one PLC can easily handle the requirements, so there is a high probability you will run across these controllers in the field. Just keep in mind that using these PLC for tasks beyond what they were designed for can be challenging, to say the least.

The specifics of the various components of PLCs vary from manufacturer to manufacturer and from model to model, but they will fit into the broad categories presented here. If you come across something outside of what was covered here or that you are unfamiliar with, I strongly encourage you to get the manual for it and learn more. Often little things like knowing how switches or jumpers work for specific settings, whether the module has fuses, or how to utilize redundant circuits can save you hours of frustration during setup or repairs. Luckily many of the manuals for equipment today come in PDF or similar format. This makes it simple to download the manual and take it with you out into the field on your smart device.

OPERATION OF THE PLC

The hardware is only half the story when it comes to PLCs. Even though more than 95% of the problems in industry are hardware related, there is no way you can fix the other 5% if you do not understand what the PLC actually does inside the box. First and foremost, you *must* recognize that the PLC only processes data. The inputs tied to the PLC provide data in when certain conditions are met at the terminal for that input, data is processed by a user-created program, and finally connections are made to the output terminals in direct relation to the data manipulation. The PLC does not "check" the inputs in the field; it just knows what is happening at the terminal screws. The PLC does not "turn on" outputs; it merely triggers connections between specific terminals and whatever is tied to common. The PLC only does what we instruct it to do via the program; it has no idea what we "want." The PLC is an industrial-grade, harsh-environment-hardened, specific-task-designed computer—and nothing more.

One of the first keys to understanding the PLC is to understand how it processes information and the sequence in which this happens, which we call the scan cycle. Some PLCs use a step-by-step scan cycle, whereas others rely on an asynchronous process. We will look at the step-by-step type first.

With a step-by-step scan cycle, the first step is to check all the inputs. For the binary inputs, if the proper amount of voltage is present, a 1 is put in the box in memory that correlates to that terminal's address. If the input is analog, a group of 1s and 0s are put into a word's worth of boxes with the corresponding address, based on the amount of voltage or current present at the terminal. Notice that at no point does the PLC "check" or "determine whether the switch is on" or anything similar.

The next step in the sequential scan sequence is to read the program from top to bottom, left to right. In doing so, the processor is looking for a path of true logic through the various input instructions. If a true path is found, then the rung, or line of instructions, is considered to be true. If no true path consists, then the rung is considered to be false. The input instructions can look at only the specified addresses for specific data values; if the processor finds what it is looking for, the rung is considered to be true. The output instructions, based on whether the rung is considered true or false, create data at the assigned destination or perform a specific action.

With some software, all the inputs will be on the left of the rung and the outputs to the far right; others are less restrictive regarding instruction placement. In Figure 10-9, the first rung includes lines that drop down to form a rectangle with an instruction in them. The function of this branch is to provide an alternate path around certain instructions. The PLC will check the branches when looking for a true path, but it will not backtrack. The only way a branch can make a true path is if reading the rung from left to right creates that path. Thus, the branch can only take the place of the instruction(s) it straddles in the truth test.

The third part of the scan cycle is to connect or disconnect output connections. If the output is digital (on/off) in nature, a 1 in the data box for that address instructs the output module to make a connection between the designated terminal screw and the common connection. If the output is analog in nature, a specific voltage or amperage is sent to the terminal screw that matches the data in the word of data for that address.

In the fourth part of the scan cycle, the system checks for errors, determines how long the scan cycle took, and handles communications and anything else on the docket not covered in the other three steps. We really do not worry too much about this portion of the scan cycle for two reasons: (1) We really have no control over what happens here and (2) as technicians, we do not make any changes to this portion of the system. We can check data boxes to see the results of this portion of the scan cycle, but that is about the extent of our interaction with it.

Before we walk through how a full scan works, let us take a look at the three most commonly used instructions in the world of the PLC. The first is

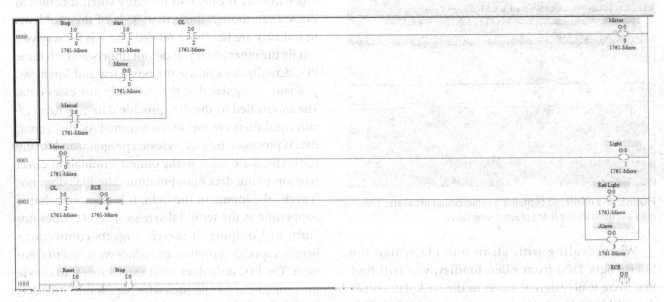

FIGURE 10-9 A program for an Allen-Bradley MicroLogix 1000. Refer to this image as you read about scan cycles and instructions to help you visualize things better.

XIC, which stands for "examine if closed": It is an input-type instruction. In Figure 10-9, on the top rung labeled 0000, the instruction labeled Stop, you can see one of these instructions. Here I:0/0 is the address for the box tied to a specific terminal. If the instruction looks in that box and finds a 1, then it will be true. If it looks in box I:0/0 and finds a 0, then it will be false. This instruction looks for a 1 in the box. That's it. That's all it does. Period. If you have taken a PLC class, you may have heard some other things in relation to the field device, but this is misleading!

The second instruction we will consider is XIO, which stands for "examine if open." It is another input-type instruction. In Figure 10-9, on the third rung down labeled 0002, the instruction labeled RCR, you can see an XIO instruction. Here O:0/4 is the address for the box tied to a specific terminal, just as we saw before. If the instruction looks in the designated box and finds a 1, the instruction will be false. If the instruction looks in the designated box and finds a 0, the instruction will be true. XIO works opposite to XIC in terms of true and false, but it, too, simply looks at data.

Notice the addresses of the boxes for the XIC and XIO instructions: One box was labeled I and the other was labeled O. Yes, I stands for input and O stands for output. Although both XIC and XIO are considered input instructions, XIO is looking at an output. This is a crucial point to understand. All input instructions look at the addressed data, *any data*! If we can assign the address for a piece of data, then the instruction can look at it. That includes inputs, outputs, status bits, portions of data words—any box of data that the system has a specific address for. Some data in the PLC is addressable only in a word format, which consists of a minimum of 16 bits or boxes of data. Data of this nature cannot be examined with the XIC or XIO instruction because they can look at only one bit or box of data.

OTE, which stands for "output energize," is one of the most commonly used output instructions. It can actually write 1s and 0s in the PLC's data tables. In Figure 10-9, on the top rung labeled 0000, the instruction labeled Motor, you can see one of these instructions. The address O:0/0 is for a bit box tied to an output terminal screw. The OTE instruction will write a 1 in this bit box if true logic comes in and a 0 in this bit box if false logic comes in. It *does not* turn on an output device or make any electrical connections;

instead, this instruction merely changes data in a data table. In the third portion of the linear scan cycle, the data in the output table is used to determine which terminals are connected to common and which ones are interrupted. That work is a function of the PLC, not the instruction.

Now that you have an understanding of the instructions, let us take a look at how the PLC processes them. For the sake of clarity, we will assume everything is digital (1s and 0s) and will look only at rung 0000, the top rung, in Figure 10-9.

- Step 1: Populate the input table based on voltage present. For the first run, the terminal screw I:0/0 (Stop) and I:0/2 (OL) have the correct voltage at the terminal, so 1s will go into their respective bit boxes. The rest of the inputs, I:0/1 (Start) and I:0/3 (Manual), do not have voltage, so 0s go in their bit boxes.

- Step 2: Read the program from left to right, top to bottom. The XIC for Stop looks for a 1, sees it in the box, and is true. The XIC for Start looks for a 1, sees a 0, and is false. The XIC for OL looks for a 1, sees a 0, and is false. Thus, no true path has been found yet. Next is the XIC for Motor. Because nothing has put a 1 in any outputs yet, the value of O:0/0 is 0, so this XIC is also false. Lastly, the processor looks at the XIC for Manual. This instruction looks for a 1, sees a 0, and so is false. There is no true path to the OTE Motor, so the false logic is present; thus, the OTE Motor instruction will write a 0 in bit box O:0/0. The processor is now done with this run; it moves to rung 0001: It does not go back to the XIC instruction that looks at the Motor box.

- Step 3: The PLC makes or breaks connections based on the data in the output table. Since O:0/0 has a 0 in the box, the connection between that terminal on the output module and the common for that terminal is opened. The PLC does not care if the connection was open or closed before this step; it simply sends the open or closed pulse as directed by the data.

Now we will take a look at a scan sequence that happens several cycles later. Keep in mind that each scan cycle on average takes 50 ms or less. That is 0.050 second or less from the end of one cycle until the end of the next cycle. A few seconds in the human world can result in numerous scans in the PLC world, especially with faster programs that complete scan cycles in approximately 10 ms.

- Step 1: Populate the input table based on voltage present. This time we have voltage at the following terminals: I:0/0, I:0/1, and I:0/2. As a consequence, there will be a 1 in each of the bit boxes for those addresses. I:0/3 has no voltage present at the terminal screw, so a 0 goes in its bit box.

- Step 2: Read the program from left to right, top to bottom. The XIC for Stop looks for a 1 in the addressed bit box, sees it, and is true. The XIC for Start looks for a 1 in the addressed bit box, sees it, and is true. The XIC for OL looks for a 1 in the addressed bit box, sees it, and is true. At this point, there is a true path to the output portion of the rung and the processor does not have to check the other two input instructions. The PLC needs only one true path to go forward with true logic as the output, so any more input instructions being true will not change things, nor will any of them being false negate the true path found. At this point the OTE Motor instruction will write a 1 in bit box O:0/0. The processor would then work on rung 0001.

- Step 3: The PLC makes or breaks connections based on the data in the output table. At this point when it looks at bit box O:0/0, the PLC finds a 1 in the box, so it signals for the terminal screw tied to this address to connect to the associate common screw. Whatever is coming into that common screw, provided nothing is damaged in the output card, will flow out of the terminal with the address O:0/0.

We will look at one more scan cycle to help you get an idea of how this all works. This scan is performed a bit later, when whoever pushed the start button and ultimately fed power to the terminal screw has released the button.

- Step 1: Populate the input table based on voltage present. This time we have voltage at the following terminals: I:0/0 and I:0/2. As a consequence, there will be a 1 in each of the bit boxes for those addresses. I:0/3 and I:0/1 have no voltage present at the terminal screw, so a 0 goes in the bit box for them.

- Step 2: Read the program from left to right, top to bottom. The XIC for Stop looks for a 1 in the addressed bit box, sees it, and is true. The XIC

for Start looks for a 1 in the addressed bit box, does not see it, and is false. The XIC for OL looks for a 1 in the addressed bit box, sees it, and is true. We do not have a true path yet, so the PLC starts to look at the branches. The XIC for Motor looks for a 1 in the Output table at O:0/0, sees it, and is true. Recall that the number in a bit box remains the same until it is actively changed by something; in this case, it had a 1 from when the start button was pushed and nothing has written a 1 in there yet, so it is still a 1. At this point we have a true path, as the Motor XIC branch provides a true path around the Start XIC that was the only false value on the main rung before the output instruction. With a true path, the OTE Motor will write a 1 in bit box O:0/0 and the processor moves on to rung 0001.

- Step 3: The PLC makes or breaks connections based on the data in the output table. At this point when it looks at bit box O:0/0, the PLC finds a 1 in the box, so it signals for the terminal screw tied to this address to connect to the associate common screw. Whatever is coming into that common screw, provided nothing is damaged in the output card, will flow out of the terminal with the address O:0/0.

As long as we do not push the stop button, which is wired "normally closed," and the overload, which is also wired "normally closed," does not trip out, the motor will start and continue to run once we push the start button. If we push the stop button, the motor will stop. Likewise, if the overload trips, the motor will stop. This assumes that everything is wired correctly and working correctly. If either the stop button or the overload does not open when it should, the motor could keep running. If someone wired in something else to the terminals assigned to those XIC instructions, then whatever is physically tied into the terminals has control. Never assume that just because the instruction is labeled Stop, it is tied to the stop button for the equipment. Only after you have tested and verified operation can you know that for sure.

You can use what we did with those three scans as a template for figuring out the basics of how a PLC program will work, applying the principles learned to predict operation. I would challenge you to run the scan cycle a few more times, playing with different scenarios such as what happens when the stop button contact no longer has power (which is what should happen when it is pushed) or what happens when the manual button is pressed. The program used for this section is one from my classes, and it has an error in operation in rung 0000 (the one we have been using). If you run the scan cycles for different scenarios, you should come across what I consider a safety concern while learning more about how easy it is to change the operation of a piece of equipment by simply changing where a branch leg joins the main branch.

Now let us look at a few more commonly used instructions. Since PLCs are designed to replace relays and there were many timer relays in use when the PLC was invented, there is an instruction that can perform this function as well. All timers have the following components:

- Preset (Pre): How long the timer will time for before it is considered done.

- Accumulation (Acc): How long the timer has been timing.

- Done Bit (DN): A bit box that is one or zero depending on how the timer uses it.

- Timer Timing Bit (TT): A bit box that has the value 1 when the accumulator is counting and 0 at all other times.

- Enable Bit (EN): A bit box that has the value 1 when true logic comes into the timer instruction and 0 whenever false logic comes into the timer instruction.

There are two main types of timers: (1) delay on energize, known as TON or timer on, and (2) delay on deenergize, known as TOF or timer off. With the TON timer, when true logic comes in, the time accumulator starts counting and the DN bit is not 1 until the preset and accumulator are equal. If false comes into the TON, everything resets, so the accumulator goes back to 0 and does not start to count again until true logic comes in. If this happens before the accumulator and preset are equal, the DN bit never gets a 1 in the box.

With the TOF instruction, the DN bit becomes 1 as soon as true logic comes in and the accumulator *does not* start to count. The timer remains in this state until false logic comes in, which is when the accumulator starts counting. While the accumulator is counting, the DN bit stays 1. When the preset and accumulator are equal, the DN bit changes to 0 and the TOF instruction stays in this state until true logic comes in and resets the accumulator count to 0 and the DN bit to 1. I have seen a lot of weird programming over the years as

students try to avoid this instruction, including some that led to problems with equipment operation. Yes, you can learn ways around this command, but you are just making more work for yourself and running the risk of causing unnecessary issues in the program.

The TT and EN bits work the same way for both the TON and TOF instructions. The TT bit has a 1 in the box whenever the accumulator is counting, and the EN bit has a 1 in the box whenever true logic is coming into the timer instruction.

Another type of timer is the retentive timer. Retentive timers can be "delay on energize" or "delay on deenergize" types. These timers do not reset the accumulator based on the logic coming in. Thus, if false logic comes into a retentive TON-type timer, the accumulator stops counting where it is, but does not rest, and the TT and EN bits become 0. When true logic comes in once more, this timer continues accumulating time from where it left off. The only way to reset the accumulator to 0 with the retentive timer is with a reset instruction, abbreviated RES. The RES instruction is given the same address as the timer. When true logic comes into the RES instruction, it sets the accumulator to 0 for whatever it is addressed to.

The PLC set includes quite a few instructions, but another set of instructions also exists that is prone to misuse—namely, the latching output instructions. Before the introduction of PLCs, mechanical relays were used that would stay on even if power was lost, causing things to fire up immediately when power returned. Since the PLC replaced the relay, the latching OTE-type instructions have provided another option.

- Output Latch (OTL): This instruction writes a 1 in the addressed bit box when true logic comes in and does nothing when false logic comes in.

- Output Unlatch (OTU): This instruction writes a 0 in the addressed bit box when true logic comes in and does nothing when false logic comes in.

These instructions look like a regular OTE instruction with either an L or a U in the middle. They are used in pairs, as one can write a 1 and the other can write a 0, but neither can do both alone.

The danger with the latching instructions arises from the fact that this is the only instruction that can put a 1 in the output bit box that will remain there when the PLC starts back up after power loss. Although some of my students prefer the latch and unlatch to the branching type of latch circuit, you have to understand the danger inherent in this instruction. Suppose a power outage occurs, and the machine powers down in the middle of grinding or cutting a part. The power stays out for a while, so the worker decides to do a bit of cleaning while waiting for the power to come back on. The worker is in the machine, cleaning things up, when the power is restored. Now, quite suddenly, the PLC has power once more, there is a 1 in the box, the connection is made, and the motor comes to life, cutting both the part and the employee without a shred of mercy or remorse. That is the danger of the latching instructions. There are times and places where we want the equipment to come back on without human input. Just make sure the latching instructions are used only for these instances and not as a catch-all as part of lazy programming.

Up to this point we have been looking at a sequential scan cycle, which produces "a last rung wins" kind of scenario. With this scheme, we can write 1s and 0s in the same box over and over, but only the last 1 or 0 is actually carried into step 3 and affects connections. Some PLCs use an asynchronous scan cycle in which the input step, program scan step, and output update step all have their own unique time tables and are happening simultaneously. With an asynchronous scan, the input table may update a few times during the program scan, meaning that XIO instructions with the same address can be true at one point in the program scan and false at another, all before the PLC reaches the end of the program. If multiple rungs are controlling the same output bit box, when the output module updates there might be a 1 in the box at one time, causing the connection to close, and a 0 in the bit box the next time, causing it to open once more. This leads to weird intermittent operation that can drive a technician crazy and is nearly impossible to find if you do not fully understand how to search through programs or how the PLC works.

Another key aspect of PLC operation is proper communications. Sometimes equipment may sit idle for hours simply because we cannot get the laptop with the program to talk to the PLC in the equipment. (See Food for Thought 10-2 for my story about this scenario.) The specifics of making the connection will differ from brand to brand, but some basic components are fairly universal. First, you will likely need some kind of software to allow the PLC and your computer to talk. Often, we use laptops to talk with PLCs in the field because a laptop is easier to lug around than a full tower desktop system. The communication software translates the data for both systems and sets up the com port parameters, which is how we determine the method of communication used. We can use an RS232 type of communication, or Data highway 485, or Ethernet, or any of several other methods that are specialized for the PLC in question. Once we pick the type of communication and set up all the needed parameters, the software handles all data transfers as directed. Some programs will automatically detect settings and do the hard work for you, whereas others leave it to you to set all the parameters.

One pitfall in this process is setting up a communication path but using the wrong cable or wrong port on the laptop or PLC. Just because a connection on the software says it is running, that does not mean data is actually flowing. The best test is to look at the PLC processor and see if the communication LED is blinking. Another good test is to delete one of the detected nodes and see if it comes back, giving you solid confirmation of communication.

Another cabling woe you may encounter is the case in which the cable you need to use has an RS232-type end and your laptop has just Ethernet or USB connections. This will generally require some form of adaptor, which can be very expensive if you get this equipment from the PLC vendor. I would recommend trying a Keyspan brand adaptor, as they seem to be fairly solid and come with software that sets them up nicely for RS232-to-USB adaptation. For other adaptations, you will have to see what you can find or else purchase something from the PLC vendor.

One last pitfall arises when you reset the path to which the PLC is listening. Because we often use touch screens and bar-code scanners that have to talk directly with the PLC on a dedicated line in a specific way, we can change the configuration of the various communication ports for the PLC processor. While this is great for getting these devices to work with the PLC, it can create a headache if you try to use one of these ports for your laptop communication with the processor. Just unplugging something from a port does not return it to the factory default. If you do reconfigure a port so that you can use it to communicate with the laptop, you will likely have to restore its settings for whatever device was hooked up to it when you are done. I have also seen cases in which someone programmed the PLC processor to look for general communication on a different port, only to discover the right cable was not available, so there was no way to change anything in the PLC. The only fix for this kind of problem is to get the right cable or to reset the PLC processor to the factory defaults, which usually wipes out the operating program.

FOOD FOR THOUGHT

10-2

One night at work I learned the hard way how important it is to learn how to set up PLC communications. We had a PLC on the floor that had the battery indicator lit up, telling us that the battery needed to be changed. The engineer working on this task powered the machine down, went to the parts room for a battery, returned about 10 minutes later, installed the new battery, and powered up the system. The battery light went away and all seemed good, until the operator tried to run the machine. While there is often a small capacitor that will maintain power for about a minute or less for the battery backup processors, none has one that will last for 10 to 20 minutes.

The engineer came into the maintenance shop and told us what happened and asked us what we thought. We told him the program was gone and would need reloaded, which is when our fun began. First, we had to find the program. We had two old desktop computers on a roll-around cart that held all the various PLC programs for the department. It took a good 20 minutes to get both systems running and then find the right program. Once we were armed with the right program,

we had to establish communication with the PLC. Over the next 2 hours, every maintenance technician in our department tried to get the PLC and computer talking, with no luck. We tried different cables, different protocols, and different settings, all to no avail.

Finally, we called a maintenance tech from a different department who seemed to know more about PLCs than anyone in our department. Once he arrived, he took about 5 minutes to figure out what we were trying to do and what we had to work with. He then began working on setting up communications. In less than 3 minutes he had the two devices talking and the program back in and was on his way back to his own department. While we were really grateful for the help, it was totally demoralizing to realize that none of us knew the first thing about establishing communications with a PLC.

The lessons from that day stuck with me, and I have since worked to learn all I can about PLCs, including setting up communications and working through the various problems each method presents. Believe me when I tell you that it does no good to have the right program or to write great code if you have no way to get it into the PLC for use.

Entire books and courses are dedicated to how various PLC programs work and what all the instructions do. One section of one chapter is not nearly enough to bring you up to speed on all the ins and outs of PLC operation. While this section is a great foundation and kickoff point for learning more, you will need to seek out a good training course to learn more. There is a good chance that you can take some courses on this subject at the same place that introduced you to this text. If that is not the case, make sure you seek out a reputable training venue. Not all training is created equal, and in the PLC field there is a lot of misinformation that tries to tie the instructions to what is happing in the real world instead of what the PLC really does—examine or alter data. This is not malicious misdirection on the part of the trainer but a function of years of people being taught about PLCs from a place of misconception. A good test is to have the trainer explain to you how the XIC works; if you get a lot of "It goes out and sees if the switch is closed" or "It checks the state of the field device," I would be wary.

HUMAN–MACHINE INTERFACES

Whether you are working with a robot, a PLC-controlled machine, or some kind of relay logic system, you need a way to control the actions of the equipment. For many years, this kind of control meant switches, buttons, knobs, thumb wheel number selectors, or something along those lines. All of these items are still used in industry, but we have moved past the simple yes/no or numerical input via binary groups to the modern touch screen, which works much like a smart phone from the operator's perspective. The input devices present on the operator panel generally depend on what we expect the operator to do and monitor. They range from simple button panels, to the somewhat standard CNC-type control panel, to the modern touch screen. In this section, we look at all three levels of operator control so you will have a better idea of what to expect in the field.

The simple control panel with buttons, switches, and selectors is still alive and well today for equipment where there is very little variation in control. For some operations, we really do not need to know every little thing going on, nor do we need to make a lot of changes in how the machine operates. For these systems, a few indicator lights, a button to control power, start/stop/E-stop buttons, and perhaps a bit of manual operation are all that is needed (Figure 10-10). For example, a factory where I once worked had a machine that pressed bearings into the part. There was no changeover between various models for the press due to the nature of the parts, nor was there a need to change the program. The press would go down as long as the operator had a finger on each of the infrared finger sweep switches; it would stop and raise when one or both of the switches were open. The fixture for the parts acted as the stop for the machine, so there was nothing to monitor or change there. A prox switch let the machine know when the ram was fully up. This machine had a simple start/stop/E-stop control panel with one alarm light if the hydraulics failed and another alarm light when the ram returned to the full up position. There was no need to add anything else to this machine for control purposes. There are still many pieces of equipment in industry like this machine, which simply do not need complex control systems. For these machines, the button controllers of old work just fine.

FIGURE 10-11 A CNC machine, being tended by a robot, and the controller for the system. A graphical representation of the machine appears on the screen, and the various buttons used for operator input appear below it. The buttons along the edge and bottom of the display screen are typically used to navigate to various menus and pages in the system.

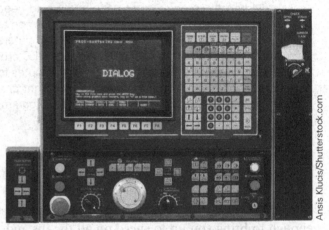

FIGURE 10-12 An older-style CNC controller. Older CRT technology is used for the display, and fewer input buttons are available for this system.

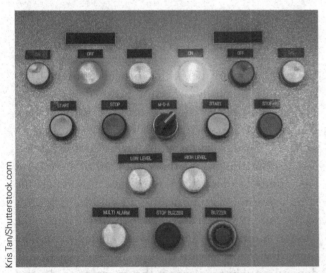

FIGURE 10-10 A simple control station with several indicator lights, a few pushbuttons, and a selector switch in the middle.

The next level up in the world of machine controls is the complex control panel, which includes some kind of a display screen, a way to enter text and numbers, and often several specific function keys that help with the operation of the machine (Figures 10-11, 10-12, and 10-13). The cathode ray tube was the standard

FIGURE 10-13 A close up of the various input buttons you may find on a CNC controller. The pictures on the buttons are intended to help the operator determine what the button influences.

display device for decades in the industrial world, but it is starting to give way to the thinner technology found in modern flat-screen monitors for computers and TVs. Most CNC-type machines have this kind of complex control panel so the operator can put in offsets, change or enter code, and move the complex equipment manually. There is a good chance you will come across this control type if you work in industry where complex machining occurs. Just as there are differences in the controllers and teach pendants for various robots, so too are the operator panels for various CNC machines different. On the plus side, once you understand the basics of using one, you can quickly learn how to use the controllers from other manufacturers.

The new player in the field of operation control is the **human–machine interface (HMI)**, which is an industrial-grade touch-screen system. The HMI comes with a processor inside, runs specific software, and is designed to communicate with PLCs and other control devices in industry. When HMIs first came onto the scene, they were relatively pricy and hard to work with, like most new control systems. Over time, their cost has dropped somewhat due to the demand for them and their mass production. The HMI has done for operator control what PLCs did for machine logic: It is replacing buttons, switches, keypads, and many other input devices. You will still find power buttons, E-stops, start buttons, indicator lights, and other such devices in use with HMI control stations, but the total number of switches has been greatly reduced (Figure 10-14). In many cases the only buttons left are those that are considered part of the hard-wired safety circuits for the equipment.

FIGURE 10-14 A few buttons are found on this control, but most of the operations are controlled by the HMI in the middle. Notice the gauges and on/off buttons on the screen.

The reason for the lingering buttons with HMI control is due to the fact that HMIs work much in the same way PLCs do: They manipulate data. With the HMI, the programmer uses software to create various graphical objects that appear on the screen, such as buttons, switches, charts, graphs, animations, tables, or other needed items. Once the graphical piece of the puzzle is complete, the next step is to tie it to data in some way. Sometimes the goal is to have the HMI pull data from the PLC or controller and display the information for the operator. At other times touching the area on the screen for the graphical object will send 1s or 0s to bit boxes in the PLC. In essence, the HMI can manipulate/display any unprotected data or display any protected data to which it has access. The versatility of the HMI system has made it a go-to device in the modern manufacturing environment.

The versatility of the HMI derives from the ease of adding in new switches or controls by simply uploading a graphical object and assigning it to the right data address. With the older technology, creating a new control meant physically going to the operator panel, drilling holes, running wires, making power connections, and physically mounting something. With the HMI, all you have to do in many cases is log into the network, open the right software program, access the correct HMI, make a few changes, and upload those changes. Voilà—the operator has a new control feature! This versatility is what makes the HMI a game-changer for the operator control side of things, much like the PLC created a sea-change in its own realm. On the downside, just as we have to be careful with PLC programs because one simple mistake could create a dangerous situation in the field, so too must we be careful with the HMI. Neither the PLC nor the HMI replaces a hard-wired E-stop when it comes to safety because it is far too easy to make a mistake in the program and negate this functionality. If you create two objects in the HMI that are linked to the same bit box in the PLC, you can get weird intermittent operation depending on which data the HMI sends and when it hits the PLC in the scan cycle. A wrong address in the HMI could lead to any and all the problems you would encounter with the same type of situation in the PLC, so it is crucial to verify any changes made to the system before hitting the big green go button and walking off.

With that said, there are things we can do with the HMI that the older button controls of old could never match. With the simple systems, all we get is a light to tell us something. With the CNC-type controllers,

Amnarj Tanongrattana/Shutterstock.com

we can get error messages and numbers on the display screen. The newer systems can go a step further and send emails to whoever needs to know and what that person needs to know about what has happened. Most HMIs have multiple user screens and all kinds of graphical elements, which means we can use graphs, gauges, tables, and so on to display data and add in all the buttons we could ever need (Figure 10-15). We can add in keyboard interfaces that pop up on the screen, negating the need for a separate keyboard attached to the system or on the controller as in the older CNC panels. Because the HMI is a large touch screen, every graphical element can be interactive, adding more levels of functionality to the system. It is this versatility that has made the HMI a commonly used control device in the modern manufacturing world. You can also find HMIs in many of the modern teach pendants, such as those used by Universal Robot's UR3 robot.

FIGURE 10-15 An operator interacting with a touch screen. Notice that there are only a few buttons and an E-stop under the screen.

There is one other way to control industrial equipment that we should consider, and that is the cell phone. Yes, you read that right—the cell phone. In my opinion, use of the cell phone as a controller is the culmination of all the advances in the controls realm made over the past several decades (Figure 10-16). Without the PLC and similar control elements, we could not have machines on the network, which is a crucial piece of this puzzle. Without the advances in networking, it would still be difficult, if not nearly impossible, to get the controller or PLC tied to the Internet, which is another big piece of this puzzle. Finally, we had to develop truly smart devices that are capable of

FIGURE 10-16 With the advances in control technology, we can now use smart phones and other such devices to monitor and trigger the operation of equipment.

working with various types of software through apps. The phones we use today in many cases are minicomputers, capable of doing much more than the average user ever tasks them with, making the cell phone similar to a portable laptop in the palm of your hand.

The specifics of cell phone control vary, as one might expect, but the basics are as follows:

- The PLC or controller is linked to the plant's network and set up to work with remote signals.
- The plant network is tied to the Internet and secured with passwords and the usual virus protections.
- An app is developed for the type of smart device intended for use, similar to what is done for an HMI.
- The app on the smart phone is linked to the plant's network via the Internet and specific equipment via code.
- Once the bugs are worked out, monitoring and control via the smart device become possible.

This almost sounds like science fiction, but it is indeed science fact! Many equipment manufacturers have built-in remote reporting that can email or message a cell phone when something goes wrong, alerting the people who need to know quickly—in some cases before the operator even knows there is a problem. Smart device control is simply the next evolution in this line of control. One savvy engineer I know set up a chemical production facility for which he could start runs from his cell phone back in 2013–2014, proving that this capability has existed for a while now.

One of the main reasons why more industries have not bought into cell phone control is the fear that someone will hack into the company's network and cause havoc with their equipment. With every level of control freedom, there is always a cost, and adding the Internet into the mix opens up the possibility of someone breaching the protective firewalls and gaining access to sensitive information or machine control, as the case may be. With the rise of the Industrial Internet of Things (IIOT), which is the process of connecting everything in the plant to the network and then the network to Internet to increase flexibility, many companies are having to really examine the risks of getting hacked from the outside and determine how they can prevent this from happening. As industry figures out how to secure its networks better and reduce the risks, we will likely see an increase in the utilization of remote control for processes where it makes sense.

If you have a chance to learn more about the operation of HMI or take some training on these interfaces' function, I highly encourage you to do so. There is a high probability you will work with HMIs in the industrial realm, and the more you know, the easier it will be for you to interact with these devices. Just as each type of PLC has its own software, so too does each type of HMI require a specific software. You may need to attend a vendor-run course or something similar down the road to learn the specifics of the software used for the devices in your facility, but having a good knowledge base will make learning about the new software easier. Also, the manuals that come with the HMIs are another great resource for information, so keep them in mind as well.

REVIEW

While PLCs are rarely used as the sole control for robots in the modern world, they are still a vital part of equipment control and something you can expect to work with, even if just as auxiliary sources of inputs and outputs for the robotic system. The versatility of the PLC has made it a common component in most machines, and thus something worth learning more about. The same is true of HMIs, which have become part of some teach pendants for robots as well as the preferred option for many operator controls. If you work with robotics for very long in the industrial realm, there is a good chance that you will need to work with one of these technologies to get the job done.

Over the course of this chapter we covered the following topics:

- **What is a PLC?** You learned what a PLC is and what it does.
- **A brief history of PLCs.** This section covered the rise of the PLC from a concept to the modern systems we use today.
- **Basic components of the PLC.** We briefly introduced the main components of the PLC and saw how they work.
- **Operation of the PLC.** This area covered the scan cycle, some of the basic instructions used in PLC programs, and thinking like a PLC.
- **Human–machine interfaces.** You learned about some of the ways to control industrial equipment as well as the benefits of using HMI systems.

KEY TERMS

Auxiliary contacts	Contacts	Force	Ladder logic
Backplane	DIN rail	G-code	Module
Bit	Diodes	Human–machine	Motor starters
Branch	Distributed control	interface (HMI)	Nibble
Byte	system (DCS)	Industrial Internet of	Optical isolation
Card	Download	Things (IIOT)	Overload module

KEY TERMS

Power rails	Proportional integral derivative (PID)	Rungs	Upload
Program mode	Rack	Safety PLC	Word
Programmable automation controllers (PACs)	Remote mode	Scan cycle	
	Run mode	Smart relays	
		Triac	

REVIEW QUESTIONS

1. What is the difference between changing a system using relay logic and changing a system using a PLC?

2. What is the difference between a PLC and a standard computer?

3. Do we still use relays? If so, why?

4. What is the difference between a contactor and a motor starter, and how does the part that makes the difference work?

5. What are smart relays?

6. What is a PLC?

7. What were some of the requirements that Dick Morley had to work to meet General Motors' specifications?

8. What was the first PLC sold on the market?

9. What is a PAC?

10. What is the difference between a relay logic schematic and a PLC program?

11. What is a DCS?

12. What are the four basic components of any PLC?

13. What is the difference between digital and analog inputs?

14. What is a bit, a byte, a nibble, and a word?

15. What is optical isolation?

16. What is the difference between a diode and a triac?

17. Which problem can we encounter when checking triac outputs? How do we eliminate this problem?

18. What is a PID module?

19. What are the different modes of the PLC processor?

20. What are forces in the PLC context, and what is the concern when using them?

21. What does the backplane do?

22. What are the three steps of a sequential scan cycle for a PLC?

23. How does an XIC instruction work?

24. How does an XIO instruction work?

25. How does an OTE instruction work?

26. What are the components used by timers?

27. What is the danger of using a latch instruction?

28. How do HMIs work?

29. What are the basic steps of using cell phones to control machine operation?

30. What is IIOT?

Reference

1. Automation Direct. (2017, August 19). *History of the PLC*. Retrieved from Library. AutomationDirect.com: http://library.automationdirect.com/history-of-the-plc/

ndoeljindoel/Shutterstock.com

CHAPTER 11

Maintenance and Troubleshooting

WHAT YOU WILL LEARN

- What preventive maintenance is

- Common preventive maintenance tasks for electrical, hydraulic, pneumatic, and mechanical systems

- The dangers of arc flash and the personal protective equipment needed to work safely around electricity

- The basics of the troubleshooting process

- How to deal with robot crashes

- Tips to help with the repair process

- Why swapping parts is not necessarily fixing the robot

- Precautions before starting a robot after repairs

215

OVERVIEW

It is a fact of the modern world that all equipment requires care and repair or it will simply stop working at some point. This upkeep can be proactive, to prevent issues, or reactive, when the machine fails and then repairs happen. Both types of upkeep occur in most industrial facilities. Finding the time for preventive measures may be difficult, but the time spent in reactive repairs is often costly from a production standpoint. The maintenance technician is specifically tasked with carrying out these activities, but often engineers are also assigned to assist with the process while looking for improvements and refinements to increase the amount of time the machine runs normally.

In our exploration of maintaining and troubleshooting the robot, we will cover the following topics:

- Preventive maintenance
- Arc flash
- Troubleshooting
- Crash recovery
- Repair tips
- Part swapping versus fixing the problem
- Precautions before running the robot

PREVENTIVE MAINTENANCE

De Legibus (c. 1240) perhaps said it best: "An ounce of prevention is worth a pound of cure" (Titelman, 2000). When it comes to repairs, proper preventive maintenance can equate to a savings in both cost of parts and time involved. **Preventive maintenance (PM)** is the practice of changing out parts and doing repairs in a scheduled fashion *before* the equipment breaks down or quits working. It is analogous to a yearly checkup at the doctor's office or changing the oil in a vehicle. The whole point is to find problems in the early stages or to prevent the problems from occurring in the first place. Moreover, since we schedule preventive maintenance before the machine breaks down, we can prevent the loss of production or functionality of the robot by doing the work at a time that is convenient, instead of in a reactionary mode when it finally breaks.

Most industrial robotic systems include a schedule of preventive maintenance tasks with the literature for the robot. This tells you how often to check or replace various components as well as when, with maintenance periods commonly being divided into daily, monthly, quarterly (every three months), semiannual (every

FIGURE 11-1 These gears turn the minor axes of the FANUC Delta robot. If you look closely, you can see the grease that reduces friction in the system—the same grease that needs added or changed from time to time.

six months), yearly, and multiple-year intervals. For instance, checking the level of a hydraulic tank would be a daily task, whereas tightening all the electrical connections would be a yearly task in most cases. Greasing the bearings and tasks of this nature tend to be performed every three or six months (Figure 11-1), whereas replacement of the wiring would fall into the three- to five-year PM category. Because of the way we divide the PM tasks, sometimes only a few tasks are called for, taking a small chunk of time, while other PM sessions require a few days of downtime due to the large list of tasks. For instance, the monthly, quarterly, semiannual, and annual tasks often come due at the same time, giving you a large list to work through. If one of the three- to five-year tasks happens to come due at the same time, you may end up adding another day or two to an already busy PM schedule. For this reason, it is helpful that we can schedule when we do the preventive maintenance.

You may be wondering at this point, "How do we determine when tasks are due?" We schedule PM based on the previous performance of equipment. At some point, the equipment runs under normal operating conditions and someone keeps careful records of part failures. Each time something breaks, wears out, loosens up, or ceases to function properly, it is recorded somewhere. Over the weeks, months, and years, this logging creates a pool of data that someone analyzes for trends or patterns of failure. For instance, if 5 out of 10 robots require replacement of the wiring harness after 5 years of normal use, then the manufacturer would recommend that all robots have their wiring replaced every 5 years. If the grease in the bearings tends to sling out or become contaminated after 3 months, then the recommendation would be to grease all the bearings every 3 months. This collection of recommendations becomes the preventive maintenance schedule provided by the manufacturer. The base information sometimes comes from other equipment that uses the same components in a similar fashion. For instance, if the same motor used for an axis of a robot is used by a CNC machine and has been proven to fail after 10,000 hours of use, then it is reasonable to assume that the same motor should be replaced in the robot after 10,000 hours of use.

The systems of the robot determine the types of preventive maintenance required. On the electrical side, electrical systems need all the connections tightened yearly (Figure 11-2). If the wires run through any of the moving parts—say, the center or outside of a robot arm—then you will have to replace them at some point. Electric motors wear out faster in harsh environments and when installed incorrectly, but have an average overall life of 13.27 years based on running hours with proper preventive maintenance (figured at 2080 working hours per year). You may need to clean or replace encoders based on where and how they are used. Another common task is checking the cooling fans for functionality and dirt buildup. You may have to pull the electronic cards from the controller and put them back in or **reseat** them to ensure proper connection. The verification of electrical systems and other sensors via voltage, amperage, and other checks is another common task in the electrical checks portion of the PM cycle. If the robot you are working with has battery backup, then expect either annual or some other PM period based on years for replacement of the batteries (Figure 11-3). Failure to comply with the battery change PM task is a good way to lose robot data, including zero points and programs, leading to hours of work needed to restore the data. The full list of electrical tasks depends on the robot you are working with and the equipment it comes with.

FIGURE 11-3 For the delta-style robots in my classroom, the black cover is where the batteries for backup are located. Make sure you know where the batteries for your robot are housed and how often they need to be changed.

Hydraulic systems require periodic monitoring of the fluid for contaminants such as dirt, water, and metal fragments, as well as periodic replacement of the oil, commonly at the one-year mark. Periodic testing of the oil for the proper consistency and chemical makeup is another oil-related PM task. Cursory checks for leaks are a daily task, whereas pulling all the covers and really looking deep into the system is typically a task placed on the semiannual or annual PM list. You need to change out the filter for the system whenever it shows a clogged condition or periodically as directed, again typically at the year mark (Figure 11-4). If your system uses **accumulators**, which are devices that store hydraulic pressure and then release it back into the system as needed, you might have to check the gas charge depending on the type. (See Food for Thought 11-1 for more about accumulators.) Some of the valves may require removal and cleaning that involves taking the valve apart and cleaning the internal components. Checking hoses for leaks and signs of wear is another common PM task for hydraulic systems. These checks are common to most hydraulic systems—not just

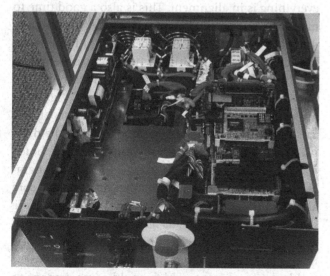

FIGURE 11-2 The internal components of a FANUC R-30iA Mate controller. A common practice is to tighten all electrical connections, like the ones in this controller, yearly to prevent loose connections and all the havoc they could cause.

FIGURE 11-4 This stand-alone power unit is used in various fluid power labs. The white canister on the left is the filter. The cooling unit is located on top of the pump motor. Both the filter and the cooling unit require yearly maintenance.

FOOD FOR THOUGHT

11-1 Hydraulic Accumulators

Some accumulators use a rubber bladder filled with nitrogen gas inside a metal housing to store pressure. As the system forces oil into the accumulator, the gas bladder is compressed and exerts pressure on the oil in the accumulator. If the system pressure drops, the pressure from the bladder forces the oil back into the system, helping with system flow and maintaining pressure. With this type of system, any loss of gas pressure in the bladder equates to a loss of stored pressure in the system. Since the accumulators create

flow when the system is overtaxed, lower pressure in the bladder could mean slow operation of hydraulic motors or cylinders, which can influence robot movement or tooling performance.

Another type of accumulator uses a solid weight with seals to force fluid back into the system. The PM for such an accumulator is to check the seals and make sure the drain for any oil that gets past the seals is working properly. No matter the type of accumulator, there will be some form of preventive check to verify proper operation.

---

those used with robots—so do not be surprised if you see a list like this for other equipment you encounter in industry.

Pneumatic or air systems have their own checks. Of course, any fluid power system requires checks for leaks, damage to hoses/piping, and filter changes, so you can expect to do this on a predetermined basis. Additionally, many pneumatic systems have lubricators that add oil into the compressed air as it passes, so as to lubricate moving parts in the system. Over time, the system depletes the oil and replacement is required. Because of the noise created by exiting air, mufflers are often needed as well, and you will want to check these for damage or clogs (Figure 11-5). Water and dirt fall out of the air in the reservoir tank of a pneumatic system, so it needs periodic cleaning and draining to ensure proper operation and prevent

damage to the components in the system. The last thing you want is dirty wet air forced through the valves and components of your system—that is a fast track to breakdowns.

For mechanical systems, the big PM task is greasing and lubrication. Unless the bearings are sealed, they will need replenishment of the grease, using the correct type of grease, at some point. Many robotic systems use white lithium grease for the bearings and gears because of its high-speed properties and the tight tolerances of the systems. For gears, the PM checks that everything is meshing properly and there is no excessive wear or broken gear teeth. For belt and chain drives, the PM makes sure that any stretching is within the appropriate tolerance, that the systems are still transmitting power without slipping, and that everything is in alignment. This is also a good time to make sure nothing is rubbing against the metal skin of the robot or contacting electrical wires or fluid power hoses, among other things. If these problems are caught early, many times the technician can minimize the damage or deal with it easily; conversely, ignoring or missing them can result in a major robot repair.

Some common tasks should be a part of the daily ritual of getting the robot ready to run. If the robot is powered down, this is a good time to check for damage to the robot, tooling, or fixtures and to make sure all the cables are securely fastened. If these tasks are done with the power on, *do not* try to tighten or adjust a loose cable! Doing so has a high probability of generating a false signal that might cause the robot to move suddenly in response, which could cause damage to the robot or equipment around the system and hurt or

FIGURE 11-5 This pneumatic regulator does not have a lubricator, but it does have a filter bowl that needs to be drained and cleaned from time to time as well as a muffler that needs attention.

even kill you if you are in the robot's work envelope. Once the system is powered up and seems ready, move each axis and check for odd noises, vibration, jerky motion, air leaks, or anything else unusual. Before declaring the robot ready to run for the day, run the position verification program, if applicable. Finally, watch the robot closely during the first few operational cycles for anything abnormal, as some problems surface only when the robot is running at full speed.

This section has presented just some of the common PM tasks; in no way is it a complete listing of everything you can expect. The specifics of your robot's PM depend on what it does, how it does it, and where it does it. Harsh environments shorten the intervals between checks or replacement, whereas light use might extend the time, depending on how many hours the robot runs. If you find a part something is wearing out more quickly than the recommended time between checks, change the time interval for replacing that part. If you have consistent problems with something not on the list, feel free to add it to your PM schedule. As the end user, the responsibility for properly maintaining the robot falls squarely on your shoulders, so make the PM schedule changes necessary to improve the system's overall performance.

ARC FLASH

A very serious condition to watch for during any electrical repair is the dreaded arc flash. An **arc flash** occurs when a short of some kind in the electrical system ionizes the air and creates an explosion complete with intense heat, shrapnel, and pressure waves. The heat produced can exceed 35,000°F (19,500°C), approximately four times the surface temperature of the sun, which is hot enough to vaporize copper and cause instant second- and third-degree burns. When copper changes from a solid to a vapor, it expands to 67,000 times its original volume, which is enough to propel shrapnel at a staggering speed of 700 miles (1130 km) per hour or more. The pressure waves created by an **arc blast** (i.e., an arc flash explosion) can exceed 2160 ft/lb^2, which is well over the 1720 ft/lb^2 threshold at which massive internal injuries and death occur. To help put this in perspective, a 50-kA arc fault generates enough force to accelerate a 170-lb (75-kg) person standing 2 feet away from the blast at a speed of 330 ft/s^2 (100 m/s^2), literally blowing that person away from the event and adding to the misery as he or she

bounces off whatever is in the way, and likely causing internal injuries such as collapsed lungs or crushed organs. If that wasn't enough, an arc flash also generates intense, blinding light and sound waves exceeding 140 decibels, enough to rupture unprotected eardrums.

An arc flash/arc blast can be considered akin to detonating the electrical equivalent of a nuclear bomb at arm's length. The intense magnetic fields generated by this event can literally pull other components into the maelstrom, adding more shrapnel and expandable materials to the mix. It happens so quickly that there is no chance to dodge it or get out of the way. One minute you are doing your job; the next minute the world explodes in your face, leaving you blinded, half-deaf or actually deaf, burned, possibly on fire if you are wearing the wrong clothing, and quite possibly dying, if you survived the initial moments of the event.

Several things can cause arc flash events, but the essence is that two power leads short between each other, possibly involving the ground as well, and generate a huge amount of current flow. The longer the flow continues, the greater the damage and the chances of an arc blast. Some of the major causes are tools or conductive parts touching two lines at once, closing into faulted lines (which pop breakers or melt fuses), and loose or damaged connections. Conductive dusts and other impurities can also create a bridge between power leads and start the disaster. Arc flashes tend to occur inside the electrical cabinet when movement occurs, such as door or cover removal/installation, switch contact opening/closing, and person-driven movement including use of test equipment or reaching in with anything conductive.

How can you avoid arc flash? The answer is simple: Kill all electrical power before opening the electrical cabinet. This sounds easy to do, but many times troubleshooting requires power checks and other testing in which the power needs to remain on. When you run into these situations, there are two points to keep in mind. First, do not use any tools on the electrical equipment besides meters and other such test equipment. Whenever you pick up a screwdriver, pliers, or wrench, that is considered live work—and the only justification for live work is life support systems or instances where shutting down the power creates a greater hazard than the risk of arc flash. There are very few instances that truly justify live work; rather, the usual excuse is that management does not *want* to shut the power down. Second, when doing live power checks, make sure you wear the proper **personal protective equipment (PPE)**. PPE is your last line of defense from the dangers around you, but it works only if you have the right gear, it is in good condition, and you use it correctly.

A document produced by the National Fire Protection Association, known as NFPA 70E, is the go-to source of information for all things arc flash. In it, you will find the specific clothing requirements for each level of danger. In addition, the NFPA has issued a two-category kit that will cover most, if not all, of the tasks performed during electrical maintenance in your area. This kit includes the following elements:

- An arc-rated long-sleeve shirt with arc pants *or* arc-rated coveralls, all at a minimum arc rating of 8 cal/cm²
- Flash arc suit hood *or* arc-rated face shield and arc-rated balaclava
- Standard PPE including hard hat, electrical gloves, leather protective gloves, leather work boots with safety toe, safety glasses, and ear plugs that go into the ear canal (Figure 11-6)

Keep in mind that if you wear flammable clothing underneath the coveralls or shirt and pants, they may ignite and cause secondary burns or simply melt into your skin.

Some tasks, such as working with the main bus bar or dealing with circuit breakers in a motor control center, will exceed the protective level of the general kit. In those instances, you should wear the clothing mentioned previously with the following modifications: You must wear the arc flash suit hood;

then, over the other clothing, you add the arc flash suit, which includes another jacket and pants to wear (Figure 11-7). Arc flash suits are rated for specific calorie event levels, with 40 and 80 being common ratings. Yes, this gear is just as uncomfortably hot as you imagine, and yes, you want to wear it for as little time as necessary. Once you have powered down the system, properly verified that no electricity is present in the system, and applied lockout/tagout (LOTO) precautions, you can remove the suit and work as you normally would. If you need to work in the suit for more than 15 to 20 minutes, you will want to watch out for heat exhaustion or other heat-related illnesses.

Unfortunately, even if you do everything correctly, have on all the right stuff, and be careful and aware of what is happening at all times, something can still go wrong and an arc blast can occur. Even with all the right gear on, you are not guaranteed to walk away from an arc blast should it occur, but the odds in your favor greatly increase when you have donned the right PPE. Without the right equipment, even a low-energy arc flash event might be enough to send you to the hospital with second- and third-degree burns on all exposed skin as well as secondary burns from clothing that ignites. An arc flash event can cause injuries requiring months or years of rehabilitation and medical costs that can exceed $1 million. It is a serious danger for anyone who works with electricity, and one that needs to be taken seriously. Can you work your whole life and never have this happen to you? Absolutely. But if it does happen and you do not have the right equipment on, the odds of that being the last day of your life greatly increase.

FIGURE 11-6 An example of the personal protective equipment required during the majority of industrial electrical tasks.

Jay Petersen/Shutterstock.com

FIGURE 11-7 A worker wearing the full arc flash suit over the other protection required. This personal protective equipment allows the individual to work with equipment that has the potential for a high-energy-release, arc blast event.

TROUBLESHOOTING

While proper preventive maintenance goes a long way toward preventing system failure, sooner or later something is bound to go wrong. At that point, we enter a **reactive maintenance** mode, in which we have to figure out what is wrong and do something to get the system back to normal running condition. Sometimes this is a simple change such as a tool change or touching up a couple of points. At other times the repair can take days, requiring the majority of the robot to be disassembled and parts to be ordered. To determine how to respond to these situations, we use **troubleshooting**, which is the logical process of determining the cause and correcting faults in a system or process. Troubleshooting is a part of our daily lives, but those who master this process and make it a part of their work skill set can earn a very livable wage fixing broken equipment and getting product flowing once more.

The specifics of each troubleshooting scenario differ from machine to machine and from event to event, but some general guidelines can help you along the process. Of course, the more you know about how a robot or piece of equipment works, the easier it is to fix it when problems arise. A misconception about how the robot should run can lead to flawed logic when you try to address the problem. While the troubleshooting process includes some intuitive elements, for the most part it is a logical approach to finding the problem and then correcting it. It is these logical steps that we will address in this section:

- Analyzing the problem
- Gathering information
- Finding a solution
- Responding to continued problems

Before you can fix the problem, you have to know what the problem is. To analyze the problem, you can talk with the operator, see which alarms are activated on the controller or teach pendant, and simply observe the equipment to see what is wrong. As we are analyzing the problem, we also begin the next step of gathering information. As we figure out what is wrong with the system, we want to look for answers to the following kinds of questions to give us clues: Where did it stop in the program? What was it doing? Is anything out of place? Did the operator report any strange sounds? Are there signs of damage? Which alarms does the system have? We can also use our senses to look, listen, feel, and possibly smell clues to the issue. Some issues we discover in this process require faster action than others. For example, if we see smoke or smell burnt insulation around the robot, we need to kill the power quickly to prevent further damage.

Another way to find more information about serious problems is to track the power or signal and find where it stops. This portion of the analysis often requires specialized equipment and a good understanding of how the system works. For instance, if we are tracking electrical power and find a component where power goes in but does not come out, we have to know whether this is normal for the current machine state. If it is, then everything is likely to be okay. If not, this might be the problem. We can do the same with mechanical power transmission as well as various data signals, if we have the right equipment. With a good understanding of the system, we can even skip over several components to check the power or signal at an easier-to-reach location. If we find what we are looking for, then likely everything between the current check and the previous one can be considered running normal. Again, this takes a good understanding of how the system works and the ability to read printouts and schematics properly. A misconception in this part of the troubleshooting process can lead to hours of lost time chasing a problem that does not exist.

Another place to check for information is the technical manuals that came with the system. You can find wiring diagrams, mechanical drawings, exploded views, troubleshooting tips, error code meanings, and other important information in these manuals that greatly facilitate the analysis portion of the troubleshooting process. Some manuals are less than helpful, and sometimes the manuals may have been lost or do not exist. In these cases, I suggest checking the Internet for pertinent resources. Many companies post their manuals online in PDF form for easy access and downloading. You can also call the company or its tech support line and see if company representatives can provide additional information. Calling the tech support line may require paying a fee if you want answers, but that cost is often cheaper than having the machine remain down for days, trying to get someone from the company out to your location, or swapping parts until something works.

Once you have enough information, it is time to deduce a solution. At this point, we look at what all the information is pointing to, determine what would cause the problem, and then do something to fix it. This solution can range from creating offsets or mastering the robot all the way up to a major disassembly and part swap repair. One recommendation is to start with the simple and quick solutions first. If you get lucky, you can save a huge amount of time; if not, at least you know what does not work with little time lost. As people become more experienced in troubleshooting they tend to go for the complex fix first, only to discover a blown fuse or something else simple was at the heart of the matter.

Unfortunately, the problem may still exist after you try the simple fix, it may get worse, or a whole new problem may arise after repairs. This happens to all troubleshooters sooner or later, but following a few guidelines can help you work through these situations.

Guideline 1: Do not get frustrated. If you allow yourself to become aggravated or upset by the fact your first attempt failed, there is a good chance you will not be thinking clearly when you try again. This can lead to missed steps, ignored information, and illogical thought processes that will impede troubleshooting. If you find yourself getting frustrated, take a short break from the problem or get some help. Frustration and anger can also arise when you injure yourself while fixing the equipment. Most people eventually end up getting hurt in the process of repairing a problem, whether in the form of a smashed finger, a cut, a pulled muscle, a tool dropped on the foot, hitting the head on something solid, or some other painful event. No matter what the cause of the injury, many of us react to these injuries with anger. This, in turn, can lead to a missed step or steps in the process of the repair, and either cause you more problems or at the very least prevent you from fixing the problem. If you find yourself slipping into this kind of negative cycle, you definitely want to do something to break the chain.

Guideline 2: Reevaluate the information, adding in what you have learned. When you try something, no matter the outcome, it provides new information. As long as you affect the problem, for better or worse, you are on the right track. With this line of thought, you must first confirm that you have not created a new problem that is masking the original problem. As long as this is not the case, you have valuable information to use in your next attempt at troubleshooting. If you made it worse, then you are on the right track, but likely need to do the opposite of what you just tried. If you made it better, but did not completely solve the problem, there are two main possibilities: (1) You are on the right track but simply did not make enough of a correction or (2) you are facing a core problem caused by multiple failures. **Multiple failures** happen when more than one component is at the heart of the problem. In some instances, multiple failures happen when one component goes bad and in the process damages other components connected to it. Last but not least, if your efforts had no effect on the problem, then you are likely on the wrong track altogether. No matter what the outcome, some new information can be gained from what you tried.

Guideline 3: Try, try again. If your first attempt at fixing the problem does not work, then you still have a problem that needs to be solved. Good troubleshooters realize that sometimes perseverance is the only way to find the core problem and correct it. This is especially true if you take an erroneous path or if you missed a step in the process on your first try. When you are working with something new—whether it is a new problem, a new piece of equipment, or a system that you have little experience with—you may find that it takes you more tries than usual to solve the problem. Complex problems often require several efforts to eliminate all the possible variables regardless of your experience level with the system. This leads us to guideline 4.

Guideline 4: "*Once you eliminate the impossible, whatever remains, no matter how improbable, must be the truth*" (Doyle, 2012). This famous quote from Arthur Conan Doyle's character Sherlock Holmes should be one that every troubleshooter remembers, as this approach is especially beneficial when there are several possible causes for a problem. In these situations, there are often simple things that we can try first before we spend large amounts of time and money on the more complex solution. The best way to organize how you work through the possibilities is to use availability, time involved, and cost to weigh your options.

We start with availability for simplicity's sake. If you have to choose between three possible solutions, and one of them requires a part to which you do not have access, then you would obviously want to try one of the other two solutions first. We also use availability to make sure we exhaust all the options available before we order an expensive part or one that will take a long time to arrive.

Once we have decided what is available, we next look at the time involved. If checking one possible solution would take 30 minutes, while the other solution would take a day to implement, we want to try the quicker of the two options first.

The last way we filter is by cost. On occasion, we may have to factor in cost before the time involved, especially if there is a risk of damaging a high-dollar item, but generally we use this consideration as a tie-breaker when no clear-cut path is evident. We would continue to filter our options by using this method and trying again until we have found the problem or have no other options left.

Guideline 5: Back to the drawing board. Sometimes you may try everything you can think of, but nothing seems to have any effect on the problem. In these cases, you may have followed a flawed troubleshooting path. Do not worry; this happens to every troubleshooter eventually (no matter what the individual might claim). When you find yourself in this situation, take all the data you have and filter it in a new way. Look for things such as other systems or components that could affect the problem. Check whether you missed some data in the analyzing and data-gathering phases. Get a fresh set of eyes to look at the problem and give you some suggestions. Look into additional resources. The simple fact is that once you have tried everything you can think of, you need to look at the problem in a new way. Albert Einstein (2012) said it best: "Insanity [is] doing the same thing over and over again and expecting different results." If as troubleshooters we continue to try what we know does not work, it would be insane to think that approach will suddenly work the third or fifth time.

Troubleshooting can be a very complex process, especially when you are working with multiple systems, problems with multiple possible causes, or multiple system and/or component failures. In this setting, there always exists the possibility that your first solution will not work. What separates excellent troubleshooters from run-of-the-mill problem solvers is what they do next. Skilled troubleshooters persevere, get help as needed, and solve the problem. It is that simple. It will take time and effort to become an experienced troubleshooter, but it can be a very rewarding and profitable career path. Most technicians take great pride in the repairs they perform and the problems they solve. Those who have honed their skills are sought out by the industries and businesses that stand to benefit from their knowledge, even in times of recession. As long as we have complex systems and equipment in our lives, there will be a need for the troubleshooter.

CRASH RECOVERY

In the industrial setting, there is a high probability that eventually the robot will make unexpected contact or **crash**. Indeed, this is one of the more common problems that technicians have to deal with. Often this happens when the robot is at full speed. The results, while spectacular, may be devastating to the various parts and systems involved. The damage from a crash varies with the energy and type of contact involved, but may include shattered parts, machine covers bent or torn, destroyed tooling, motors slipping, broken belts, gears with lost teeth—the list of possible damage goes on and on. No matter the type of robot involved in the crash, we can use the following basic approach to deal with the situation:

1. Determine why the robot crashed.
2. Get the robot clear of the crash or impact area.
3. Determine how to prevent another crash.
4. Check the alignment of all equipment involved in the crash.
5. Determine what to do with the parts involved in the crash.

Step 1: Determine Why the Robot Crashed

This action sounds simple and is an obvious first step, but in the heat of the moment many technicians and operators have overlooked this step. Before you move the robot, before you deal with the damage, before you do anything, determine what caused the problem to begin with. Once you move the robot and start doing things to react to the crash, there is a good chance you will lose important information about what specifically happened. Think of the crash like a crime scene and imagine that you are a CSI tasked with determining what happened. Take pictures, look at the crash from different angles, and check programs, paying special attention to the line on which the program stopped for each machine and system involved. Do whatever you need to gather enough information so you understand exactly what happened.

In cases where you cannot gather enough data to draw a definitive conclusion about what happened, do the best you can. Sometimes there is no clear-cut explanation for why the robot hit something. For example, intermittent problems that occur randomly with no discernible pattern will make it difficult to find the root cause of the problem. When you are unsure of the *why*, record as much information as possible and save it for future reference. Unfortunately, sometimes the system has to fail, and even crash, a few times before we can get enough information to find the root cause. Just do the best you can with the information you have available. Do not forget to talk with other operators or technicians who have experience with the system, as they may be able to give you some insight or tips on possible causes.

Step 2: Get the Robot Clear of the Crash or Impact Area

Once you have gathered the necessary information, it is time to get the machine(s) untangled. Different systems will require different steps to accomplish this disentanglement. Sometimes the robot becomes stuck or hooked on something in the crash area, and you may have to use tools to free it. In these cases, try to minimize any additional damage by applying the right amount of force in the right ways, instead of going straight to using a sledgehammer at full force. Some systems will let you reset the alarm and then move the system manually. In this case, make sure you move the robot in the proper

direction and keep all initial movements small to minimize any potential secondary damage. This is where having a good working knowledge of the robot's movements comes in handy. You can also use the right-hand rule as long as you are in the World frame and remember what each finger represents.

Some systems have manual brake release switches or buttons to allow free movement of the robot (Figure 11-8). If you plan to use this capability to free the robot from the crash, there are some very important points to keep in mind. Once you release the brake, there is nothing to hold the robot in place. For the minor axes of the robot, this may not be a critical concern, but releasing the major axes of the robot can result in a sudden and dangerous collapse of the arm. Use the axis number (discussed in Chapter 3) to make sure you are releasing the proper axis.

FIGURE 11-8 This Panasonic robot includes four white pushbuttons and a red pushbutton on the side, near the base, with a black background. These break release switches help with crash recovery of the Panasonic system.

Also, always double-check that you are using the switch for the axis you want and *not* the one next to it. During a classroom demonstration of the brake release method for some of my students, I inadvertently pressed the brake release for the axis next to my intended axis, and the robot arm suddenly fell downward under its own weight at the third axis. Luckily, I was off to the side when this happened, but the arm caught the edge of my safety glasses and knocked them across the room. If I had been over a few inches, I would have had a nasty knock to the head and likely a trip to the emergency room. Yes, even instructors make mistakes—but we share those mistakes so you do not have to recreate the learning event.

In a worst-case scenario, you may have to physically remove motors or remove power and turn gears by hand to move the robot. This action should be your last resort and not the first plan of attack. If you cannot clear the system alarms to move the robot manually, check the repair or maintenance manuals for a way to release the axes of the robot for movement. Often clearing alarms that do not reset easily requires going to a rarely seen menu and /or setting, but be aware that the majority of the systems provide some form of alarm override.

To complete the second step of crash recovery, you have to understand the operation of the robot in question. This is yet another reason why taking the time to understand your robot is time well spent. These machines are truly marvelous in terms of what they can do and how flexible they are, but they require a great amount of knowledge on the user's side to work properly.

Step 3: Determine How to Prevent Another Crash

If step 1 was the CSI investigation, then step 3 is the arrest. During this step, we make the necessary modifications to the programming or operation to prevent future crashes. In some instances, step 3 and step 4, checking alignment, are combined into one mammoth step with the goal of preventing future damage to the robot system. This could require changing the position of a point, putting a pause in the timing of the program, adding an input to the robot system, replacing a worn-out or faulty sensor, tightening bolts in a fixture, removing things that do not belong in the work envelope, and so on. The main point of this step is to ensure that the problem will not happen again, causing even more damage.

In the case of intermittent or unclear problems, it may be difficult to determine a course of corrective action to take. If you have to run the system without a corrective action, try to minimize any potential for damage. When you start it back up, watch the system run with your hand on the stop button and reduce the operating speed if possible. Do a few dry runs where no part is present, if that is an option. Do anything you can to minimize the potential damage while checking the system for symptoms to help fine-tune the diagnosis. Unfortunately, you may simply have to let the system run for a while before any new information surfaces. If this is the case, use your best judgment and do what you can within the constraints you have to work with.

Step 4: Check the Alignment of All Equipment Involved in the Crash

As mentioned previously, the unexpected contact of a crash can cause physical damage to the machine and cause the various axes of the robot to move. This creates a difference between where the robot actually is and where it thinks it is. When this happens, all the various points in a program become affected and the programs do not perform properly. In some cases, the difference may be small enough that there is no apparent change in the robot's operation until we examine the program as a whole or run the alignment check program. As a rule, the tighter the tolerances required of the robot, the less room there is for error or axial slip.

When you determine there is error in the system, you need to take some form of corrective action. The appropriate action depends on the robot, the type of tooling it is using, and the amount of error. Many systems allow for corrective offsets as long as they do not exceed a set amount, usually somewhere less than 1 inch. Some types of tooling slip within their holders to prevent damage. In these cases, there is often a fairly quick and simple procedure for realigning the tooling to get back to business. This is another one of those instances where you need to know how your system works and how best to deal with positioning errors of that system. (Have you noticed a trend yet?)

When checking alignment, do not forget to check the fixtures as well. If the error is in a fixture and not in the robot, you can cause yourself a lot of grief by changing the position of the robot instead of fixing the alignment of the fixture. Yes, once you align the robot to the new fixture position, the system will function properly—but what happens if you change fixtures and go back to the proper fixture alignment? Suddenly the robot is out of position once more, but there was no crash to blame.

Another mistake to avoid is realigning the program instead of the robot. This is the quick fix and a common action for operators in a hurry. This approach will get the system turning and burning once more, but what happens when someone changes the program? Once again, we have positioning problems with no clear-cut cause. If this situation progresses for a long time and the operator adjusts multiple programs in response, and then something happens that warrants the realignment of the robot that should have occurred in the first place, can you guess what happens next? Yes, the programs that were changed in those series of adjustment snow need changed back. Remember, you may combine steps 3 and 4, but you still have to meet all the criteria of each.

The best hope for normal operation is to correct any damage or misalignments caused by the crash instead of just treating the symptoms. Any shortcuts taken at the time of the crash to get the robot up and running are likely to haunt the operation with points in the wrong position here and odd alarms or operation there until someone finds the original root cause and fixes it. At that point, we will lose even more operation and production time because of the need to remove all the various symptom fixes from the robot to get it back to normal operation. This unhappy possibility is why you need to take your time with the crash recovery steps instead of rushing to get the system back up and running.

Step 5: Determine What to Do with the Parts Involved in the Crash

Before we restart the robot and let it continue with the tasks at hand, we need to figure out what to do with any parts involved in the crash. This may seem an odd step to some and not apply to every robot crash, but for industry it is a key step. With lean manufacturing and the desire to reduce inventory, a company may produce only the number of parts needed by the customer, with no extras. If this is the case and the robot just destroyed one of those parts, what happens to the customer's order? Yes, the customer does not get the full order because there are no longer enough parts to make enough of whatever it is the customer wants. Just as you would be mad if your favorite restaurant forgot part of your order, so too do customers become upset when they do not receive all the products requested. The fix for this problem is usually fairly simple: Use the proper in-house procedure and let the responsible party know you need another part.

Tech Note

Lean manufacturing is a cost-saving set of operational procedures that has gained a great amount of popularity in industry. The key principle is to remove all possible wasted time, space, and motion by placing similar operations close together and streamlining the production process. Another aspect of this approach is time manufacturing, in which the company produces parts ordered instead of anticipating parts that might be ordered and stocking them in a warehouse or other storage area.

Sometimes we can save crash-involved parts by finishing out the process. In such a case, the operator needs to manually step through the program up to the point of the crash and then start the system in automatic mode or finish the part in manual mode, depending on the process and robot in question. This is why we take care of the crash-involved parts before beginning normal operation, as it is easier to run the system manually before we start normal operation as opposed to stopping the system at some later point and trying to get everything back in position. If you are unsure how to save a salvageable part, ask someone with more experience for help.

Once you have determined whether the part is salvageable or junk, make sure you follow your employer's guidelines. If it is junk, report it, get a new one, and carry on as directed. Make sure to put junk parts in the proper area for disposal or recycling instead of just somewhere handy. I have seen many parts placed "somewhere handy" remain there for days, weeks, or even longer. Most importantly, do not try to hide the damaged part. Eventually it will come to light that a crash or damage happened; if you try to hide the damaged part, it will appear as if you know you did something wrong. This will likely lead to repercussions you would rather not experience, ranging from a reprimand to loss of your job.

If you salvage the part, pay extra attention to the part's quality. No one wants a part that looks good, but is junk in disguise. Some parts might initially seem salvageable, but in the process of reworking become scrap with all the responsibility that entails.

Once you have completed the five steps of crash recovery, you are ready to fire up the system and let it run. I recommend doing at least one manual run from start to finish, if possible, before switching to automatic mode. This gives you a chance to confirm that everything is working correctly and that you did not miss any other problems with the system. Your employer may have some specific guidelines that apply to these five steps, and those guidelines may require some actions or steps not covered here. If you are unsure or if questions arise anywhere along the crash recovery path, talk with someone who has more experience and can help you out.

REPAIR TIPS

When you are ready to begin repairs, there are some things you can do to make the process easier. The tips in this section come from my years of fixing equipment, and several of them were paid for in blood and bruises. Ideally, you will learn from my experiences and save yourself some pain and trouble along the way.

"One picture is worth a thousand words," according to Fred R. Barnard—a statement that I have found to be very apt when it comes to maintenance. Before you remove a part or unwire a system, take your cellphone or other device with a camera and snap a few pictures. This will give you a visual roadmap back to the original system configuration and help with any questions that might arise during the repair. I have used this approach many times to help me remember how the system was wired, where everything went, how something looked before I took it apart, how hoses or cables were routed, and so on. This technique has saved me many hours of trouble, especially when

days, weeks, or months have separated the point when I started a repair and the point when it was finished (Figure 11-9). These pictures also help anyone who has to finish what you started or if you need to explain what is going on with a system.

Keep all the parts together. To fix most equipment, you need to remove covers, take off brackets, remove parts, and in general windup with a pile of stuff that needs to go back on the machine once you have finished the repairs. Keeping everything together makes the assembly process much easier. When deciding where to place the parts, be aware of things such as areas where people are still working, openings that parts could fall into, areas where parts could roll under the machine, places where the parts might fall

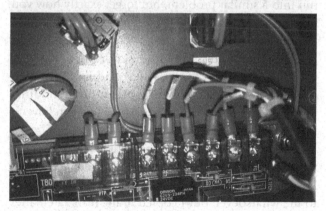

FIGURE 11-9 I took this series of photos before unhooking a FANUC R-J2 controller to move our lab robot from our old building to the new campus. It was nearly a year later before I had the chance to hook the system back up, but these pictures allowed me to complete the task as if it had been only a day or two.

on you or others, and anything else that could cause problems. For small parts such as nuts, bolts, screws, and washers, you may want to get a small container to hold them so they will not roll away. Another trick is to lay out the parts in the order they came off the robot. By organizing parts in a linear fashion, you create an easy-to-reverse timeline indicating what goes on next.

Get help *before* you need it, not *after* you have created an emergency. Many times I have seen maintenance technicians bite off more than they could chew and get into dangerous situations because they would not ask for help. There is no trophy for taking foolish risks just to prove you can do it all by yourself. If you need help lifting something, ask. If you are not sure how to proceed with a task, see if a coworker has experience and can help. If you need someone to use a forklift or crane to remove or install a heavy portion of the equipment, make sure the person knows how to run the equipment. The only prize for trying to be a one-person show when two would be better is injury, scars, and possibly death (Figure 11-10).

When you have someone help you, make sure everyone is on the same page. One of my favorite parts from the *Lethal Weapon* film series is when Danny Glover and Mel Gibson are arguing over is it "One, two, three, go" or is it "One, two, and go on three." While this is one of the running jokes of the movies, it brings to light a problem that happens when people

Photo Courtesy of FANUC America Corp.
www.fanucamerica.com

FIGURE 11-10 While this robot is not small by any means, take a moment to look at the tooling it uses during operation. Now imagine trying to hold on to this tooling while simultaneously removing the bolts holding it in place. In situations like this, you definitely need help from your coworkers.

are working together. If two people are lifting a heavy motor, but they are not working in unison, there is a good chance for injury and accidents. If one person is about to lose his or her grip and does not tell the partner, then disaster may follow. If both people are thinking that they will go to different places with the heavy object or if they have different ideas of what various hand signals mean, then mayhem is almost certain to occur. The whole point of teamwork is to work as a team, not a group of individuals all trying to do their own thing. Take into consideration the physical differences as well. With a two-person lift, if one partner is more than 6 feet tall and can lift 300 lb and the other is less than 5 feet tall and can lift only 100 lb, then there is a good chance the team lift will fail. Carrying anything at an angle puts more weight on the person on the lower end. If the shorter person drops whatever it is due to muscle fatigue or the weight being more than he or she can handle, then the taller person may hurt himself or herself trying to catch the load or end up under it. These are all considerations when working as part of a team.

Take good notes. Often during repairs, we learn better ways to do things, track down information that did not come with the equipment, or make other discoveries that can help later. This information can be very beneficial six months or a year later when you run into a similar problem, but forget exactly how you fixed it. This is one of the main purposes behind the technician's journal: It is the perfect place to put technical sheets found on the Internet or in parts boxes, those pictures you took earlier, and anything you found helpful or learned during the troubleshooting process. The more you save today, the less you have to search for tomorrow.

I remember copying the binder of my mentor when I first started my industrial maintenance career. It took a few hours and I had stuff that made no sense to me whatsoever, but I copied it all. Years later, after my mentor had moved on to another company, I still used that book and found from time to time a nugget of knowledge that meant nothing to me the day I copied it but was invaluable when I finally understood its purpose. Of course, I added to this book over the years from my own experiences. When I left that company to pursue teaching, I passed down my binder of knowledge to one of the newer techs, in hopes that it would serve my coworker as well.

Break the mold. We tend to get stuck in a rut and do things the same way over and over again because that is the way we learned to do it. Just because that is the way it has always been done, it does not mean it is the best way! Sometimes a fresh set of eyes can see a different path that can save time, money, and frustration. You may need to create your own special tools to do what you want, but do not let that stop you. You will likely get resistance from those who are stuck in the rut, but again, do not let that stop you. Sometimes your idea will fail. Sometimes your idea will make matters worse. But more importantly, sometimes you will find a new way that makes it all worthwhile. When talking with your peers about trying something new, the argument that "This is the way we have always done it" is *not* a valid reason that rules out trying something new. Valid arguments would be that something is preventing your plan from working, your plan could cause unnecessary damage to parts or equipment, you would have to make major modifications to the equipment, and other things of this nature.

These are a few of my favorite tips to help smooth the way when you begin repairs, and I highly encourage you to find other tips and tricks that help. Talk to people who have experience with the equipment and see what they recommend. See if you can persuade someone with years of experience to teach you about the finer arts of maintenance. Search the Internet for information from reputable sources. In this digital age and with all the social media resources available, information is easy to gather for those who are willing to put a little effort into the search. When you discover a gem of knowledge, glean a new insight, or create a new process, make sure to record it in your technician's journal to preserve it for future use.

PARTS SWAPPING VERSUS FIXING THE PROBLEM

When it comes to repairs, replacing broken parts with working parts is a big part of the process, but it is not the end of the process. In our modern world where many things are disposable, it is common practice to replace whole assemblies of parts instead of trying to fix the single part or component of the assembly that is causing problems. Years ago, maintenance technicians would pull a control card from the robot controller,

find the specific damaged component or components, replace them, and then reinsert the same card into the controller. Today, technicians pull the damaged card and put in a brand-new card. Sometimes they send the old card back to the manufacturer for repair, but many times we simply toss or recycle the bad card. This failure to perform deeper troubleshooting has led to a trend in which technicians swap out bad parts for good and then stop their troubleshooting, assuming they have repaired the robot. Yes, the robot is ready to run once more, but have they truly solved the problem?

In truth, everything faults out for a reason and simply changing parts may not be enough to remove the root cause of the problem. If the motor burns out because the robot is picking up too heavy a part, then changing the motor does nothing to fix the problem and merely buys a bit more time (Figure 11-11). If a loose wire is sparking and causing a current problem that damages components on a control card, then replacing that card does nothing to address the loose wire. Sometimes part swapping addresses the symptoms rather than the cause of the problem. To truly fix the robot or other equipment, we have to figure out why it failed in the first place. Perhaps the part in question really did fail, in which case swap-and-go is the right path. For everything else, try to answer the question, "Why did it fail?" Ignoring the root cause can result in repeated damage to certain components and a possible cascade effect in which the second or third time the problem arises other components and systems are damaged as well.

FIGURE 11-11 You will likely have to change out a servo motor at some point. Make sure you determine whether the motor had just reached the end of its life or if something caused it to wear out prematurely.

If you cannot find a bigger reason why, make sure you perform all reasonable checks for functionality and record your results. If possible, compare these results to those for another robot of the same type and make and determine whether the system is running normally. If you do not find any smoking guns or obvious reasons, make sure to pass along the information you learned to others who work on the machine, so those involved will have this background information should the problem happen again. Sometimes it takes two or three failures before the root cause is determined, similar to the way a doctor may need to run a series of tests and try a couple of medicines before truly diagnosing the patient's problem. This deeper level of analysis is what separates the novice technician from the experienced maintenance person.

PRECAUTIONS BEFORE RUNNING THE ROBOT

Once repairs are complete, there are a few things you need to do to ensure everything is ready to go before powering up the robot and immediately thereafter. These simple steps verify that the system is ready to go and can help you avoid the prospect of turning around and coming right back to a now incorrectly working robot that you just declared functional. My own experience has taught me that it is easier to do things right the first time than to go back and explain why the robot you just fixed is down again.

Check the system for tools, spare parts, and foreign objects before putting on the covers. You would be surprised at the amount of damage a forgotten wrench or screwdriver can cause when you fire up the system. Once the covers are on, these foreign objects remain unnoticed until they become a source of damage to the system that requires repair.

Make sure everyone is clear and all the covers are in place. This is part of the LOTO process, but it is important to reemphasize here because you really do not want to be the person who started the robot with someone in harm's way. Having all the covers on before start-up assures your own safety. It protects you electrical arcs, flying parts, rotating equipment, and other dangerous events that may occur

after a repair, especially if you were unable to find the problem's root cause. Even if all the locks are off the lockout, make sure everyone is clear of the work envelope before turning the power back on. The robot could start unexpectedly, with all the attendant ills that could happen. Make sure any operators or other people around the robot know you are powering up the system as well. This prevents them from being startled and possibly hurting themselves on something else.

After power-up, check for any alarms or unusual action by the system. Alarms let you know that the system's diagnostic functions detected issues and irregular actions and can indicate deeper problems or improper repair of the system. Once you verify the system is ready, place it in manual mode and move each axis to confirm proper operation. If you have a simple program, such as a homing program or alignment check program, try running it to see how the robot responds. If the robot stopped at a specific point in a program before repairs, load that program and step through it manually to see if it functions correctly. Also, verify that all the safety equipment and sensors of the robot are working. Sometimes sensors are damaged either by the breakdown of the robot or by accident during repairs.

The last, and perhaps most important, check is to load the proper program and start the robot back into normal operation. While it is running, keep your hand on the E-stop or stop button so you can halt the system quickly should any signs of problems arise. Murphy's law of maintenance says that those who fail to check proper operation in automatic mode are doomed to receive a call back to the robot. I learned this the hard way during my years of fixing machines: Many times everything seemed fine during start-up and all the checks, but the problem returned during normal operation. I always spend some time watching the robot run normally before I truly call anything repaired. After a few cycles of the program with no apparent issues, I usually turn the machine over to the operator and tell that person to let me know if anything goes wrong.

Another great benefit of these steps is the fact you have a front row, center seat for any problems that arise. There are times when no amount of data from the system or descriptions of what happened from the operator can replace your own experience with the system. As a technician, you view the system differently than someone who just runs or programs the systems, which is why seeing the event can give deeper insight into the root cause than just hearing about it. Do not be afraid to spend some time with the equipment after start-up just watching the system for those special clues that mean something to you. This is time well spent that could save you hours of work later.

REVIEW

There are many other areas we could drill down into when it comes to maintenance and troubleshooting, but this chapter provides a good overview of what maintenance entails and some things to watch for. Nothing compares to time in the field, but you need to have a basic understanding before you start. The maintenance and engineering fields are lucrative trades and rewarding career choices for those who like to dig into the deeper workings of the robot.

As we explored repairing the robot, we hit on the following topics:

- **Preventive maintenance.** Preventive maintenance includes those tasks we perform to prevent the robot from breaking down and increasing operational time.

- **Arc flash.** This section provided a basic introduction to this dangerous situation and is meant as a primer that encourages you to learn more.

- **Troubleshooting.** You got a look at the basic steps of the troubleshooting process to start your journey into the technical field of your choice.

- **Crash recovery.** This section described a methodical, step-by-step approach to dealing with robot crashes.

- **Repair tips.** I shared some tips and tricks I have picked up over my years working in various maintenance roles.

- **Part swapping versus fixing the problem.** We discussed the difference between changing parts and truly fixing the machine.

- **Precautions before running the robot.** This was about what to do before powering the robot up and immediately thereafter.

KEY TERMS

Accumulators	Multiple failures	Reactive maintenance
Arc blast	Personal protective equipment (PPE)	Reseat
Arc flash		Troubleshooting
Crash	Preventive maintenance (PM)	

REVIEW QUESTIONS

1. What is preventive maintenance?

2. How do we determine when PM tasks are due?

3. What are some of the common electrical PM tasks?

4. What are some of the common fluid power PM tasks?

5. Why should you never tighten a cable when the robot is powered up?

6. What are some of the dangers of an arc flash?

7. What are some of the events that can cause an arc flash?

8. What is the commonly recommended PPE for arc flash, and where would you check to see if anything has changed?

9. What is reactive maintenance?

REVIEW QUESTIONS

10. What is troubleshooting, and what are the main steps in this process?

11. What are some of the things we do during the data-gathering phase of troubleshooting?

12. What are the five guidelines to keep in mind if the troubleshooting solution does not work?

13. What are the five steps in crash recovery?

14. What are the benefits of taking pictures before you start a repair?

15. What is the benefit of taking good notes during the repair process?

16. What is the potential problem with just swapping out parts and not looking for the root cause?

17. What are some of the precautions to take when repairs are finished and the robot is ready for a trial run?

Reference

1. Titelman, Gregory. *America's Popular Proverbs and Sayings.* New York, NY: Random House Reference, 2000.

Image courtesy of ABB Inc.

CHAPTER 12

Justifying the Use of a Robot

WHAT YOU WILL LEARN

- Why we use robots instead of humans to perform some tasks

- Whether robots are taking humans' jobs

- What a SWOT analysis is and how it is used

- How to figure the return on investment for a robotic system

- How precision and quality figure into justifying the use of a robot

- How consumables are better used by robots

- How to justify using robots that can go places where humans cannot or should not go

233

OVERVIEW

By now you should have a good base knowledge of the parts of the robot, how they work together, how to make the robot run, and some of the things needed to keep the robot running. This is a good time to round out your knowledge with an exploration of the specifics of why we use robots. We have touched on the *why* of using a robot in several chapters along the way, but there are some considerations and things we have yet to explore.

With that in mind, we will look at the following topics:

- Robot versus human labor
- SWOT analysis
- Return on investment
- Precision and quality
- Use of consumables
- Hazardous environments

ROBOT VERSUS HUMAN LABOR

Robots are the perfect solution for handling many of the Dull, Dirty, Difficult, or Dangerous tasks that people perform every day—these qualities are known as the four Ds of robotics. If we damage the robot, we can get replacement parts to repair it, whereas a person requires medical care if hurt. A robot does not feel pain; it knows only what its sensors tell it and even then, the information is purely data. When a person receives damage, there is pain and suffering involved, not to mention the possibility of scars, surgery, rehabilitation, and permanent loss of mobility or function. Robots do not have emotions, so they have no fear of doing a dangerous job. Likewise, robots do not get bored by dull tasks or frustrated by difficult tasks (Figure 12-1). Robots simply continue to perform the task until told to stop or something breaks, whereas a person may become fed up over time and quit.

The same lack of emotion that gives the robot a leg up in the four Ds also helps to justify their use in other situations. Robots are methodically constant where we mere mortals are not. A robot performs its tasks in the same manner until something mechanical changes or we modify the program, and this consistency translates into efficiency and improved quality. In contrast, humans have emotions and—as much as we like to deny it—our emotions do color our actions. We have Monday mornings where we would rather be anywhere but work. We have Friday afternoons where all we can think about is the upcoming weekend. We get sick

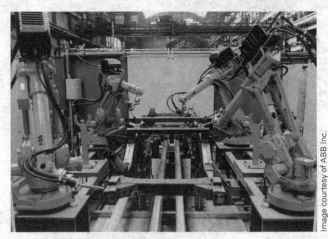

FIGURE 12-1 Repetitive jobs, such as those found on assembly lines, are perfect tasks for robotic systems.

Image courtesy of ABB Inc.

enough to be off our game but not sick enough to stay home. If we are unhappy because the boss chewed us out or because we are having problems outside work, our heart will not be in our work. If we are excited about something, we tend to let our thoughts wander or find excuses to leave our post and tell others about the exciting news. In other words, our own emotions work against us in the work setting, creating distractions that influence the quality and quantity of our work. While we may be off our game, the robot is steadily doing its job and being productive (Figure 12-2).

Wages are another factor that comes into play when we talk about using a robot versus a person. A midsized

FIGURE 12-2 A robot's consistency often outperforms even the fastest human workers due to the fact a robot can work 24 hours a day, 7 days a week, at the same pace.

Image courtesy Yaskawa America, Inc. Motoman Robotics Division

robot costs approximately 72 cents per hour to run when you figure in costs such as electricity, maintenance, programming, and other running costs. (This operational cost does not include the initial cost of the system or cost of damaged parts. We will talk more about the initial cost when we discuss return on investment later in this chapter.) At the writing of this book, the federal minimum wage in the United States was $7.25 per hour. In other words, an employer can run a robot for 10 hours for the same amount it must pay a human worker at minimum for 1 hour of that person's time. Now factor in the reality that most factories pay well over minimum wage, plus the cost of the various benefits that employees receive, and you can see the cost savings achievable through robotics (Figure 12-3). In many cases, this type of cost savings is what allows a manufacturer in the United States to compete with companies in countries such as Mexico, Africa, China, and others where a large number of workers are willing to work for low pay.

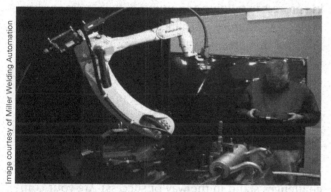

FIGURE 12-3 Part of the human support staff needed to keep the robot operating and productive.

While talking about the cost of human labor versus running the robot, I want to take a moment to address the concerns about lost jobs for human workers when robotics is adopted in a plant. Over the years I have heard workers, students, employers, and others debate or bemoan how the use of robots leads to fewer jobs in the industrial world. At times when jobs are tough to find, as the recession of 2007–2009 and the recovery following the official years of that recession, discussions about the use of automation happen frequently and sometimes become rather heated. When we talk about this issue in my class, I make the following points.

- *If the business is not profitable, it will not stay in business.* Yes, the use of robots may eliminate some positions at a company and lead to fewer total people employed, but this is better than the business closing its doors and no one having a job there. Moreover, the use of automation

and robotics may mean the difference between a company staying where it is or moving operations to another country.

- *Every robot needs a support staff.* Every robot in industry needs someone to run it, someone to program it, and someone to fix it at the very least. Usually three different people complete these jobs. If a company has a large number of robots, it is a safe bet that it will need multiple employees for each area.

- *Many tasks are better suited to humans than robots.* Robots are great at black-and-white tasks but still struggle with the gray areas of decision making. It is these gray areas in which people excel and that explains why we still need people in the automated factory. The robot can load the machine while the operator tweaks programs, checks quality, double-checks order quantities, or performs any of the other thinking tasks necessary in modern industry. Someday we may figure out a way to write programs that deal with the gray areas better, but until then we need the decision-making capacity of humans.

There are other justifications for using robots that fall under the four Ds of robotics, but we have looked at many of the common reasons that led to the increase in robot use. As we continue to advance the field of robotics, we will find new and exciting ways for the robot to perform tasks that people would rather not tackle. When we free the operators from the mundane tasks of industry, we give them the time needed to perform functions such as monitoring quality, adjusting programs, and figuring out better ways to perform tasks. Moreover, the more we use robots in the industrial world, the greater the number of good-paying technical jobs for people interested in robots—like you!

SWOT ANALYSIS

SWOT stands for strengths (S), weaknesses (W), opportunities (O), and threats (T). **SWOT analysis**, then, is a technique used to compare all of these factors to help with important financial and business decisions. It is a step-by-step process that is a great tool for making important decisions, such as whether to use automation or more people. We can break the analysis into the following steps:

- Determine the focus.

- Do research.

- List company strengths.

- List company weaknesses.
- List opportunities.
- List threats.
- Establish priorities.
- Develop a strategy and execute it.

The first step in any planning or analysis process is to define the desired outcomes, or goals. This step is where you set two or three objectives that you want to reach as a company, which then guide the various types of data collected and analyzed in the following steps of the process. If the end goal is not yet clear or is a work in progress, then specific questions could be used instead. For example, perhaps the goal is to add enough automation to free up an hour of the employee's day or to add automation to tasks where people are suffering from repetitive motion injuries. Questions might be something along the lines of "How much could we save by using robots to paint the parts?" or "Is it possible to fully automate the welding line?" Whether you use goals or questions in the step of the SWOT analysis, there has to be some focus to the data gathered or you run the risk of just doing busy work gathering a bunch of information that has no clear use.

During the research phase, you learn all the ins and outs of your company's business operations as well as how the company compares to its competitors. This is a good time to get data on the cost of raw parts, labor costs, profit margins, number of hours spent manufacturing each part, and anything else that is beneficial for understanding the physical and financial sides of your business. When you look into your competitors, it is unlikely you will be able to glean all of the same data, but you can get general cost of goods sold, market share, information on new products about to hit the market, patents, and other information to help with comparisons.

The next step is to list the strengths of your company. These include advantages your company has over its competitors and things your company does better (or best). Unique resources or materials purchased at a lower cost fit into the strengths category as well. Another component to think about here is the strengths perceived by the general public. Does the average consumer trust your product because your company is more than a hundred years old? Perhaps you have a top rating of some kind, which translates into buyer confidence. Your facility's location or the skill level of the employees may also factor in here. For instance, Missouri is a state with a low cost of living state; consequently, employers can pay employees less in St. Louis

than they would in higher-cost cities like Detroit and the employees can still earn a good living. This drives down the overall cost of production and leads to happier workers, as they can take care of their families.

Next, we gather all the weaknesses. There are downsides to every business, and this step is where we log those disadvantages for the company in question. This could include factors that are costing the company sales or market share, not owning the patents or intellectual property on which the product is based, or difficulty in getting raw materials. On the people side of this equation, we can include high turnover rates, lack of skilled workers, low-quality production, and any image problems from the customer's perspective. The whole point of this step is to identify those factors that might be a problem in the process so they can be properly weighted when making the decision.

Next, it is time to find all the opportunities involved with the goals or question at hand. This can range from new technology to increase the company's manufacturing process to training opportunities to increase workers' skills. New products or changes in social patterns would fall into this area as well. Perhaps another company is preparing to leave a particular market, opening up chances for the company to obtain new customers and new production avenues. Generally, the opportunity side of things comes from outside the company and should not include things listed as threats in the next step.

As with any venture or change, there will be threats to success, and that is what we document next. Which challenges stand in the way of success? Are your competitors doing what you want to do already, or are they doing it better than you can for a while? Are there changes in technology or the related standards and specifications that could prove problematic? Does the company already have financial issues to deal with, such as low cash flow or high debt? During this step, we look at the negative side of the coin in a realistic, not fearful, way. It is easy to imagine all kinds of dark outcomes, so make sure everything is based on solid fact and not the stuff of nightmares.

Now that we have our SWOT lists, it is time to put things in order of importance or priority. The best strengths and opportunities go at the top of the list, while the lesser factors filter to the bottom. For the weaknesses and threats, we put the ones with the worst impact at the top and the lesser evils toward the bottom. This creates an easy-to-see hierarchy of information for the final step of the process.

The last step of SWOT analysis is to take all this information, filtered through the initial questions or goals, and create a strategy. In this step, we consider which strengths and opportunities are compatible with what we want to do, while simultaneously which threats and weaknesses we need to overcome to be successful. At this point, you may discover that the original goal or action in question is not a good idea at this time. If that is the case, take this opportunity to identify a new goal or objective that will get you closer to your original goal. If the goal or objective offers too little return given all the challenges that must be surmounted, you might be off abandoning ship on that idea and looking for something else that would be a benefit.

We have looked at only the basics of the SWOT analysis in this section; there is much more to learn for those interested in this subject. There may be a team in your company that is good at doing these analyses and all you need to do is provide them with the information for the SWOT portions. If you run into trouble along the way, training in SWOT analysis is available as well as companies that will do the analysis for your company, so do not be afraid to look for help when needed.

RETURN ON INVESTMENT

Return on investment (ROI) is a measure of the amount of time it takes a piece of equipment to pay for itself. We calculate ROI by taking the operating cost of the equipment and subtracting it from the amount of money that equipment saves. Once we know the amount of profit made using the machine, we divide the total cost of the equipment by this number to determine how long it will take for the equipment to pay for itself. To figure the ROI for a robot on a parts basis, we would determine how many parts the robot needs to make to pay for itself and then turn that result into a time scale using some correlation between parts produced and the time needed to produce them. If we are comparing cost savings with the robot to human labor costs, then we would simply figure the per-hour savings and use that number to get the total hours needed to pay for the robot. Once we have that figure, we simply divide by the number of hours the robot runs per year or week to figure out the time for payback. The following examples illustrate ROI calculations.

Example 1

For this example, we will figure the ROI for a new robot that costs $107,500, which includes the cost of shipping and installation. For cost savings, we will use $0.50 per part after subtracting the operating cost of

the robot, at a rate of 50 parts per hour. Lastly, we will use 2040 hours per year, as this is the standard 40-hour week multiplied by 51 weeks. Remember, the robot does not get sick days, paid vacations, or holidays, but we do factor in a week of downtime due to the facility being closed for holidays.

First, take the per-part cost savings of $0.50 and divide it into the total cost:

$$\frac{\$107,500}{\$0.50} = 215,000 \text{ parts}$$

Next, take the total number of parts needed to pay for the robot, and divide it by the rate of 50 per hour to determine the total number of hours needed:

$$\frac{215,000}{50} = 4300 \text{ hours}$$

Lastly, divide the 4300 total hours by the 2040 hours for a year we figured earlier:

$$\frac{4300 \text{ hr}}{2040 \text{ hr}} = 2.11 \text{ years (just over 2 years)}$$

Example 2

Let us look at another example in which we use the same data, but instead of a per-part cost savings, we will figure the ROI versus the wages of a person. The new robot still costs $107,500, which includes the cost of shipping and installation. For cost savings, we will use the $0.72 per hour operating cost and assume an hourly wage for our fictional worker of $19.75 when benefits and wages are added together. Lastly, we will use 2000 hours per year, as this is the standard 40-hour week multiplied by 50 weeks: Our worker gets a week off each year for vacation and all the holidays mentioned previously. To make a straight comparison in this example, we will not give the robot an extra week of work relative to its human counterpart—though this would be a week of extra profit because the robot is producing parts during that week while the company pays the worker who is *not* producing parts.

First, we take the $19.75 hourly wage of the human worker and subtract from that the $0.72 cost of hourly production for the robot:

$$\$19.75 - \$0.72 = \$19.03$$

Next, we take the total cost of $107,500 and divide it by $19.03 to see how many hours the robot must run to pay for itself:

$$\frac{\$107,500}{\$19.03} = 5648.975$$

We will round this up to 5649 hours so we are working with whole hours.

Lastly, we divide the 5649 hours by the 2000-hour working year for an employee at this company:

$$\frac{5649 \text{ hr}}{2000 \text{ hr}} = 2.82 \text{ years (almost 3 years)}$$

These are rough figures that do not factor in real work events such as machine downtime for preventive maintenance, unexpected plant closures, programming, or breakdowns. Any parts used to maintain or repair the robot would lengthen the amount of time for payback, as this would be an added cost. If anything changed over this time, such as the cost of electricity or having to buy new tooling, that would need to be included as well. The reality is that payback would likely take longer than our math shows in both cases when you add these unpredictable factors in, but for a general idea of the amount of time until the robot is profitable, these figures are sufficient. Figures 12-4, 12-5, and 12-6 show the initial costs for some robotic systems.

FIGURE 12-5 The Panasonic-Miller welding robot PA55 has an initial cost of approximately $55,000.

FIGURE 12-6 The Panasonic-Miller PA 102S has an initial cost of more than $100,000.

Another factor affecting ROI is how long the robot will run—that is, its life expectancy—after it pays for itself. Some systems come with a designated useful life from the manufacturer that suggests how long you can expect the robot to run before it should be replaced. This is just an estimate, however, and there are numerous accounts of robots running long past the time stipulated by the manufacturer. For example, the first electric ABB robot has been running for almost 40 years. The longer a system runs after it has paid for itself, the more money it makes for the company. Most companies want their equipment to have an ROI time of 2 years or less, so they can get the most out of the equipment before they need to replace the equipment or make large investments in repairs and preventive maintenance.

FIGURE 12-4 The FANUC M-1iA training system has an initial cost of approximately $50,000.

In some cases, the cost of the system far outweighs the return on investment. In those cases, we have a few options. First, we can forget about the robot and continue to do the work the way it is currently done. Second, we can try to find a cheaper robot that will perform the task. Third, we can figure out a new way to use the robot. Remember, robots work in the same manner over and over again, with little human input. This means that with some thought and preparation, we can set up the system in such a manner that it can run all night while the human workers are away. Perhaps there is a way that the robot could do several jobs instead of just one or two, thereby increasing the amount of savings the system creates. In the next two sections, we will cover some of the other areas that figure into the cost savings of a robot and, therefore, offer more ways to boost the equipment's ROI.

The simple truth of the matter is that if we do not find a way to justify the cost of a robot, there is a very small chance of industry using it. Companies are in business to make money, and things that cost money and provide no tangible return rarely get funding. You can argue until you are blue in the face about the value of something, but until you can show the math supporting that value, you are likely just wasting everyone's time. If the ROI is more about avoiding injuries to people, make sure to figure out a way to show this return mathematically in cold, hard, monetary facts in addition to arguing that it is the right thing to do. This approach gives you the best chance for justifying the purchase of a robot.

PRECISION AND QUALITY

As discussed previously, robots have the ability to reach an exact point repeatedly with only the smallest of errors, often within 0.0003 to 0.005 inch. This level of precision is difficult for a human to match once, much less repeatedly. For jobs requiring this level of precision, such as placing micro-sized components on an integrated circuit board or precision laser fabrication for aerospace, we need either a dedicated machine or a robot that is up to the task. When you add the robot's speed to this level of precision, it is clear why some tasks are better suited to a robot than to a human.

This level of repeatability also translates into better quality of work. Since a robot can follow the same path repeatedly with only the smallest amount of error, you simply need to edit the program until you get the quality of part you want and then let the robot run. People can and do produce quality parts all the time, but there is a greater risk that they will do something different from time to time that could compromise quality. When we catch bad parts inside the facility, it equates to lost time and materials. When bad parts make it to the customer, it can hurt future sales when the customer shares that experience. Imagine what you would have to say about the manufacturer if you bought a new TV or cell phone and turned it on for the first time, only to discover that the new device was dead. It does not take too many of these types of stories spread by customers to decrease the sales of a product, which is one reason why the quality possible with a robot is so attractive to industry.

Continuing with this train of thought, robots give operators with low skill levels or experience the ability to produce parts that require a great amount of precision and technical skill. For instance, if someone with 20 years in the welding field creates a program to weld parts the same way that person would do it by hand, then anyone who knows how to run the robot can create parts with proper welds. This creates situations where companies need only a few workers with the years of experience and training to program the robots, while operators with a lower level of training run them (Figure 12-7). In a time when employees with many years of experience are retiring, this has already become the operating philosophy

FIGURE 12-7 Welding thick metals (also called heavy plate) requires a great amount of skill and experience when done by hand. When we program a robot to perform welding, we give operators who are new to heavy plate the chance to make parts with quality welds while they learn all the specifics of this type of welding.

Image courtesy of Miller Welding Automation

at more than one company. The company can train the operators in the basics of running the robot and quality inspection in a week or two and have quality parts flowing down the line. This leaves their experienced workers free to write programs, make changes as needed, and perform other tasks to improve the overall quality of the products.

Precision and quality savings also figure into the overall ROI of a robot. Each time a company scraps a part, it costs money. Each time a bad product leaves the factory, there is a good chance for a decrease in customer satisfaction and potential loss of sales. Lack of precision requires larger spaces for various components and raises the overall cost of the product. Therefore, by increasing the precision of processes and part placements while ensuring these actions occur in the same manner every time, the company reaps cost savings that figure into the overall cost per part savings associated with the robot. Given the ongoing improvements in robots and their precision levels, some companies are replacing worn-out dedicated machines with precision robotic systems and saving money from day 1, as the robot is cheaper than the cost of a new dedicated machine. As long as the length of service is comparable, there really is no need to worry about the ROI in this instance.

USE OF CONSUMABLES

On top of its quality and precision benefits, many times a robot offers savings in terms of the amount of raw materials used in the production process. The same factors that allow the robot to increase quality and precision in production also tend to optimize the use of materials, thereby minimizing waste (Figure 12-8). In many applications, subtle changes in the speed of the process translate into large dollar savings via reduced use of materials and higher quality. For example, when human workers paint a car, they typically use some form of spray device and sweeping motions to ensure the entire vehicle is covered. If they move the sprayer too fast, the paint may be thin in certain areas and not stand the test of time. If they move too slowly, they end up putting on a thicker coat that wastes paint and could lead to cracking or other problems down the road. If we use a robot in this application, we can program a constant motion rate that equates to the optimal paint thickness, saving on

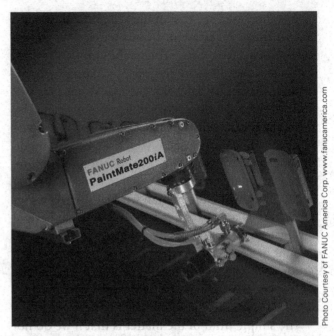

Photo Courtesy of FANUC America Corp. www.fanucamerica.com

FIGURE 12-8 This FANUC robot is hard at work painting parts as they pass by on the conveyor. Notice how the parts get a nice, even coat of paint, while the conveyor system under them is free from overspray or waste.

the amount of paint used while increasing the overall quality of the finished product.

Robots also have the ability to work with materials that are hazardous to humans. Many chemicals and materials can be used to minimize costs in the manufacturing process that would cause harm or death to humans. For human workers to interact with these materials, they often need to wear personal protective equipment (PPE)—that is, various items worn on the body to negate the hazards of a work task or area, such as safety goggles, chemical-resistant gloves, special clothing, and other safety equipment. The PPE has to fully protect the worker from the hazards of the task, can be less than comfortable for the wearer, and represents an added cost in the manufacturing process. In addition, if the PPE fails, it exposes the worker to a hazard of some kind. All of this equates to added risk and cost above and beyond the cost of the materials used. Since a robot is a machine and not a person, many times these materials present no added danger to the robot. In the cases where a chemical or material could damage the robot, often a simple change in the construction of the robot is all it takes to mitigate the hazard. This is another of those dangerous tasks that robots do for us, but it can also deliver cost savings for the company.

The precise use of consumables was the driving force behind the first industrial robot built by the DeVilbiss Company in 1941. To this day, industries continue to look for ways to use an ounce less here or a coat less there to increase their overall cost savings by reducing the amount of materials used. Every dollar saved in this way adds to the ROI from the robot.

HAZARDOUS ENVIRONMENTS

One of the great benefits robots have over people is the fact that they are machines, rather than organic systems. This gives the robot the ability to not only survive but also continue to work without issue, in environments that would be physically taxing, dangerous, or even deadly to their human counterparts (Figure 12-9). Sometimes this is all the justification we need for using a robot, especially in situations where the ROI math does not add up. Even if we overlook the moral obligations of the company and its managers, there are expenses associated with

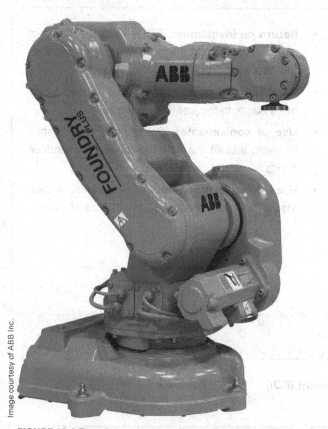

Image courtesy of ABB Inc.

FIGURE 12-9 Foundry operations often have multiple areas where human habitation is hazardous if not downright lethal. The use of robots in these areas makes good sense!

injury or death of employees. On-the-job injuries can cost a company thousands, if not millions, of dollars, so avoiding even one of these events can mean the difference between a company making a profit or going bankrupt. Even in cases where the injury to the human worker is relatively minor and the medical expenses are not that high, the employer still has to find someone to do the work while the injured employee recovers. In addition, the injury could increase the company's worker's compensation insurance premiums. Collectively, this equates to increased expenses. While this factor is not typically included in the ROI math, some cases warrant the addition of these figures in the calculations.

In instances where worker death is highly probable, there should be no debate: Does the company hire someone knowing there is a fair chance that person could die while performing the work or does it purchase a robot that it can replace? The obvious answer is "Get a robot!" Even if the company's management has no concern for human life, the cold, hard facts about the costs should make the decision clear. A death on company property, due to working conditions, means an immediate inspection by the **Occupational Safety and Health Administration (OSHA)**. OSHA's job is to ensure that all employees have a safe and healthy working environment. If someone dies on the job due to work-related factors, OSHA immediately inspects the facility and investigates the death to figure out why it happened and to prevent anyone else from being hurt or killed. The company might be forced to pay thousands of dollars in fines, not to mention the costs associated with the lost production time if OSHA shuts the plant down until corrective actions are complete. There is also the possibility that the victim's family will file a wrongful death lawsuit, which is another big expense in terms of legal fees and possible damages awarded to the family. Insurance rates for the company are likely to increase, which means more money flowing out of the company's pockets. Moreover, we cannot forget the impact a work-related death would have on the moral of the workers at the facility and all the costs that could have. Low moral generally decreases workers' productivity, reduces the quality of parts produced, and leads to increased **turnover** (employees leaving the company), which necessitates hiring and training new workers. Taken together, these factors should convince even the most hard-hearted company that no job is worth a worker's life when a robot can take the risks instead.

Robots can go into some hazardous environments that people simply cannot tolerate. Even with our impressive array of current technology and equipment, it is still impossible to send a person below a set depth in the ocean because the pressure would crush the human body. To gather information or do research at these depths, we need a machine. Likewise, without protective equipment and oxygen, humans cannot work in certain areas with toxic or low-oxygen atmospheres. Current technology can shield people only from certain levels of radiation, and even then only for a finite amount of time. We have miles of buried pipeline that are too small for adults to fit through that require regular inspection. In all these instances and many more, the only good option is to use a robot to perform the tasks that humans cannot.

No matter why it is hazardous to humans, it generally makes good business sense to use a robot in a dangerous environment. With the maturing of robot technology and the advances in sensors over the last decade, the old argument of "A robot isn't able to do it as well as a person" has gone out the window for most of these applications. While there is a cost associated with robot repair and replacement, it pales in comparison to the pain and suffering that humans incur when performing the same task. It is this reasoning that has put robots into use in repetitive-motion jobs, under the sea, cleaning the Chernobyl nuclear site, destroying buildings, inspecting pipes, and performing a multitude of tasks in industry.

REVIEW

The justification for using a robot depends on the *where*, *what*, and *why* for using a robot in the first place. In a perfect world, we would use robots because they are a good fit and would not have to look at budgets or cost in making this decision. We live in an imperfect world, though, where we have to justify the use of robots before they are adopted. This chapter by no means covers all possible justifications for using robots, but you should have a better understanding of why we use robots and some of the financial considerations that go along with this decision.

During our exploration of why to use a robot, we covered the following topics:

- **Robot versus human labor.** We considered the differences between using a robot and using a person to perform tasks from a justification standpoint.

- **Return on investment.** ROI is all about the robot paying for itself, and the math that goes along with this payback.

- **Precision and quality.** This section added a few more points to the justification toolbox.

- **Use of consumables.** Robots can save money by using less of the materials needed to produce things.

- **Hazardous environments.** The dangers associated with a job can help to justify the use of a robot.

KEY TERMS

Occupational Safety and Health Administration (OSHA)	Return on investment (ROI) SWOT analysis	Turnover

REVIEW QUESTIONS

1. What are the four Ds of robotics?

2. How does damage to a robot compare to damage to human workers?

3. How do people and robots compare in terms of the effects of emotions on job performance?

4. How does a robot's operating cost compare to the human labor cost?

5. What are the three main points to remember when talking about the loss of human jobs to robots?

6. What are the eight steps in a SWOT analysis?

7. How do we figure ROI?

8. What are the options if the cost of the robot outweighs the return?

9. How does using a robot help save on consumables such as paint or welding wire?

10. What are the downsides of PPE for human workers?

11. What are the costs associated with a worker's injury on the job?

12. What are the potential costs and impacts of a worker fatality for the company involved?

13. List some of the places robots can go that humans cannot.

Appendix A

	Flow Charting Symbols
	Terminator: Used at the beginning and the end of the program. Place Start in the beginning terminator, End in the ending terminator.
	Process: A step or group of steps in a program.
	Alternate process: An alternate to the normal process step.
	Predefined process: A subroutine that is used more than once in the program.
	Data: Inputs to and outputs from a process.
	Delay: A delay or hold. The time or hold condition should be specified in the symbol.
	Manual input: Data or signals that are manually entered into the process.
	Manual operation: A process step that is not automated.
	On-page connector: A jump from one point to another on the same page. An alpha or numeric designator should be entered in the symbol at matching connections.
	Off-page connector: Used when connecting to a point of the flow chart that is on another page. An alpha or numeric designator and the page number should be entered into the symbol at matching connections.
	Summing junction: A logical AND connection.
	OR: A logical OR connection
	Flow line: Indicates the direction of program flow. The arrow points toward the next block in the program.
	Decision: Indicates a Yes/No decision. An arrow enters the block at the decision, and two arrows—one for yes, the other for no—leave the block.
	Decision: Indicates a multiple choice decision. One arrow enters the block at the decision, and an arrow for each choice leaves the block

Appendix B

RECOMMENDED TOOL LIST

This is a recommended tool list based on the author's experience. It contains only general, commercially available hand tools. Most of these tools are in metric dimensions, as most industrial robots use metric hardware. Any special tools required by the robot manufacturer should be obtained from the manufacturer. Always read the manufacturer's maintenance instructions before performing any maintenance on a robot.

Ratchet and extensions, ¼-in drive
Ratchet and extensions, ⅜-inch drive
Ratchet and extensions, ½-inch drive

Torque wrenches

- 0–100 lb-in.
- 0–50 lb-ft
- 50–300 lb-ft

Hex drive bits

- 3–6 mm ¼-inch drive
- 5–14 mm ⅜-inch drive, short and long
- 6–20 mm ½-inch drive, short and long

Torx bits, ⅜-foot drive

Socket sets, metric and standard, ⅜-inch and ½-inch drive

Allen wrenches, metric and standard
Phillips screwdrivers, #1, #2, #3, 6–8 inches long
Screwdrivers ⅛-inch tip, ¼-inch tip, ⅜-inch tip,
6–8 inches long
Wire cutters, 6 inch
Needle-nose pliers, 6 inch

Adjustable wrenches

- 6 inch
- 8 inch
- 12 inch

Adjustable pliers

- 6 inch
- 8 inch
- 12 inch

Pry bars

- 10 inch, 2 each
- 16 inch, 2 each

Tap and die sets, metric

- 3–8 mm fine thread
- 6–20 mm coarse thread

Hammer

- 12 ounce
- 16 ounce
- 24 ounce

Metric reamers, 6–16 mm
Slide hammer and adapters for pulling dowel pins
Soldering iron, 15–30 watt, and solder
Crimp-on connectors and crimping tool
Heat-shrink tubing and heat gun

Appendix C

Torque Conversion Formulas		
Convert From	**To**	**Multiply By**
ft-lb	in.-lb	12
ft-lb	Nm	1.356
in.-lb	Nm	0.113
Nm	in.-lb	8.85
Nm	ft-lb	0.7375
kgfm	Nm	9.81
kgfcm	Nm	0.098
kgfm	Nm	7.233
kgfcm	Nm	0.07233

Appendix D

Bolt Torque Chart: UNC/UNF Threads				
Diameter (Pitch)	Grade 5 Dry	Grade 5 Lube	Grade 8 Dry	Grade 8 Lube
¼ (20)	8.0	6.3	12.0	9.0
¼ (28)	10.0	7.2	14.0	10.0
5/16 (18)	17	13	24	18
5/16 (24)	19	14	27	20
3/8 (16)	30	23	45	35
3/8 (24)	35	25	50	35
7/16 (14)	50	35	70	50
7/16 (20)	55	40	80	60
½ (13)	75	55	110	80
½ (20)	85	65	120	90
9/16 (12)	110	80	150	10
9/16 (18)	120	90	170	130
5/8 (11)	150	110	210	160
5/8 (18)	17	130	240	183
¾ (10)	260	200	380	280
¾ (16)	300	220	420	310
7/8 (9)	430	320	600	450
7/8 (14)	470	350	670	500
1 (8)	640	480	910	680
1 (14)	720	540	1020	760

Torque values are in ft-lb.

	Class 4.8				Class 8.8 or 9.8				Class 10.9				Class 12.9			
	Lubricated		Dry		Lubricated		Dry		Lubricated		Dry		Lubricated		Dry	
SIZE	N·m	lb-ft	N·m	lb-ft	N·m	lb-ft	N·m	lb-ft	N·m	lb-ft	N·m	lb-ft	N·m	lb-ft	N·m	lb-ft
M6	4.8	3.5	6	4.5	9	6.5	11	8.5	13	9.5	17	12	15	11.5	19	14.5
M8	12	8.5	15	11	22	16	28	20	32	24	40	30	37	28	47	35
M10	23	17	29	21	43	32	55	40	63	47	80	60	75	55	95	70
M12	40	29	50	37	75	55	95	70	110	80	140	105	130	95	165	120
M14	63	47	80	60	120	88	150	110	175	130	225	165	205	150	260	109
M16	100	73	125	92	190	140	240	175	275	200	350	225	320	240	400	300
M18	135	100	175	125	260	195	330	250	375	275	475	350	440	325	560	410
M20	190	140	240	180	375	275	475	350	530	400	675	500	625	460	800	580
M22	260	190	330	250	510	375	650	475	725	540	925	675	850	625	1075	800
M24	330	250	425	310	650	475	825	600	925	675	1150	850	1075	800	1350	1000
M27	490	360	625	450	950	700	1200	875	1350	1000	1700	1250	1600	1150	2000	1500
M30	675	490	850	625	1300	950	1650	1200	1850	1350	2300	1700	2150	1600	2700	2000
M33	900	675	1150	850	1750	1300	2200	1650	2500	1850	3150	2350	2900	2150	3700	2750
M36	1150	850	1450	1075	2250	1650	2850	2100	3200	2350	4050	3000	3750	2750	4750	3500

Metric Bolt Torque Chart

Glossary

A

absolute optical encoder. A device that adds enough emitters and receivers, usually four or more in total, to give each position of the encoder its own unique binary address.

accumulator. A device that stores hydraulic pressure and releases it back into the system as needed.

acyclic. A data transfer that happens only when specifically requested, often as a one-time event.

adaptor. A device that changes the style of the pinout so that communication or power can flow.

algorithm method. Use of a mathematical formula to analyze a sample part and then compare newly processed image data to the initial calculation result.

alternating current (AC). Electrical power where the electrons flow back and forth in the circuit.

American National Standards Institute (ANSI). The organization that administers and coordinates private-sector voluntary standardization systems.

amp-hour (Ah). The number of amperes deliverable over a length of time.

amperage. A measure of how many electrons, or how much electricity, is flowing through a system; measured in units of amps (A).

ampere. The unit of measurement of electron flow; $1\ A = 6.25 \times 10^{18}$ electrons passing a point in 1 second.

analog. A range of voltage or current values scaled to mean something in the system.

AND. A logic function that requires two or more separate events or data states to occur before the output of the function is triggered.

android. A synthetic organism designed to imitate a human.

angular grippers. Fingers that hinge or pivot on a point so as to move the tips outward to release parts or inward to grip parts.

anthropomorphic. A descriptor for motions that look very organic and lifelike.

arc. Part of a circle.

arc blast. The pressure wave created by an arc flash explosion, which can exceed 2160 ft/lb², which is well over the 1720 ft/lb² threshold at which massive internal injuries and death occur.

arc flash. An event in which a short in the electrical system ionizes the air and creates an explosion complete with intense heat, shrapnel, and pressure waves.

area scanning. An application in which the imaging device includes an array of charge-coupled device (CCD) or complementary metal-oxide semiconductor (CMOS) elements, giving the system the ability to take a picture of an entire area or object all at once.

arithmetic instruction. An instruction used to perform arithmetic operations.

artificial intelligence (AI). The ability of a computer program to make decisions when there is no clear-cut right answer or learn from previous events.

automata. Devices that work under their own power and are capable of being humans' equal in functionality, design, and purpose.

automation. Machines that work largely on their own, performing tasks in industry and providing many of the wonderful products we enjoy today.

auxiliary contact. A device used to latch circuits and maintain power after a pushbutton is released, power other devices at the same time as the main device, or provide voltage signals as needed.

axial diffuse lighting. A lighting scheme that uses a light source set at 90 degrees to the part; a mirror in line with the camera reflects the light straight down to the part, and then allows the returning light to pass through to the camera.

axis. Each part of the robot that has controlled movement.

B

backlash. The distance from the back of the drive gear tooth to the front of the driven gear tooth; it represents loss of motion in the system.

backlighting. A lighting scheme that provides the greatest possible contrast, as the light source is placed opposite to the camera, with the object between the two.

backplane. A piece of hardware that provides a path for power and communication between all the various modules or parts of the programmable logic controller (PLC).

ball screw. A screw consisting of a large shaft with at least one continuous tooth carved along the outer edge, along with a nut or block that moves up and down the length of the shaft.

base. The area where a nonmobile robot is mounted to or bolted on a solid surface for stability, or where a mobile robot's platform for a manipulator is mounted.

bellows gauge. A gauge that uses thin-walled tubing that can expand or retract in reaction to pressure, with the bellows portion often being augmented by an internal spring.

binary address. A unique sequence of 1s and 0s that the controller can understand and use.

bit. Each 1 or 0 in the processor's memory.

blunt-force trauma. An impact that does not penetrate the skin; the proper action depends on the level of the injury.

bourdon tube gauge. A gauge that uses a thin-walled, slightly elliptical, cross-sectioned tube bent in a C shape, which is tied directly to the system to read pressure.

branch. An alternate path around certain instructions.

branching. A programming structure that causes a program to jump to another point in the program; it is classified into one of two broad types: conditional branching and unconditional branching.

bright field lighting. A lighting scheme in which a large portion—if not almost all—of the light from the illumination source is reflected from the object back into the camera.

bus topology. A linear bus; a topology that consists of one main cable (trunk cable), with terminating resistors attached at each end of that cable.

byte. A group of 8 bits.

C

calibration. A process that ensures a precision-based system performs properly and provides for any adjustments needed.

call instruction. An instruction used to call other programs from a main program; it can be considered a type of unconditional branching instruction.

capacitive level sensors. Sensors that work on the same basic principle as the capacitive proximity switch, with one exception: Instead of the fluid acting as the second electrode, the liquid is the dielectric.

capacitive proximity switch. A switch that generates an electrostatic field and works on the same principle as regular capacitors. The switch provides the electrostatic field, air is the insulating material or the dielectric, and the object being sensed provides the conductor that completes the capacitive circuit.

capsule gauge. A gauge designed for measuring low or very low pressures and vacuums.

card. An electronic unit found on a flat platform that connects with the system in some fashion to provide for communication and power draw.

cathode ray tube (CRT). A glass envelope that contains an electron gun and a fluorescent screen in a vacuum, and that includes some means of accelerating and deflecting electrons.

cautionary zone. The area where one is close to the robot, but still outside of the work envelope or reach of the system.

cell. A logical grouping of machines that aid in the work flow required to turn raw parts into finished goods.

center of gravity. The part of a robot where the mass is considered to be centered.

character recognition. The ability to read written or printed letters, numbers, and symbols.

charge-coupled device (CCD). A device made up of pixels constructed of metal-oxide semiconductor (MOS) capacitors that convert the light photons falling onto them into electrons.

circuit. The path through which the electrons flow.

circular motion. The formation of arcs and full circles as described by no less than three points.

closed-loop control system. A system that has some means of verifying proper operation, usually in the form of encoders or other position-sensing devices that confirm commands were executed as directed.

closed loop. A system that sends out a control pulse—whatever it may be—to initiate movement; it then receives a signal back that confirms movement and sometimes identifies the direction and distance of that movement.

clustering. Region growing; a means of sorting data by finding elements that have a similar value and growing the clusters outward from there.

CMOS. Complementary metal-oxide semiconductor; a combination of p-type and n-type metal-oxide semiconductor field-effect transistors (MOSFETs) used to create logic gates and other solid-state circuits.

coaxial cable. A cable consisting of a single copper wire conductor surrounded by a layer of insulation, which in turn is surrounded by a braided metal shield that is connected to ground, which in turn is covered by an outer jacket.

collaborative robots. Robots that are designed to work with humans, instead of separately away from humans; they have safety systems that limit the danger to humans by carefully monitoring their surroundings and often slowing down when humans are nearby.

color detection. Use of photoelectric sensors to analyze the light coming into the receiver and determine the color or a color difference in parts.

common industrial protocol (CIP). An information transmission method that creates a common language among protocols so information can flow between them without the hassle and loss of time associated with translating everything into a different format.

communication. The portion of the vision system that is responsible for transmitting the refined data from the processor to the robot controller or other control system so it can respond appropriately.

compound gears. Two or more gears on the same shaft, often made from one solid piece of material.

computer numerical control (CNC). Computers based on microchips that became strong and small enough to use in industry, thereby replacing the punch cards and magnetic tapes of old with programs and computer code.

connector. A device designed to fasten together and make electrical connections through cable wires being attached to its numbered internal pins.

consistency. The ability to produce the same results or quality each time.

contactors. They tend to have one to three main contacts and one or more lower-amperage contacts.

continuous mode. A mode that once started continues until the operator stops the process or it is complete.

controller. The "brain" of the robotic system; the part of the robot responsible for executing actions in a specific order under specified conditions.

coordinate system. A system used to specify points in space for robot movement.

crash. When the robot makes unexpected contact with an object.

cycle. One complete AC sine wave from zero to positive to zero to negative and back to zero.

cycle time. The time taken by a robot to complete a program.

cyclic. A data transfer process that happens at set intervals.

D

3D printing. Creation of a three-dimensional design, layer by layer, in some medium such as ABS (acrylonitrile butadiene styrene) plastic, hardened resin, glued powder, heat-fused metal powder, or even concrete.

dangerous. A descriptor for environments that includes all things hazardous and any other condition that has a high risk of injury or illness for humans.

danger zone. The area the robot can reach or the work envelope; the area where all the robotic action takes place.

dark field lighting. A lighting scheme in which most of the light from the light source is reflected away from the camera, with the lighting usually at an angle of less than 45 degrees or focused off the object.

data structure. A collection of related items used to control a process, often arranged in a table format.

DC brushes. Devices made of carbon that transfer electricity from the power wires going into the motor to the rotating portions of the motor.

dead man's switch. A trigger- or bumper-type switch on the back of the pendant, which is required to move the robot manually.

degree of freedom (DOF). Each axis or way that the robot can move.

delta robot. A robot that has three or more arms coming down to a central point to create a unique motion type.

deterministic. A descriptor for a system in which the user can set the minimum rate of data transfer within given parameters and the system will keep the timeline that has been set.

diaphragm gauge. A gauge that uses a thin membrane, often made of metal in a capsule arrangement, to move the indicator assembly.

difficult. A descriptor for environments that feature heavy tasks, which includes all the tasks in industry that humans struggle to perform.

diffuse. A descriptor for a system consisting of photoelectric sensors that house both transmitter and receiver in one unit but in which only a small amount of returned light is needed to activate the sensor.

diffuse dome lighting. A lighting scheme sometimes referred to as "cloudy day illumination." With this technique, a light source is placed inside the surface of a partially reflective dome, with the camera focused on the object through a hole in the middle.

digital. Data that can have only one of two states: 1 or 0, on or off, yes or no.

DIN rail. A mounting system that consists of a specially designed metal (aluminum) rail, often 35 mm in width, to which users can clip components to provide stability.

diode. A solid-state device that allows current to flow in only one direction.

diopter. A unit of refractive power; 1 diopter = 1/1 meter of focal length.

dip switch. A small on/off-type switch used to set options or parameters.

direct current (DC). A one-direction flow of electrons.

direct drive. A system in which the rotating shaft of the motor is connected directly to the part of the robot it moves.

dirty. A descriptor for processes that produce dust, grease, grime, sludge, or other substances that people would rather not get on them.

distributed control system (DCS). A system in which autonomous controllers are distributed throughout the equipment but there is a central operator supervisory controller.

download. A mode in which the user sends a program from a laptop to the robot's programmable logic controller (PLC).

driven pulley. A pulley connected to the output of the system.

drive pulley. A pulley connected to the motor or force.

drive system. The way force is transmitted from motors or other input systems to the output or portion of the system that completes tasks.

drop cable. A cable that includes several taps along its length, allowing for node connections.

dull. A descriptor for the tasks that are repetitive in nature and often require little or no thought.

duplex. A communication system in which transmissions flow in both directions, thereby overcoming the problems of simplex transmission; it is the preferred transmission method for networks.

E

e-stop. A device that stops the function of the machine and often kills the power to multiple areas of the equipment.

eddy current. A flow of electrons created by a magnetic field moving across a ferrous metal item; this current then generates its own magnetic field.

edge and region statistics. Comparison of data describing unique features of an object, usually selected by the user, to a newly processed image to determine if it is indeed the object and how it is oriented.

edge detection. A data sorting process in which the processor looks for sharp differences in the light values between elements or pixels and then uses the elements nearby to confirm that it has, indeed, found an edge.

electricity. The flow of electrons from a place of excess electrons to a place of electron deficit.

electromotive force (EMF). A technical term for the voltage of a system.

emergency. A set of circumstances or a situation that requires immediate action and often involves the potential for or events that have caused injury to people and/or severe damage to property.

encoders. Built-in devices that provide feedback about the motor's rotational position.

end-of-arm tooling (EOAT). The devices, tools, equipment, grippers, and other tooling at the end of a manipulator that a robot uses to interact with and affect the world around it.

exoskeletons. Wearable frames with motors and electronics embedded that allow people with reduced or lost mobility to walk again.

expanded metal guarding. Metal that is perforated and stretched to create diamond-shaped holes with 0.25-inch pieces of metal around it.

external axis. The axis of motion that often moves parts, positions tooling for quick changes, or in some other way helps with the tasks of the robot.

F

ferrous metals. Metals that contain iron.

fiber-optic cable. A cable whose central core is made of glass or plastic.

fieldbus. A generic term for a number of industrial network protocols designed to provide real-time control of various devices and work with all the common network topologies.

fingers. The part of the gripper that moves to hold parts.

fixture. A device that holds a part in place for various industrial processes by clamping or holding the part in some manner.

float sensor. A sensor based on something a liquid can lift to determine level.

flow. The motion of a fluid; it can be measured with some type of sensor.

fluid. A substance in which the particles can move past one another, conform to the shape of their container, and continually deform under an applied shear stress.

fluorescent lights. A type of light produced by using high-voltage electricity to excite gases inside a glass tube coated with a phosphor material, which gives off light when struck by the ultraviolet radiation released by the excited gases.

force. A condition in which the user sets the state of data for an input or output, ignoring the actual field condition for inputs and the data result of the program for outputs.

four Ds. Situations that are Dull, Dirty, Difficult, or Dangerous.

four Hs. Conditions that are Hot, Heavy, Hazardous, and Humble.

frame. A set of three planes at right angles to each other; where the three planes intersect is the origin of the frame.

friction. The force resisting the relative motion of two materials sliding against each other.

fringe benefits. Benefits such as health and dental insurance, retirement plans, life insurance, Social Security contributions, and anything else for which the company pays part or all of the cost on behalf of the employee.

full bright field lighting. Use of flat array lights, axial diffuse lighting, or diffuse dome lighting to shine light directly down on an object and return the maximum amount of light to the camera.

G

G-code. A specific programming method that has proved very effective for motion control and is relatively easy to learn.

gantry. Typically a simple two- or three-axis machine designed to pick up parts from one area and place them in another area.

gantry base. A linear base with a finite reach.

gear train. Two or more gears that are connected together.

general station description (GSD). Detailed information about a device that is readable by the configuration software.

global function. Variables, subroutines, and other code or data accessible by any program created on a robot.

global positioning system (GPS). A system that determines a geographical position on the basis of the amount of time it takes to receive signals from three or four separate satellites in orbit around the earth.

grayscale. Any measurement of an element or pixel that is between zero, which is considered the full absence of light, and the maximum output, which is considered full light.

grippers. A type of end-of-arm tooling that applies force to secure objects for maneuvering.

grounded point. A point somehow connected to the earth.

ground wire. The wire that provides a low-resistance path to earth to protect people and equipment in case of electrical shorts.

guards. Devices designed to protect humans from the dangers of a system.

gun time. The amount of time during which current flows through the spot-welding gun tips.

H

hall-effect sensor. A sensor that responds to magnetic fields and can produce either a digital on/off-type signal or a varying analog signal, depending on the specific sensor and application.

halogen light. An evolved form of incandescent lights that uses a tungsten filament inside a pressurized, gas-filled bulb made of high-silica glass, fused quartz, or aluminosilicate.

harmonic drive. A specialized gear system that uses an elliptical wave generator to mesh a flex spline with a circular spline that has gear teeth fixed along the interior.

hertz (Hz). A unit of measure defined as sine wave cycles per second.

high-intensity discharge (HID) light. A type of light that uses an arc tube filled with gas to vaporize mercury, sodium, metal salts, or other substances. When current flows between two tungsten electrodes, it ionizes the gas and creates heat and pressure inside the tube.

hot spot. A high-intensity light spot.

hub. A repeater that sends a signal to all the connected nodes.

human–machine interface (HMI). A device (touch screen) that displays information via interactive graphics and is often tied into the system controller to cause actions of the robot.

hydraulic power. The use of a noncompressible liquid that is given velocity and then piped somewhere to do work.

I

idler gear. One or more extra gears added to a system to change the direction of rotation or to cover large gaps between the drive and driven gears.

imaging device. The heart of a vision system, in which light is captured and turned into some form of transmissible electrical signal.

imaging system processor. The portion of a vision system that takes the raw data from the sensors, usually in the form of voltage or current, and converts it into a digital signal that other controllers can use one frame or picture at a time.

impact. The point at which the robot contacts an object in its movement path.

incandescent light. A type of light in which electricity passes through a tungsten filament or wire, which in turn produces heat and light.

incremental optical encoder. An encoder that consists of a disk that has either holes for light to pass through or special reflectors to return light, an emitter, a receiver, and some solid-state devices for signal interpretation and transmission.

inductive proximity switch. A switch that uses an oscillating magnetic field to detect ferrous metal items; an oscillating field goes on and off.

industrial automation. Equipment that completes processes with minimal human assistance, which became readily available and competitive in the latter part of the 1980s.

Industrial Internet of Things (IIOT). The process of connecting everything in the plant to the network, and then the network to the Internet, with the goal of increasing flexibility. With this scheme, the company must assess the risks of getting hacked from the outside and determine how it can prevent this type of attack from happening.

industrial robot. An automatically controlled, reprogrammable, multipurpose manipulator programmable in three or more axes, which may be either fixed at a place or mobile for use in industrial automation applications.

infrared radiation (IR) temperature sensor. A sensor that detects the heat of an object by measuring the infrared radiation created; infrared radiation is a long wavelength of light below the human scale of vision.

input. Data that come into the robot, often from devices such sensors or registers and other data storage places.

integration. The process of combining equipment and design elements to create a functioning system or work cell.

International Standardization Organization (ISO). An organization that creates standards that are voluntarily used by industry members to prove levels of quality and safety in their products.

J

jaws. The part of the gripper that moves to hold parts.

joint frame. A type of robot that moves only one axis at a time while in manual mode, with the positive and negative directions being determined by the setup of each axis's zero point.

joint motion. A point-to-point motion in which all the axes involved usually move at the speed of the slowest axis, with no correlation between the separate axes involved.

jumper. A connection between two or more settings pins.

K

kinematics. The branch of mechanics that uses math to describe motion without considering either the cause of that motion or the masses involved.

kink. A sharp bend that causes damage to a fiber-optic cable.

L

ladder logic. The programmable logic controller (PLC) equivalent of an electrical schematic, in which the physical components of a relay logic schematic are replaced with symbols that represent instructions for data sorting in the PLC.

laser light. Light that has been intensified through a process of optical amplification.

laser photo eye. A technology that works like the infrared photo eye except that the light emitted to a concentrated beam, allowing for greater distance of travel before the returned signal becomes too diffused for sensing purposes.

laser welder. A device that uses an intense beam of light to create the high temperatures needed to melt and fuse two pieces of metal together, without the need for the traditional welding gun.

lens. The part of a camera that focuses the incoming light directly on the charge-coupled device (CCD) or complementary metal-oxide semiconductor (CMOS) elements of the camera.

level sensing. Determining how much fluid is present in a container, usually with the goal of preventing overfilling or other less-than-desirable conditions.

Leyden jar. A device for storing static charge and the predecessor of the modern capacitor.

light-emitting diode (LED). A special type of semiconductor with a p–n junction that emits light when current is applied, via energy released as photons when electrons from the n-type material cross the boundary and fill holes at a lower energy level in the p-type material.

light curtain. A technology that protect workers by creating a vertical or horizontal light barrier that detects any disruption.

light level detection. Use of a photoelectric sensor that can tune out or ignore the background, so as to differentiate between parts on a background or even to detect transparent objects.

light source. A device that provides a specific type of light that strikes the object in a specific manner to facilitate taking an image.

limit switch. A switch that senses the presence or absence of a material through contact with a movable element attached to the end of the unit.

linear motion. A type of motion in which the controller moves all the axes involved at set speeds to create straight-line motion between two points.

line scanning. 1D scanning. The imaging device uses a single line of the charge-coupled device (CCD) or complementary metal-oxide semiconductor (CMOS) element to gather the picture.

local area network (LAN). A network that brings Internet connectivity to areas where it would otherwise be unavailable, as the information is passed along the network from node to node as needed.

local function. A function in which only one program can access the data.

lockout. A mechanism that holds the power source in the blocked or de-energized state.

lock out/tag out (LOTO). A system in which all equipment have a means of shutting off and locking out the various power sources of the machinery.

looping. A programming structure that causes a program to repeat a series of instructions either for a specified number of times or until a specific program condition is met.

M

macro. An adaptation from the computer science world, in which a specific series of instructions invoke a definition that generates a sequence of instructions or other outputs.

magnetic flowmeter. A device that measures the change in voltage induced in a passing fluid. Such a system requires the fluid to conduct electricity, so it does not work with all fluids.

major axes. The axes responsible for getting the robot's tooling into the general area of use.

manipulator. What the robot uses to interact with and affect the world around it; they come in all shapes and sizes.

master/slave communication. A type of communication in which the controller (the master) asks for data from the nodes (the slaves) and the nodes respond only when directed to do so.

mechanical relay. A mechanical or solid-state device in which a coil of copper wire is used to generate a small magnetic field that moves parts inside the relay to open and close contacts.

media access control (MAC) address. The unique address for each device.

mesh. Where the gears meet up to transmit force.

mesh topology. A network consisting of multiple connections, where each node is connected to all or several nodes, with the ability to share information along those connections.

metal halide lamp. A type of high-intensity discharge (HID) lamp that is used in photography applications. It produces the greatest amount of light in the HID family, with its output being close to daylight frequencies.

metal mesh. Thick wire welded and/or woven together to create a strong barrier that is easy to see through.

MIG welder. A welding system that uses electrically charged wire to fuse metal together.

minor axes. The axes responsible for the orientation and positioning of the robot's tooling.

mobile base. A robot that uses a system to move the manipulator to various locations so it can perform its functions.

module. A removable device designed to serve a specific purpose or function.

motor encoder. A device that directly monitors the rotation of a motor shaft and turns that information into a meaningful signal.

motor starter. A contactor with an overload module used to start a motor.

muffler. A device that slows the air as it passes through this device's interior, reducing the sound and spreading the air that was flowing in a focused direction into a circular pattern.

multi-drop. A scheme in which one transmitter can talk to multiple receivers but not all of them.

multi-point. A scheme in which one transmitter can talk to any receiver.

multiple failures. A failure event in which more than one component is at the heart of the problem.

N

NAND. The opposite of the AND command; a logic condition in which all the input conditions must be false before the output is triggered.

NC. "normally closed," indicates that there is connection between two points when there is no power to the coil.

negative temperature coefficient (NTC). A term used to define when there is an increase in temperature that causes a decrease in resistance.

network. The connection of two or more devices for the purpose of information sharing.

neutral wire. The second wire attached to the larger of the two prongs on a plug, which provides a return path for the electrons and allows for a complete circuit.

nibble. A group of 4 bits.

NO. "Normally open"; indicates that there is no connection between the two points when there is no power to the coil.

node. A point tied into the network, such as a controller, a smart sensor with a processor, a communication module, a smart switch that helps with information transfer, a computer, a programmable logic controller (PLC), or anything else that can send and receive data.

noise. The mathematical difficulties that arise when trying to properly calculate the torque required at the start of motor motion for robot movement.

noise. Voltage induced into the line by magnetic fields or radio waves. It can create false signals and alter data transmissions.

NOR. A logic filter in which all input conditions must be false before the output is triggered.

normal force. The force that an object pushes back with when acted on by a force.

numerically controlled (NC). A descriptor for machines that used punch cards (like the Jacquard loom) or magnetic tapes (similar to VHS or cassette tapes), with position and sequence information being used to control the motions and actions of the machine.

O

object dictionary. A structure that holds all accessible values within the system. It includes the index, name, and type of each variable in the system, as well as descriptions, structures, data types, and relationships for device addresses; these values are organized in such a manner that the information can be broadcast over the network as needed.

Occupational Safety and health Association (OSHA). An agency that ensures all employees have a safe and healthy working environment.

octet. An 8-bit unit of data.

odd-shaped parts. Parts with unique shapes and proportions.

Ohm's law. The current through a conductor between two points is directly proportional to the potential difference across the two points; $E = I \times R$.

open-loop control system. A system in which commands are issued to various systems, such as stepper motors or control valves, and it is assumed that the robot did what it was told.

open loop. A system that works on the assumption that the control pulse—whatever it may be—activates the motion system so that the robot performs as expected.

optical isolation. Creation of a dead zone where the only connection between the internal electronics and the outside connections is light.

OR. A logic condition in which if at least one of two or more events happens, then the output of the function is triggered. If more than one condition is true, the output of this logic command is triggered as well.

output. A function that causes an action or writes data somewhere.

overload module. A device designed to allow the large in-rush of current needed to start a motor but to change state if the flowing amperage exceeds the set value for more than a few seconds.

P

packet. The quantity of data sent from node to node.

parallel grippers. Robot fingers that move in straight lines toward the center or outside of the part to close and grip or open and release the part, respectively.

parallel robot. Another name for the delta-style robot configurator in which multiple arms meet at a point to move the tooling.

partial bright field. A lighting scheme in which one or two light sources are set at an angle greater than 45 degrees to the object, so that most of the light transmitted is reflected into the camera lens.

payload. How much weight the robot can safely move.

peer-to-peer (P2P). A type of communication in which each node can talk directly to the others without having to go through a third node, such as the main controller, and can act as sender or receiver of data.

personal protective equipment (PPE). The last line of defense from the dangers around you; it protects the wearer when that person has chosen the right gear, the gear is in good condition, and it is used correctly.

photoelectric proximity sensor. A switch that detects levels of light, senses objects via reflected or blocked beams, and detects colors via an emitter that sends out a specific wavelength of light and a receiver that looks for this wavelength to return after it has interacted with the surrounding environment.

photo eye. A device that emits an infrared beam that strikes a shiny surface or reflector, which then reflects the light back to a receiver in the unit.

pick and place operation. The process of picking up items from one area and placing them in another.

piezoelectric pressure transducer. A transducer that uses crystals that generate electricity in reaction to pressure to generate an analog signal.

pinch point. An area in the danger zone where a person needs to watch out for the robot, any tooling used by the robot, and any place where the robot could trap the person against something solid.

pitch. The minor axes of motion, analogous to the up-and-down orientation of the human wrist.

plunger pressure gauge. A gauge that uses a plunger exposed to system pressure, a bias spring, a pointer, and a calibrated scale to measure pressure.

pneumatic power. A type of power that resembles hydraulic power but uses a compressible gas instead of a noncompressible liquid.

polarity. The positive and negative terminals on DC components that ensure proper electron flow through the part.

polling. A communication scheme in which the master (requesting unit) sends a signal asking the slave (sending unit) if it has any data to send.

positive temperature coefficient (PTC). A term used to define when there is an increase in temperature that causes an increase in resistance.

power rails. Two vertical lines with the one on the left side being the main power in and the one on the right being the return path in a 120-V or DC system or the second power line in a 220-V system.

precision. Performing tasks accurately or exactly within given quality guidelines.

presence sensor. A sensor that detects when a person is inside an area.

pressure relief valve. A valve that either vents pressure or returns fluid to the tank whenever the pressure exceeds a set point.

pressure sensor. A sensor that detects the presence or absence of a set level of force.

pressure switch. A switch that reacts to system pressure and, at some determined point, changes the state of contacts inside the system.

preventive maintenance (PM). The practice of changing out parts and doing repairs in a scheduled fashion before the equipment breaks down or quits working.

producer/consumer. A communication scheme in which a device "produces" (sends) information with specific identifiers attached and other devices "consume" (receive) the data that matches the identifiers they are looking for.

program. A scheme that makes decisions based on a system of logic filters and commands, and then activates various outputs based on the pertinent instructions.

programmable automation controller (PAC). A combination of programmable logic controller (PLC) operation and a PC-based control system.

programmable logic controller (PLC). A specialized system that runs one type of code very efficiently in industrial environments to monitor signals coming in, filter information through instructions entered by the user, and then send power out to activate various pieces of equipment.

programming language. The rules governing how the user enters a program so the robot controller can understand the commands.

program mode. A mode in which the user can upload or download data, but the processor does not scan the program and the I/O does not update with changes.

proportional integral derivative (PID). A mathematical process used to vary an output to maintain a specific set point.

proprietary network. A network in which the specifics of the protocols are typically a closely guarded secret; such a system is often incompatible with instrumentation and devices built by other companies.

protocol. A set of rules governing how data is sent, which voltages and currents are involved, how the data is read, and how packets are addressed, among other things.

proximity switch. A device that generates an electromagnetic field and senses the presence of various materials on the basis of changes in this field (instead of physical contact).

proximity switch. A solid-state device that uses light, magnetic fields, or electrostatic fields to detect various items without the need for physical contact.

publisher/subscriber messaging. A communication scheme in which nodes talk on a one-to-one or one-to-many basis; these capabilities are helpful for coordinating multiple-axis motion.

punch card. A rigid card that contains holes in a specific pattern that can control the functionality of equipment; this technology would survived until the late 1970s.

R

rack. A metallic shell into which various modules fit, with guide rails to keep everything straight.

random-access memory (RAM). Memory that allows information to be entered, changed, and deleted.

reactive maintenance. An action taken to figure out what is wrong and do something to get the system back to normal running condition.

read-only memory (ROM). Memory that allows the data stored there to be removed once it has been written.

receiver. A device that receives light (or other input) and looks for specific light information; the internal process/control takes the incoming signal and either turns it into information to send on or changes the state of contact(s) within the sensor.

Recommended Standard (RS). A standard that designates specific pin connections and connector types to aid in the transmission of data.

reduction drive. A drive that takes the output of the motor shaft and alters it via mechanical means.

register. A structure that allows the programmer to use arithmetic to store and manipulate data.

relay. A mechanical or solid-state device that, when control power is applied, opens or closes internal connections.

relay logic. A control system that utilizes relays to create logic gates for function control. By activating certain relays at certain times, the system creates logic gates for sorting information. The commonly used gates are AND, OR, NOR, NAND, and XOR.

remote center compliance (RCC). A simple approach to tooling that allows the tooling to respond to parts that are not always in the same position.

remote mode. A mode in which the operator can use a laptop to upload or download data into the system.

repeatability. The ability to perform the same motions within a set tolerance.

repeater. A device that takes the signal from a segment and amplifies it for the next segment.

reseat. An electronic card is pulled from a controller and put back to ensure proper connection.

resistance. The opposition to the flow of electrons in a circuit; the reason why electrical systems generate heat during normal operation.

resistance temperature detector (RTD). A detection system based on the principle that a linear increase in resistance occurs when a metal is exposed to heat.

retroreflective. A descriptor for a photoelectric sensor in which the transmitter and the receiver are placed in one assembly; such a system often uses something like a bicycle reflector to return the emitted light to the receiver.

return on investment (ROI). A measure of the amount of time it takes a piece of equipment to pay for itself.

right-hand rule. A way to determine robot motion while using a Cartesian-based frame that relies on the thumb, forefinger, and middle finger.

ring topology. A network in which all the nodes are connected in series to form a continuous loop or ring.

Robotic Industries Association (RIA). An organization with the goal of driving innovation, growth, and safety in manufacturing and service industries through education, promotion, and advancement of robotics, related automation technologies, and companies delivering integrated solutions.

robotic integrator. A company, usually other than the robotic system manufacturer, that specializes in selecting, adapting, installing, and programming robots for whatever application(s) its clients need.

robot program. A list of commands that run within the software of the robot controller and dictate the actions of the system based on the logic sorting routine created by the user/programmer.

roll. The rotation of the wrist in a robot.

root mean square (RMS). A mathematical average of the sine wave.

router. A smart device that resembles a switch; its connections can be either wired or wireless, depending on the application.

RS232. A communication standard that was introduced in 1962 and can still be found in industry today, though it is no longer the most popular method of data transmission.

RS422. A communication standard in which one transmitter can talk to multiple receivers but not all of them.

RS423. A communication standard that is similar to RS232, but adds half-duplex operation and transmission speeds up to 100 Kbps out to 40 feet (12 meter), which dwindles to 1 Kbps at the maximum range of 4000 feet (1200 meter).

RS485. A communication standard that is the current favorite and is the next step up from RS422. It can handle true multi-point communication, in which a transmitter can talk to any receiver.

rung. In a schematic, a line drawn from the left-hand side to the right, placing various components in the electrical sequence of the path, thereby mapping out the logic of the circuit and showing what has to happen to get power flow from left to right.

run mode. A mode designated by an extreme switch or key position in which the processor scans the program and manipulates data.

S

safety factor. The margin of error built into a process.

safety interlock. A system in which all the safety switches must be closed or "made" for the equipment to run automatically.

safety PLC. A programmable logic controller that contains two processors that are identical in nature; they are set up so that if the primary processor fails, the secondary processor seamlessly picks up where it left off.

safe zone. An area where a person can pass near the robot without having to worry about making contact with the system.

sail flowmeter. A flowmeter that works on the same basic principle as a target flowmeter, but instead of the fluid striking a round disk, the fluid drags along a triangular or rectangular flag.

scan cycle. The sequence in which the PLC processes signals coming in, the program, and the activation of output connections.

SCARA. Selective compliance articulated robot arm; a system that provides a unique motion type by blending the linear movement of Cartesian geometry robots with the rotation of articulated geometry robots.

seed pixel. A starting point picked by a system based on mathematics.

semiconductor temperature sensor. An integrated circuit (IC) temperature sensor; a device consisting of semiconductor diodes that rely on temperature-sensitive voltage rather than current characteristics to determine temperature.

sensing array. An organized group of sensing elements, which gathers information about the elements' contact with objects.

servo motor. A continuous rotation-type motor with built-in feedback devices called encoders, which provide feedback about the motor's rotational position.

shielding. Wires taking the form of wire mesh or metal foil that is connected to ground.

shock. A state in which electricity enters the body of a person at the point of contact with the electrical system.

simplex. A communication scheme in which data transmissions flow in only one direction, and that lacks any way to verify whether the data was received; it is rarely used in today's networks.

single-phase AC. A current in which one sine wave is provided to the system via a single hot wire and returned on a neutral wire.

single ended. A communication scheme in which one transmitter sends signals to one receiver.

singularity. A condition in robotics in which there is no clear-cut way for the robot to move between two points.

sinking signal. An input that provides the negative connection or ground to an input module.

slide rule. A 2-feet-long ruler that includes a logarithmic scale and makes complex math much easier to perform.

slippage. The loss of output power due to a belt slipping on the pulley.

smart relay. A relay that performs multiple timing functions and may offer the ability to control how outputs act, almost like limited programmable logic controller (PLC) systems.

solid-state device. A unit consisting of a solid piece of material that manipulates the flow of electrons.

solid mount base. A system in which the robot is firmly mounted to the floor or other structures using bolts and fastening systems.

solid state. A descriptor for those components that control power without moving parts.

sourcing signal. An input that provides the positive connection or DC voltage (usually 24 V) to an input module.

spot-welding. Use of a device that resembles a large copper clamp to fuse metal together by passing current from one tip to the other, through the material fused.

sprocket. A device with teeth designed to fit into the links of the chain instead of pulleys, and a chain, usually made of metal, that connects the drive sprocket to the driven sprocket.

spur gear. A gear made by cutting teeth into the edge of a cylindrical object.

staging point. A position that gets the robot close to the desired point but still a safe distance away, allowing for clearance and rapid movement.

star topology. A type of network in which a central connection point, known as the hub node, has connections (often hard-wired) leading to each separate device on the network.

step mode. A mode in which the robot advances through the program one line at a time, with the operator being required to press a button on the teach pendant keypad before the robot reads the next line of the program and responds.

stepped drum. A cylinder with high and low spots of varying lengths around the outside that trigger switches during the drum's rotation.

stepper motor. A motor that moves a set portion of the rotation with each application of power: The more steps per rotation, the finer the position control.

structured lighting. A variation of partial front lighting that typically relies on a laser. Specifically, a fiber-optic line light is used as the light source to capture object features or measure specific points.

subroutine. A sequence of instructions grouped together to perform an action that the main program accesses for repeated use.

switch. A device that knows the address of each connected node and sends the packet only to the designated node, instead of doing a network-wide broadcast.

SWOT analysis. A technique used to compare four types of factors—strengths, weaknesses, opportunities, and threats—when making important financial and business decisions.

synchronous belt. A belt that has teeth at set intervals along its length.

T

tactile. Having the ability to sense pressure through physical contact.

tap. A trunk cable that carries data and power, and that allows physical and electrical connections.

target flowmeter. A flowmeter that places a disk or similar shape in the fluid and uses the force of the fluid flow to deflect the target and generate a signal.

teach pendant. A handheld device, usually attached to a fairly long cord, that allows people to edit or create programs and control various operations of the robot.

template matching. Matching that involves comparing the processed data to the data stored from a previous image.

thermistor. A resistor made of platinum, nickel, cobalt, iron, or oxides of silicon, which experiences a consistent change in resistance with a change in temperature.

thermocouple. A type of temperature sensor based on the principle that a small DC millivoltage is generated when the junction of two dissimilar metals is heated.

thermography. The process of turning thermal data into colored images.

three-phase AC. A type of alternating-current power in which three sine waves are located 120 degrees apart electrically.

through-beam. A type of photoelectric sensor that separates the transmitter and the receiver into different units, placed opposite each other.

token. A bit of code that allows a node to transmit information.

tool center point (TCP). A point of reference based on the tooling and some important point related to that tooling.

tool frame. A system that establishes the location and orientation of the end-of-arm tooling for the robot software and the direction of motion for the x-, y-, and z-axis; it also affects payload and inertia calculations.

tooling base. The part that attaches to the robot and holds the mechanisms to move the fingers.

torque. The rotational force generated by the motor; the limiting factor when determining the robot's payload.

tourniquet. A tightened band that restricts arterial blood flow to wounds of the arms or legs in an effort to stop severe bleeding.

transmission. Two or more gears connected together.

transmitter. A device that sends out a specific type of light.

tree topology. A network that combines aspects of a bus network and multiple star networks, usually comprising three levels of devices.

triac. A solid-state device designed like two diodes in antiparallel orientation, with a trigger connection to start current flowing.

troubleshooting. A logical process of determining the cause of a problem and correcting faults in a system or process.

trunk cable. The main cable in a bus topology network.

turbine flowmeter. A flowmeter that uses a propeller- or turbine-type assembly placed inside the fluid to generate a signal via a magnetic pickup sensor positioned nearby, but outside of the fluid.

turnover rate. How often workers quit a job or leave an employer.

U

ultrasonic sensor. A type of sensor that is similar to a photo eye or photoelectric proxy switch, except that it emits and receives sound instead of light.

ultrasonic welding. A welding process that uses intense pressure and vibration to fuse materials.

upload. A mode in which a program can be pulled from the programmable logic controller (PLC) to the interface device, generally a laptop.

user frame. A specially defined Cartesian-based system in which the user defines the zero point and determines the positive directions of the axes, including the option of straight-line motions moving at an angle instead of along the flat lines of the world frame.

V

vacuum. A suction device that uses less than atmospheric pressure to secure parts.

vacuum gripper. A gripper that works by creating a pressure that is less than atmospheric pressure around it.

ventricular fibrillation. A condition in which the heart quivers instead of actually pumping blood.

voltage. The force that drives electricity through the system: The greater the voltage, the greater the driving force. It is measured in units of volts (V).

voltaic pile. A battery consisting of alternating disks of zinc and silver or copper and pewter separated by a paper or a cloth soaked in either salt water or sodium hydroxide.

W

weave motion. Straight-line or circular motion that varies from side to side in an angular fashion while the whole unit moves from one point to another.

welding guns. Tube-like tools that people or robots use to direct the electrically charged wire used to fuse metal during welding operations.

wheeled base. A base that consists of wheels, a drive system, and some form of navigation system so the robot can find designated points and avoid obstacles along the way.

word. Data that consists of multiple bytes, with the number depending on the design of the system.

work cell. A logical grouping of equipment in which the processes flow seamlessly from one operation to the next, with the goal of improving quality and reducing per-part production times.

work envelope. The area the robot can reach or where all the robotic action takes place.

world frame. A Cartesian system based on a point in the work envelope where the robot base attaches.

X

xenon light. A type of light that uses tungsten electrodes and an arc tube filled with xenon gas to produce intense light.

XOR. A logic command that is similar to the OR command, with the exception that only one of the conditions can be true. If more than one condition is true, the output does not occur.

yaw. The side-to-side orientation of the wrist in a robot.

Z

zero-energy state. The process of removing all power from equipment so there is no active or latent power.

Index